CHALLENGES OF THE NEW WATER POLICIES FOR THE XXI CENTURY

PROCEEDINGS OF THE SEMINAR ON CHALLENGES OF THE NEW WATER
POLICIES FOR THE 21ST CENTURY, 29–31 OCTOBER 2002, VALENCIA, SPAIN

Challenges of the New Water Policies for the XXI Century

Edited by

Enrique Cabrera & Ricardo Cobacho
Institute for Water Technology, Polytechnic University of Valencia, Spain

A.A. BALKEMA PUBLISHERS   LISSE / ABINGDON / EXTON (PA) / TOKYO

UIMP

C.O.I.I. Castellón

INSTITUTO TECNOLÓGICO DEL AGUA

Copyright © 2004 Swets & Zeitlinger B.V., Lisse, The Netherlands

All rights reserved. No part of this publication or the information contained herein may be reproduced, stored in a retrieval system, or transmitted in any form or by any means, electronic, mechanical, by photocopying, recording or otherwise, without written prior permission from the publisher.

Although all care is taken to ensure the integrity and quality of this publication and the information herein, no responsibility is assumed by the publishers nor the author for any damage to property or persons as a result of operation or use of this publication and/or the information contained herein.

Published by: A.A. Balkema, a member of Swets & Zeitlinger Publishers
www.balkema.nl and www.szp.swets.nl

ISBN 90 5809 558 4

Printed in the Netherlands

Table of contents

Foreword — VII

Acknowledgements — IX

Introduction

Situation and trends of water management. The Mediterranean approach — 3
E. Cabrera & R. Cobacho

Political aspects

Towards the water policies for the 21st Century: a review after the World Summit on Sustainable Development in Johannesburg — 17
J.J. Bogardi & A. Szollosi-Nagi

Water policies in Europe after the Water Framework Directive — 39
B. Barraqué

The risks and benefits of globalization and privatization of fresh water — 51
G. Wolff

A structure for water administration in the 21st Century. The case of Australia — 99
B. Martin

Practical aspects and sustainable water management

Effects of climate change on hydraulic resources — 117
L.W. Mays

Demand management of water in a regulatory environment — 141
D. Howarth

Sustainable water management in agriculture — 157
E. Fereres & D. Connor

Sustainable water management in cities — 171
B. Chocat

The future of desalination as a water source — 187
J. Lora, M. Sancho & E. Soriano

Economical aspects

Water pricing: a basic tool for a sustainable water policy? — 209
A. Massarutto

Author index — 243

Foreword

New water policies for the 21st Century, the subject of the book you have in your hands, represent one of the main challenges that forthcoming generations will have to face. Because of the population growing and the look for higher standards of life, limited renewable water resources are becoming more and more pressed. Traditional water policies cannot provide adequate answers to the new arising problems, and the new Water Framework Directive that Brussels promoted in December 2000 is an evidence of it.

A Water Framework Directive that will have a significant influence in countries, as it is the case of Spain, in which large hydraulic infrastructure works have been built during the last century throughout a wide policy of subsidies. This water policy from the supply perspective needs now to be balanced by a demand water policy or, in other words, water development must be deeply complemented by water management.

My Organization, aimed to work for the society development, finds sound reasons to promote the publication of this book. First of all, contributing to a wider knowledge of all these new water policies; and secondly, emphasizing the fact that such policies will require a better water management, a field in which the professionals integrated in the Chartered Institution of Industrial Engineers of Castellón that I chair, will make a relevant contribution in a next future.

Javier Rodríguez-Zunzarren
Dean of the Chartered Institution of Industrial Engineers
Castellón

Acknowledgements

This book contains the lectures presented at the International Seminar 'Challenges of the New Water Policies for the XXI Century', hosted and sponsored by the Universidad Internacional Menéndez Pelayo (UIMP). The Seminar took place in Valencia (Spain), in late October 2002, and was attended by more than one hundred delegates.

First, we must acknowledge and thank Dr. Vicente Bellver, the UIMP Director (Valencia site), and his collaborators Mr. Luis Moreno and Ms. Mabel López for their support and help prior and during the Seminar. UIMP sponsorship allowed assembling a very good panel of lecturers, with complementary backgrounds, that fully covered the main issues concerning the forthcoming challenges of water policies. Second, we express our gratitude to all these professionals, because they have made possible this publication and last but not least, we want to thank Mr. Javier Rodríguez-Zunzarren, Dean of the Chartered Institution of Industrial Engineers (Castellón site), for his sponsorship in this publication.

The Institute for Water Technology, the editors' institution, carried out the organization and technical direction of this International Seminar, as well as the preparation of the texts, while the presentation and production of the book has been performed Mr. Janjaap Blom and his colleague Mr. Martin van der Zwan at Balkema Publishers. We thank them both for their great professionalism and understanding.

Enrique Cabrera
Ricardo Cobacho
Editors

Introduction

Situation and trends of water management – The Mediterranean approach

E. Cabrera & R. Cobacho
Instituto Tecnológico del Agua, Universidad Politécnica de Valencia, Spain

ABSTRACT: The 21st Century is going to witness a radically different water systems management from the one we have known all along the recently finished century, which has been characterized by the construction of numerous and enormous, though necessary, hydraulic infrastructure works. Today, three facts have set up a new scenario: firstly, water, subjected to high requirements due to the competition of different uses, is a very vulnerable resource. Secondly, it should be kept in mind that the most profitable investments have already been made. Those still to be undertaken imply strong investments that would set the price of water at an uncompetitive value, if the principle of cost recovery demanded by the new EU Water Framework Directive is implemented. And finally, it cannot be ignored the society's growing environmental awareness, brought about by the evidence of the progressive deterioration of our environment which is demanding a sustainable management.

1 INTRODUCTION

Throughout the 20th Century, water policies have been mainly focused on water management from the supply side. In short, such policies have consisted on supplying higher and higher amounts of water to those requesting it. This (and it should be remarked from the beginning) was somewhat right until a few decades ago. For that reason, society, in relation to water, has been mainly interested engineering of surface and underground hydrology, systems applied to an optimum use of the available water resources, water transfers between basins, analysis and regulation of natural currents, maritime and coastal engineering, transportation of sediments, hydroelectric power stations and, finally, the great transformation of dry lands into irrigated lands. As a result, hydraulics engineering has focused its efforts towards these aspects. Therefore, until recently, water policy was associated almost exclusively to civil engineering.

In this context, urban water supply and sewage systems are nothing but lesser works, nothing comparable to colossal dams such as the Hoover Dam in the Colorado River in the USA which supplies water to most of the Nevada desert and Arizona and, particularly, the city of Las Vegas. Nothing compares either to the Aswan Dam on the Nile River whose long tail of backwater, in the early 70s, forced the authorities to raise the Abu Simbel temple (300 km away) to a height of over 50 meters. Rameses II had this temple built four millenniums ago. The Aswan Dam's environmental impact is so great that, in order to compensate for it, the project was presumptuously and demagogically sold as Egypt's barrier against hunger. As we said before, these pharaonic works have made civil engineering to occupy the most important position in the field of water.

However, great changes occurred by the end of last century. The main one was produced in the technological field in which modern computers have enabled engineers to make calculations that were unthinkable in the past, consigning the large physical model laboratories to a secondary position. Moreover, the development of instrumentation in general, and of electronics in particular, has increased the possibility of accurately and quickly measuring the different physical magnitudes involved in the phenomena under analysis, insofar as the mathematical models representing physical systems can be calibrated from the measurements actually carried out. This first change has revolutionized the way by which classical problems are dealt with, but in no way relegates civil engineers to a secondary position. It has just modified their framework for action.

Our main concern here is a second change whose origin is undoubtedly very different from the one derived from the only technological progress. On the contrary, that origin is economic and social, that is, a crucial one in modern society. Thus, it appears as a counterweight, as a "this is as far as we can get" in the face of the practically unlimited possibilities rendered by technology in the service of great works. This turning point officially started in 1987 when the Brundtland Commission published its conclusions (Brundtland et al., 1987). There, emphasis is made on the need to utilize most effectively the planet's natural resources, as long as such development does not put the future in jeopardy, i.e., that it "can be sustained throughout time".

It is important to highlight the decisive influence this report has had (and it is still going to have) for the future of hydraulics engineering. The sustainability of the natural resource through time has implied the updating of many issues related to the field of water that were, until a few decades ago, completely unknown. It has already been unanimously accepted that water management must be carried out in an integrated and global way, without neglecting environmental, ecological and biological aspects. The field of water has become much more multidisciplinary and demands cooperation and coordination (Lopardo, 1995; Kobus, 1997). This has clearly been pointed out by the American Society of Civil Engineers, ASCE, (ASCE, 1996) when it states that "*the hydraulic engineers have to think bigger and broader*". That important reference will be dealt with in detail further on.

This new framework, and the self-evident fact that the available water resources are subjected to an ever increasing pressure as a result of the growing competition existing among the traditional uses of water (mainly irrigation, those derived from the increase of the population and its living standards, and those employed in industry and during leisure time) are making water stop being an exclusive social asset to turn it, most of the time, into an economic asset as well, thus having remarkable consequences on everything related to water policy.

The sustainability of this resource is based on a time-stable balance between water supply and demand. In fact, it is important to note that supply management has experienced a period of outstanding splendor all throughout the 20th century, mainly in developed countries. This has taken place in such a way that Rouse (Rouse, 1987) has called this period "*the last golden age of hydraulics*". As a consequence, the works with the greatest technical feasibility and economic productivity have already been built. Current conditioning environmental factors, neglected in the past, either make the building of new works more expensive or just prevent them from being made. Actually, all these facts restrict extraordinarily the possibilities of continuing the increase of water supply. This limitation makes it compulsory to rationalize water uses giving demand management a great importance nowadays. The recent appearance of this new philosophy of water management makes all those possibilities suitable for exploitation, or at least for exploration.

Some objective facts speak for themselves. For instance: the changes that have significantly affected the research programs of the US Bureau of Reclamation, which is the institution that has fostered most of the 6,372 dams of that country (Cobacho, 2000). Burgi (1998) briefly states: "*As public values have shifted from emphasis on water resource development to management of western waters, the Bureau's contemporary hydraulic research program has also changed from water development to water management*". This change of trend is being strongly imposed on most of the developed countries so that water management is ceasing to belong exclusively to the civil engineer. Sociologists, town planners, geographers, biologists, geologists, chemists and a wide range of engineering specialists have something to say or contribute to the complex and multidisciplinary field of water.

In short, the policy of supply management carried out to its logical conclusion is already showing clear signs of exhaustion in developed countries because, among other things, most of the needed works have already been built. And this has generated its counterpoint: demand management, whose main goal is not to supply all the water requested, but to begin rationalizing its use.

In order to better understand how much has been synthesized and to highlight the significance of the turning point marked by the Brundtland report in 1987, it seems appropriate to look over Man's relationship with water throughout history. For such journey, there are excellent and passionate historical reviews, since, as Levi says (Levi, 1995), there possibly does not exist another field of engineering that can show such a long and rich history as hydraulics engineering. And that, undoubtedly, is because water is part of Man's own life and he has neither been able, nor is able or will ever be able to do without it. And such fact leads to an unbreakable bond.

2 WATER RESOURCES MANAGEMENT AND ITS EVOLUTION THROUGHOUT HISTORY

Humanity, and also plants, have always needed water to live. As a consequence, irrigation has been practiced for 30,000 years (Bonnin, 1984), meanwhile the actions aimed at facilitating human consumption (for example family storage tanks of rain water) are reported several millennia before our era. With time, such individual tanks were substituted by works to bring water from the springs to the consumption points. A good example of this is found in Rome, two thousand years ago, where a system of aqueducts was designed with a striking daily capacity of transport: $600,000\,m^3$. That implied a consumption of more than $500\,l/(capita\,day)$, unimaginable amount today, even if the distribution system was managed in a rather competitive way. In any case, the whole exciting world of water supply in ancient times can be seen in the above mentioned work (Bonnin, 1984) and, also more condensed in Garbrecht (1987).

The Antiquity has left a few examples of pressure water transport set to play, as we will see, the leading role in hydraulics engineering of the future. This is a very outstanding fact concerning the objectives of the present article due to the difficulties found when developing techniques and materials that could be watertight to important pressures. In addition, the difficulty of controlling properly the changes in patterns (for instance, water-hammer) was one of the reasons for the pressure hydraulics to be hardly present in the Antiquity. All in all, the inverted siphon – attributed to King's David son – is known; probably the oldest known device used to supply Jerusalem with water. Later on, in the 2nd century, it is clear that siphons supplying the Acropolis of Pergamo (lead pipelines of 22 cm diameter and 8 cm thickness) could withstand 20 bars pressure.

Also in Ancient Rome, the final lines of water supply networks were under pressure, though fairly slight (it barely reached 10 meters of water column). On the other hand, and as the only well-known pumps in the Antiquity were those of positive displacement (Raabe, 1987) – it is not until the 17th century that we find the first centrifugal pumps – all the known pressurized pipes are siphons or pipelines that get loaded, but always with a favorable geometric slope that generates, as in the open channel transport, the flow of water. The culture of pressure hydraulics was lost almost completely, and not developed again until many centuries later.

Due to the difficulties found to transport water under pressure, hydraulics of open channel is much better known. Man tends to imitate the water transport that occurs in nature. Technologically it is much less complex, since it is only necessary to follow the slopes. On the other hand, the simple regulation problems outlined are solved very easily, and, in short, the materials required for the construction of a channel are, due to their simplicity, much better known. The Romans left aqueducts throughout all the land they conquered, authentic art works nowadays. They managed to transport large flows of water, which was not possible with pressure pipelines. Everything related to open channel hydraulics was familiar for them. Even, since river floods were relatively frequent – prophet Jeremiahs reports them in the Bible – it is not strange that the first river channelings were developed several millenniums before our era.

Although Archimedes' screw was already known, it is the treadmill, or the water wheel, what prevails in the Antiquity in everything concerning the elevation of water by a hydraulic machine. The first reference to its existence is found in the 3rd century BC (Raabe, 1987). It was very popular during the Middle Ages, and it was not only used as a water rising pump, but also as an engine to get the energy of a river current available for human activities.

The Arabs used it very frequently and, in fact, the word treadmill in Spanish (noria) comes from the Arabic word (al-na´ura), that means wonder. The water wheel allowed to rise water to the required level, from which it was released and flowed due to gravity. For these reasons, almost all the systems devised for water transport and supply were based on the principle of open channel, what contributed, during this period, to develop the understanding of this technique even more. Water transport by means of a pressurized tube, from a lower level to a higher one, is not feasible yet in the Middle Ages.

Another evidence of the good knowledge on open flow hydraulics is the regulation of the rivers by means of dams, a well-known technique from the Antiquity. According to Schnitter (1994) the first work of this type is found in Memphis (Egypt) and it dates back to the year 2,600 BC. This reference describes the development of dams throughout history, nearly all of them devoted to facilitate irrigation, although some were also built with other objectives such as urban water supply, flood control and even for energy purposes.

Regarding urban water supply after the Roman splendor, the supplied volume is limited to a few cubic meters that satisfy the most vital necessities. It should be kept in mind that, during many centuries, most of the people live in small towns. Such a so low standard of comfort, as opposed to the Romans' lifestyle, explains the great number of limitations that constrained water supply in medieval cities. The example of Paris is quite representative, since it is one of the most influential cities in the modern era. In 1553, there is a supply of only 300 m^3/day for a total of 260,000 people (Thirriot, 1987a). This equals the ridiculous amount of 1 l/(capita day). Even in 1669, already at the beginning of the Age of Enlightenment, there is only 1,800 m^3 for a total of 500,000 people, which represents 4 l/(capita day). And this situation is even worse in other European cities. For example, Lisbon with 80,000 people, in 1740, could only supply 560 m^3/day which is equivalent to 7 l/(capita day) (Thirriot, 1987b). And Madrid, which throughout the 18th century could only supply 3,600 m^3/day for the whole population (Paz Maroto & Paz Casañé, 1969).

Thus, urban water supply is hindered by the lack of knowledge and available technical resources regarding the transport of water under pressure. In fact, it is not until the work *Hydrodinamica* by Daniel Bernouilli, published in Strasbourg in 1738 (Rouse, 1963) that their basic principles are established. At the same time the foundations for the turbo-pumps are laid. Leonardo contributes with the first ideas, although it is not until two hundred years later, in the 17th century, that the first radial impeller by Papin will be born. Two centuries later, first industrial production of these radial pumps began, featuring a snail-shaped diffuser, which took place in the USA by the Massachusetts Pumps in 1818 (Mataix, 1975).

This knowledge would have been fruitless without the appropriate pipes to transport pressurized water. In 1672 the first cast iron pipes appeared, manufactured by Franzini (Paz Maroto & Paz Casañé, 1969), which will enable cities to lay first water supply networks, very similar to our current systems. It was a remarkable achievement, since the inadequacy of the supply in large cities at the 18th century was the cause of a mortality rate higher than the birth rate – cholera and typhus. However, these towns continued growing due to the arrival of people from rural areas (Steel, 1972). With no further purification treatment, which was not possible yet, open channel water supply was not very advisable. Moreover, in case of war, the high degree of vulnerability of a water transport accessible to everybody was more than evident. Regardless of the land characteristics, it was urgent to bury all the pipes which could only be achieved by means of pressure water transport.

Therefore, urban water supply, already an imperious necessity at the 18th century, starts to be a reality as soon as metal pipes, able to transport pressurized water, are available at a reasonable cost; and at the same time the hydraulic power available (flow times pressure) for the turbo-pumps start to be significant. The first urban supply in the USA, carried out with pressure pipes, dates from 1754, in Bethelehem, Pennsylvania (Griegg, 1986). The first water purification filter was installed in 1829 in London, while the first water chlorination system was carried out in the same city in 1908 (Steel, 1972).

However, urban demand requires modest amounts of water, much lower than those needed for irrigation, for which open channel hydraulics prevailed until some decades ago, when sprinkler and leak irrigation appeared. The sanitary problems of the urban supply are not found in irrigation.

Thus, open flow hydraulics is promoted as well as the necessary hydrology to evaluate the availability of resources and the appropriate flood and drought planning. Large hydraulic works aimed at transforming dry lands into irrigated lands and to generate electric power prevail in hydraulic engineering in Spain, which at that time was isolated from abroad. *"Watering is eating and, therefore, the power lays in watering"*, states Joaquín Costa at the end of the 19th century (CICCP, 1975). A statement which was necessary and absolutely right at that time.

The prevalence of water management as *"water development"* during the whole 20th century has been enormous. One of the most prestigious books of this century, the *Handbook of Applied Hydraulics*, in its 3rd edition of 1965 (Davis & Sorensen, 1965), devoted 14 out of its 42 chapters (one third of the total) to the most representative works in water development: dams, and up to 7 additional chapters to hydroelectric power stations, the great infrastructure linked to dams. In contrast, different approaches of water management are dealt with in less than 10 chapters.

Nowadays, things have changed considerably. The modern equivalent to the handbook mentioned above, entitled *Hydraulics Design Handbook* (Mays, 1999), published by the same publishing company (Mc Graw Hill) back in 1999, does not deal specifically with dams in any of its 24 chapters, however environmental hydraulics or risk analysis, which in the past were disregarded are dealt with in detail.

This prevalence of *water development* has been extremely marked in countries like Spain, where two singular facts converge: a temporary isolation that led to have a self-sufficient agriculture, and an irregular climate that favors extreme hydrological situations, that is, floods and droughts. Both facts lead to an enormous regulatory effort compelling to have a number of reservoirs higher than usual in other developed countries. Thus, it is not surprising that our country has the largest number of dams in the world, and that, depending on the indicator used, it can even be considered as the world leader. In our opinion, it should be noted that the right thing was done.

The excellent historical revision carried out by the International Association of Hydraulic Research, IAHR, in 1985 (IAHR, 1987), in accordance with the classical trend described here, appeared in the same year of publication of Commission Brundtland's report which sets the turning point in the tendency followed until then by water management and hydraulics engineering. Indeed, there is a world of difference between what Kennedy thinks in 1985 (Kennedy, 1987) in the closing report of the above mentioned historical revision by the IAHR (published two years after the celebration of the event) and what was presented by the ASCE some years later (ASCE, 1996). This Association begins by describing hydraulics as environmental, what is not an irrelevant detail. The work by the Brundtland Commission and all its later development, including among others the Agenda 21 of the 1992 Rio Summit (UN, 1992), deeply influences the Society towards the evident necessity of changing the direction.

Let us analyze these two points of view on hydraulics engineering, so close in time and so distant in terms of approach. Back in 1985, Kennedy, who focused exclusively on traditional water management systems, considered that hydraulics on year 2000 should:

– Develop measures economically feasible, which allow the reduction of water losses by filtration and evaporation, in reservoirs as well as in the transport systems.
– Extend the reservoirs life by means of a better knowledge of sediment hydraulics.
– Be able to design and produce high power pumps and turbines free of vibration and cavitation problems, thus the materials used for bearings and mechanical elements in general should be improved. It is necessary to avoid the problems that have affected many of the current large facilities.
– Develop decision support systems for the correct water allocation among the different users.
– Develop numerical methods that allow performing trials with models in a less expensive and more efficient way.
– Improve prediction models for future river behavior, once it has been established a significant modification of their pattern (dam or transfer). And not only from the perspective of quantity, but also from quality and transport of sediments.
– Develop and improve maintenance and operation strategies for current large hydraulic systems.

As we see, Kennedy had the idea of projecting hydraulic engineering towards the same direction as in ancient times. However, the ASCE, through the Task Committee on Hydraulic Engineering Research Advocacy (ASCE, 1996) keeps a very different concept. Environmental hydraulics in its way toward the 21st century cannot continue making the same mistakes, since up to now:

– Research and education have not been adequately structured.
– Researchers do not connect appropriately with the social needs that policy makers keep emphasizing.
– Training programs on hydraulics are not adapted to the needs of the labor market.
– Hydraulic engineers should adopt a more broad-minded and forward-looking approach.

The document of the ASCE has an enormous significance as it represents the most influential association of civil engineers in the world. Furthermore, it seems unquestionable that the clarifying self-criticism of the US Civil Engineering can already be seen in Spain through some of the most outstanding researchers, even when the administration – which has already recognized in the White Paper on Water (MMA, 1998) the need to introduce changes – is very reluctant to it, and policy makers, always fearful to the unknown, are not inclined either to it, because it is difficult in the short term to value the consequences that can be derived from such changes. For these reasons, in this country it seems reasonable to think that the solution will appear after a crisis caused by a drought, or by the impossibility of controlling the water pollution because of the current lack of control on consumption.

3 THE NEED FOR A NEW WATER MANAGEMENT SYSTEM

Clearly, in view of the above mentioned, the new scenario is producing a significant change in hydraulic engineering towards a new policy, a change that, on the other hand, has just begun. History, tradition, decision-making centers, even, the most developed doctrine bodies, encourage the traditional management. However, it is generally accepted the need to manage water in a very different way from it was done in the past (Beard, 1994), and it is evident that 21st century hydraulic policies will be characterized by the necessary balance between supply and demand. Undoubtedly, demand management will have in this 21st century the relevance that supply management received in the past.

Furthermore, nobody should consider that there must be a competition between both approaches. Conversely, foundations for a comprehensive water management system lay on their balanced compensation. That was the objective presented in a course organized in this same context, the International University Menéndez Pelayo (Cabrera et al., 2002). However, the supply management has been up to now, and especially throughout the 20th century, the only approach, while its complement, demand management, has just come out into scene. Whereas some countries are already very diligent in adopting this kind of changes, others, on the contrary, keep feeling more influenced by their past. That is the case of Spain.

All in all, there are lots of evidences. According to a study of the DGXII in charge of Research in the European Union, EU (EC, 1996) the most important challenges for research and development in the European Society regarding the field of water are:

– Fight against pollution.
– Wise use of water.
– Fight against water scarcity.
– Prevention and management of crises: floods and droughts.

The new EU Water Framework Directive, WFD (EP, 2000), aimed at the guidelines described in this article to ensure the sustainability of water management, will establish a new policy on water. It is certainly symptomatic that the new WFD states at the beginning that *"water is not a commercial product like any other, but rather it is a heritage that should be protected"*, a perfectly justified statement in view of the increasing citizen demand on fresh water (from rivers, lakes and coast waters) to be used for drinking, hygiene purposes and all the multiple uses it has.

The idea behind the previous statement can be perfectly identified with the wide range of problems associated to the use of water, and the subsequent multidisciplinary approach it implies. Among others, we can emphasize the over-exploitation and pollution of groundwater and surface water (rivers, lakes and regional seas), floods and water scarcity, destruction of the aquatic ecosystems and wetlands, erosion and desertification. Undoubtedly, a very different scenario from the situation just some decades ago.

In order to try to solve problems like these, the draft of the new WFD is laid on five basic foundations:

– Protection of surface and groundwater.
– Strict control on spills and the quality standards of water.
– Management of water basins.
– Development of the social stakeholders' participation in the water policy.
– Introduction of a new pricing policy.

From a cultural point of view as well as from the direct repercussion it has on the citizen's economy, the establishment of a pricing policy based on the full cost recovery principle may be the most controversial change. Overall, in the Spanish irrigated lands (and particularly, the historical ones) water is strongly subsidized. As to the urban, industrial and leisure uses, prices are very variable, even when in general they are far from recovering the costs as well as from applying rigorously the principle "the polluter pays", both criteria clearly stated in Article 9 of the WFD.

In Europe, there are countries where the recovery of costs is already almost completely fulfilled. Thus, Figure 1 (EC, 2001) shows the percentages of the Current Water Prices (CWP) and the Full Cost Recovery (FCR), related to the average family incomes. As it can be seen, the average price of water in Spain implies 0.4% out of the total income for an average family. Four countries out of the eight represented in Figure 1 are close to achieve the Full Cost Recovery, while the another four (Spain among them) will have to increase their prices significantly to reach it.

Undoubtedly, for a pricing policy adapted to achieve the full cost recovery and a strict control of spills, the exigency for complete economic and judicial responsibilities are essential conditions to enhance savings and a wise use of water. In fact, the countries in which these policies are applied (Figure 1) are much more efficient in using and managing water. We also find evidences in Spain (Corominas, 2000): since it is very well-known the higher efficiency in the use of groundwater for irrigation, compared to surface water, simply because in the first case the recovery of costs is

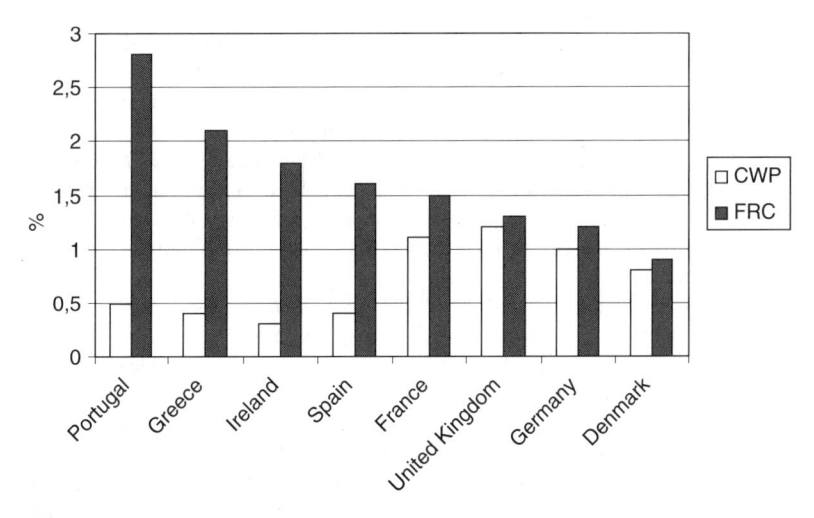

Figure 1. CWP and FRC related to average family income.

greater than in the second one. On the other hand, it should be kept in mind that among the main objectives of efficiency in management, the reduction of the environmental impact and the increase of the resource's capacities should be counted. This fact, in spite of being a fundamental principle of all sustainable water management, consciously or unconsciously, is usually ignored by Spanish water policy.

The recent 2002 Johannesburg World Summit, in which the EU has assumed a leading role, has clearly confirmed the guidelines for the future, though it has been carried out in a much wider context, since water problems are very different throughout the planet and they should not be generalized. But what is unquestionable is that the main guidelines are the same. A good synthesis can be found in the closing speech of the German Secretary of Environment (Trittin, 2001) at the international conference on continental waters held in Bonn in December 2001, as a preparation for the Johannesburg Summit. Four important points were there emphasized:

- Efficient water management is the key element to fight against poverty and to effectively achieve a sustainable development.
- Efficient water management depends to a great extent on a structure of modern and efficient management.
- The best way to achieve an efficient water management is to decentralize it, since the local users are the most interested in maintaining the availability of their water resources in the long run.
- All the stakeholders involved should participate actively in the process.

In Germany, the pollution during the late sixties, that was caused by an enormous industrial development, acted as a generator for the necessary change in water management. They did not have the history or the social pressure from a sector like agriculture that demands a special consideration, but they did have a high environmental pressure. A pressure that is now fully affecting Spain, where the inertia and the vested interests are slowing down a change in direction that, sooner or later, will certainly arrive.

4 THE GREAT CHALLENGES IN WATER MANAGEMENT FOR THE 21ST CENTURY

There is no doubt that the current scenario is totally different from the one drawn by the contemporary policies. And it is also clear that the wise Mediterranean people has managed to face water management challenges at any moment throughout history. Many examples could illustrate this fact, but within the context of this International Seminar, it seems evident to highlight the *Tribunal de las Aguas*, created by Abderramán III (Giner, 1997) back in the 10th century. This internationally renowned court, is even used today in covers of documents from prestigious international organizations (GWI, 2002), such as the Global Water Partnership that claims the dialogue as the most effective way of solving the water crisis the current world has outlined.

The objective of this International Seminar is to review some of the main problems of water management for the 21st century and to draft, from this analysis, potential lines of action, not easy to set forth when dealing with a natural resource with such a sensitive approach from traditional users and, very especially, those in the Mediterranean area. The contents have been divided into two large sections. First section takes into consideration the guidelines and trends of the current policies, while the second one deals with the aspects to remember, within the current framework, aimed at a sustainable management of water.

The first block starts with a review to the recent Johannesburg Summit regarding water policies worldwide; the next stage is to translate this to a closer context: the EU, taking into account the implications of the new WFD, and finally posing two recent topics: the possible consequences of the increasing privatization of water services, and the changes required in water management in order to effectively face an integral methodology, considered in a holistic way, instead of assuming the traditional fragmentation to which our history has taken us.

The second section deals with some specific aspects that should be taken into consideration within a sustainable water management. This is the reason to start off with the climatic change, unquestioned subject, and its relationship with the availability of resources. Next, and within this context, we analyze the importance of a new ingredient that any sustainable water policy should have: demand management. In third place, we will study, within their two main contexts, the two most outstanding uses of water, agriculture and cities, to conclude by considering desalination as an alternative source for water resources. The different pricing policies, from the approach of a sustainable use of water will also be studied. And, after this general overview, we will briefly refer to each one of these topics:

- *"Towards the water policies for the 21st century: A review after the world summit on sustainable development in Johannesburg"* is maybe the best framework to continue the developments of the subsequent chapters. In this first topic, the importance of water as basic factor for the survival and the development of Humanity is analyzed, as well as the hazards that threaten its availability nowadays: pollution, overpopulation, climate change, etc.

    Likewise, we make a review of the last three large international forums in which this problem has been dealt with (World Summit on Sustainable Development, Johannesburg, 2002; International Conference on Freshwater, Bonn, December 2001; and 2nd World Water Forum, The Hague, March 2000) detailing the trajectory followed by the agreements already signed and, especially, describing the actual situation of its practical application at the present time, as well as specific cases in developed and developing countries.

- Within a strictly European environment *"Water policies in Europe after the Water Framework Directive"* approaches a landmark in the history of the EU: the regulation of all water policies of the member countries under an only one directive: the WFD.

    This chapter aims at clarifying the content and objectives of such WFD, the main points onto which it is articulated, and the problem that impels it. At the same time, it stresses its commitment, clearly favorable, towards policies of demand management.

- *"The risks and benefits of globalization and privatization of fresh water"* develops, in great detail, one of the processes in which services of water supply are currently involved in a large number of countries: privatization, a process that is favored by the economic globalization and the ease for multinationals to expand their areas of business throughout the five continents.

    This chapter develops a precise definition of terms and introduces cases and general guidelines towards the future, stating clearly that this is not a simple question. It cannot be generally concluded with a favorable or opposed position, as there are many factors to keep in mind in each situation.

- After the overview and proposals presented in the previous chapter, *"A structure for water administration in the 21st century. The case of Australia"* shows a paradigmatic case of our days: the current regulation in everything regarding water management in Australia.

    During the last years, a process to separate official competences related to water management has taken place in Australia, complemented by a clear and progressive regulation in all them, whose conclusions can serve as guidelines to other countries willing to start reforms in this field.

- In dealing with more practical aspects of sustainable water management, *"Effects of climate change on hydraulic resources"* presents the circumstances derived from climate change that such management will have to face in the future.

    From the merely hydrological effects of climate change to its implications for water management itself, it analyzes alternatives and scenarios for Europe as well as for North America.

- The United Kingdom is an outstanding example of private water management, though subjected to a rigorous public regulation. *"Demand management of water in a regulatory environment"* describes in detail water management in that country, starting with the explanation of a paradox: a country with so many water resources is, at the same time, one of the most advanced in implementing demand management policies, studying and analyzing its results.

    Therefore and in a progressive way, the role of the different management bodies, the tools available and, last but not least, the instruments to assess the scope of the expected results are here explained.

- The chapter *"Sustainable water management in agriculture"* discusses the expected future increase in water demand for agricultural irrigation: the challenge of obtaining larger yields with the same or even lower volumes of water.

    Likewise, the efficiency measures to be considered under different circumstances are presented, keeping in mind the possible environmental impact as a consequence of their implementation or not.

- Concluding in a certain way with the perspectives of water uses in a urban environment, the chapter "*Sustainable water management in cities*" starts from a fact slightly questioned: the high maintenance cost of the current systems in the developed countries, and the underlying incoherence when intending to export such a model to developing countries.

   Four different scenarios are analyzed here to evaluate the future sustainability of the current urban systems, showing the need for a political and social support (not only technical support) and defining the necessary institutional arrangements in time.
- "*The future of desalination as a water source*" deals with a topic that at the moment is being subjected to discussions in all forums on water policy: the availability of new water resources from desalination of brackish or sea water.

   This chapter questions the, so often too clear, cost of desalinated water according to a series of factors on which it depends. At the same time, it presents its future possibilities depending on the different current tendencies and the circumstances of the expected scenarios.
- Lastly "*Water pricing: a basic tool for a sustainable water policy?*" is fully devoted to the crucial problem of the costs recovery through the price charged to the customers and, more important, the different alternatives that it provides.

   However, this chapter does not represent a simple description of possibilities, but rather it suggests a financial structure to carry out the analysis of this pricing; concluding with two clarifying examples on what was previously presented.

5 CONCLUSION

The increasing pressure on water resources, the need to use them so that they benefit the largest number of citizens, and the need for a higher respect to the environment, that is to say, the desire of the present society to establish sustainable water policies has led to a new scenario that has nothing to do with the situation decades ago. It is certain that our ancestors designed the water policy their time demanded and very especially in Valencia, where its millennial *Tribunal de las Aguas* constitutes a magnificent example, internationally acknowledged as exemplary response to the problem outlined.

Undoubtedly, the increasing industrial and economic growth, i.e., the modern consumer society is putting natural resources in general and water in particular under some extraordinarily high pressures that demand a new policy, still to come in Spain. That new policy should be participated by the whole society, which is the only way for policy-makers to establish such significant changes. And thus, improvements would progressively be introduced, at the same time that the awareness and education of society increase. Otherwise, that change could unavoidably turn out to be excessively risky, abrupt and uncertain, which is obviously the most undesirable perspective.

ACKNOWLEDGEMENT

This article is framed within the R&D project "Efficient Urban Water Management: Development of a Comprehensive Methodology for its Diagnosis and Improvement" (REN2002-03196), which is currently in progress with the financial support of the Spanish Ministry of Science and Technology.

REFERENCES

ASCE Task Committee on Hydraulic Engineering Research Advocacy. 1996. Environmental hydraulics: new research directions for the 21st century. *Journal of Hydraulic Engineering*. ASCE. April 1996. pp. 180–183.

Beard, P. 1994. *Remarks of Daniel P. Beard, Commissioner of the U.S. Bureau of Reclamation before the International Commission on Large Dam*. Durban. South Africa. Noviembre. 1994.

Bonnnin, J. 1984. L'eau dans l'antiqueté. L'hydraulique avant notre ère. *Collection de la Direction des Etudes et Recherches d'Electricite de France*. Editions Eyrolles. Paris. 1984.

Brundtland, G.H. et al. 1987. *Our common future. Report of the World Comission on Environment and Development*. Oxford University Press, 1987.

Burgi, P.H. 1998. Change in Emphasis for Hydraulic Research at Bureau of Reclamation. *Journal of Hydraulic Engineering*, July 1998, pp. 658–661.
Cabrera, E., Cobacho, R., Lund, J. 2002. *Regional Water System Management. Water Conservation, Water Supply and System Integration*. Ed. Blakema. The Netherlands.
Corominas. 2000. El papel económico de las aguas subterráneas en Andalucía. *Proyecto Aguas Subterráneas*. Fundación Marcelino Botín. Madrid.
Cobacho, R. 2000. *La gestión de la demanda en el contexto de una nueva política integral del agua. Su aplicación al suministro urbano*. Tesis Doctoral. Universidad Politécnica de Valencia.
CICCP. 1975. Política hidráulica. Misión social de los riegos. *Publicación póstuma de los discursos de Joaquín Costa*. Colegio de Ingenieros de Caminos, Canales y Puertos. Madrid. 1975.
Davis, C.V., Sorensen, K.E. 1965. *Handbook of applied hydraulics (3rd edition)*. Mac Graw Hill. New York. 1965.
EC. 1996. *Task Force Environment Water*. European Commission DGXII-Science. Brussels. 1996.
EC. 2001. *Common Strategy on the implementation of the water framework directive*. European Commission. Brussels, 2 May 2001.
EP, European Parliament. 2000. Establishing a Framework for Community Action in the Field of Water Policy. *Official Journal of the European Communities*, 22 December 2000.
Garbrecht, G. 1987. Hydrologic and hydraulic concepts in antiquity. *Hydraulics and Hydraulic Research. An Historical Review*. pp 1–22, published by the IAHR and edited by Günter Garbrecht. Ed. Balkema. Rotterdam. The Netherlands.
GWI, Global Water Partnership. 2002. *Dialogue on Effective Water Governance*. GWP Secretariat, SE-105 25 Stockolm, Sweden.
Giner. 1997. *El tribunal de las Aguas de Valencia*. Giner Boira, V. Fundación Valencia III Milenio. Valencia. Spain.
Griegg, N.S. 1986. *Urban Water Infraestructure. Planning, Management and Operations*. John Wiley & Sons. New York. 1986.
IAHR, International Association of Hydrualic Research. 1987. *Hydraulics and Hydraulic Research. An Historical Review*. 1987. Editor: Garbrecht G. Published by Balkema. Rotterdam. The Netherlands.
Kennedy, J.F. 1987. Hydraulic trends towards the year 2000. *Hydraulics and Hydraulic Research. An Historical Review*. pp. 357–362, published by the IAHR and edited by Günter Garbrecht. Ed. Balkema. Rotterdam. The Netherlands.
Kobus, H. 1997. Desafíos en la hidráulica en el Siglo XXI. *Revista Ingeniería del Agua*. Volumen 4, n° 1, pp. 27–38.
Levi, E. 1995. *The science of Water. The Foundation of Modern Hydraulics*. American Society of Civil Engineers. 345 East 47th Street. New York, New York 10017–2398.
Lopardo, R. 1995. La formación del Ingeniero Hidráulico para el Siglo XXI. *Revista Ingeniería del Agua*. Volumen 2, n° 4, pp. 67–76.
Mataix, C. 1975. *Mturbomáquinas hidrúlicas*. Editorial ICAI. Madrid. 1975.
Mays, L. 1999. *Hydraulic Design Handbook*. Mc Graw Hill.
MMA, Ministerio del Medio Ambiente. 1998. *El Libro Blanco del Agua en España*.
Paz Maroto & Paz Casañé. 1969. *Abastecimiento y Depuración de Agua Potable*. ETSICCP. Madrid. 1969.
Raabe, J. 1987. Great names and the development of hydraulic machinery. *Hydraulics and Hydraulic Research. An Historical Review*. pp. 251–266, published by the IAHR and edited by Günter Garbrecht Ed. Balkema. Rotterdam. The Netherlands.
Rouse, H. 1963. *History of Hydraulics*. Ed. Dover. New York.
Rouse, H. 1987. Hydraulics' latest golden age. *Hydraulics and Hydraulic Research. An Historical Review*. pp. 307–314, published by the IAHR and edited by Günter Garbrecht. Ed. Balkema. Rotterdam. The Netherlands.
Schnitter, J. 1994. *A History of Dams. The useful pyramids*. Ed. Balkema. Rotterdam. 1994.
Steel, E.W. 1972. *Abastecimiento y Saneamiento urbano*. Editorial Gustavo Gili, Barcelona. 1972.
Thirriot. 1987a. Pouvoir politique et recherche hydraulique en France aux XVIIe et XVIIIe siècles. *Hydraulics and Hydraulic Research. An Historical Review*. pp. 251–266, published by the IAHR and edited by Günter Garbrecht. Ed. Balkema. Rotterdam. The Netherlands.
Thirriot. 1987b. L'Hydraulique au fil de l'eau et des ans à travers les XVIIe et XVIIIe siècles. *Hydraulics and Hydraulic Research. An Historical Review*. pp. 251–266, published by the IAHR and edited by Günter Garbrecht. Ed. Balkema. Rotterdam. The Netherlands.
Trittin, J. 2001. A first concrete framework for action has been set up. Water supply and development go hand in hand. *http.//www.bmu.de/English/topics/speeches/speech_trittin_011207.php*.
UN. 1992. United Nations Conference on Environmental and Development, Rio de Janeiro. Chapter 18: Protection of the quality and supply of freshwater resources. *Agenda 21*. June, 1992.

Political aspects

Towards the water policies for the 21st century: a review after the World Summit on Sustainable Development in Johannesburg

J.J. Bogardi & A. Szollosi-Nagi
Division of Water Sciences
UNESCO – International Hydrological Programme, France

ABSTRACT: Water is recognized as a key factor in sustainable development. Its direct links with human life society, poverty, natural disasters and ecosystem functioning render water a prime concern for the 21st century. Climate change, population growth, deterioration of water quality, scarcity and competition for this resource are the key challenges facing humanity. Therefore, comprehensive policies, based on common ethical principles and political will are needed to prepare humankind to face the "looming water crisis". Three major international conferences and their respective declarations are analyzed to comment the evolutionary process shaping the public perception, producing agreement on principles and concepts towards sound water resource policies, strategic plans and integrated water resources management. While convergence on objectives, targets and principles can be documented, the "water world" is still at the beginning to adopt water policies in conformity with sustainable development at adequate scales and to put them into practice. Securing the financial base, the governance framework, the necessary human and institutional capacities and knowledge base are the first steps to achieve the political objectives.

1 INTRODUCTION

The World Summit on Sustainable Development (WSSD) – popularly, Rio+10 – was held in Johannesburg between 26 August and 4 September 2002 (WSSD, 2002).

The popular media representation of this mega event reflects certain weariness with summits, declarations and public commitments. Many critics pre-conclude that targets set in Johannesburg will not be achieved and they are quick to refer to the un-kept promises of the United Nations Conference on Environment and Development, held in Rio de Janeiro in 1992, to prove the point. Yet, human society has not produced more efficient means to reach political consensus, to formulate and adopt policies, other than large, international conferences, summits and meetings. In the series of the serious attempts to secure the sustainable future of the planet and its inhabitants, Johannesburg is thus the last event. Therefore, it is worth analysing, irrespective of the short time passed since its resolutions were negotiated and a declaration published, what impacts we may expect as the consequence of this summit.

The present paper reflects on the influence of the WSSD on water policies for the new century.

Irrespective of the primordial role of water in sustainable development and in the sustainability of life and biodiversity, the WSSD cannot be classified as an international event focusing on water. Out of the 170 paragraphs of the Plan of Implementation for the World Summit on Sustainable Development (WSSD/PI), less than 30 refer explicitly to water or water-related activities such as management, water supply, irrigation and sanitation. Thus, the real impact of the Johannesburg Summit on emerging water policies for the 21st century can only be estimated as part of the sequence of water-related global events in recent years. In fact, in the international context it is more precise to speak about principles rather than policies as these later ones have to reflect national, climatic and cultural aspirations and constraints. Without going back too far in time, the water-relevant contribution of WSSD should certainly be seen as embedded between the 2nd World Water Forum (2nd WWF, celebrated in The Hague, March 2000), the International Conference on Freshwater (held in Bonn, December 2001), and the forthcoming 3rd World Water Forum (3rd WWF, Kyoto, March 2003). This last event will ultimately show how far the Declaration and Plan of Implementation of the WSSD influence water resources management worldwide.

The WSSD, while a full United Nations (UN)-led inter-governmental conference and summit, showed clear signs of the development, or better, search for new ways of communication between stakeholders and their respective representatives. The gradual opening of the intergovernmental dialogue to involve other stakeholders (like Non-Governmental Organisations, NGOs) can be measured by the presence of several thousand non-governmental participants and the multitude of side events, organised by various combinations of NGOs and Intergovernmental Organizations (IGOs) and government partners.

It is therefore not astonishing that next to the official declarations, Johannesburg can be associated with a large number of stakeholder statements (some much more explicit than the official ones), as well as by the emergence of the so-called type 2 (non-negotiated, yet endorsed) partnerships involving governments, IGOs, NGOs, the private sector and universities. The following paper embeds thus the WSSD into the dynamic reality of the "water world" at the turn of the millennium.

In this paper, the term "water world" will be used to describe the microcosmos of those concerned and dealing with water. In many respects it is not an exaggerated statement that we are all citizens of this "water world".

After a short review of the prevailing issues and the responses formulated by the Ministerial Declaration of the 2nd WWF, the outcome of the Bonn conference as well as the "water" component of the Water-Energy-Health-Agriculture-Biodiversity (WEHAB) position paper will be analysed, prior to an overview of the Johannesburg Declaration on Sustainable Development and the WSSD/PI. How far would the WSSD influence water resources policies? How do they develop? The paper will attempt to assess the state-of-the-art and outline the development tendencies.

Water policies are very much shaped by international programmes and initiatives. Therefore, a sub-component of the UN-wide World Water Assessment Programme (WWAP), the UNESCO-Green Cross International (GCI) "From Potential Conflict to Co-operation Potential: Water for Peace" (PC→CP:WfP) project will be presented to illustrate how to prepare the ground, secure the knowledge-base to translate conference recommendations like the Johannesburg statements and principles into practical water policies and actions.

2 THE "WATER WORLD" AT THE TURN OF THE MILLENNIUM: AN OVERVIEW OF THE 2ND WORLD WATER FORUM

How can we characterise the "water world" at the turn of the millennium? The most comprehensive picture emerges from the Long Term Vision for Water, Life and the Environment (World Water Vision, WWV for short; Cosgrove & Rijsberman, 2000), which was presented to the 2nd WWF in March 2000 in The Hague. This event, with more than 5500 registered participants and 80 parallel sessions, was a real landmark. The Forum was the culmination of a, so far, unprecedented, massive public awareness-raising event, with its associated Ministerial Conference and its Statement (2nd WWF, 2000).

Since the 2nd WWF, water is high on the international political agenda. It is driven by the urgency to face and counteract the perceived water crisis, not to tolerate any longer that close to 3 billion people have no adequate sanitation, that more than 1 billion people have no safe water supply, that well over 3 million people, mainly children, die annually in water-related diseases. This sentiment was underlined in the Forum by statements like: "we know enough, we have enough data, stop talking, act now".

The prevailing intellectual environment of the "water world" entails deep contradictions. On the one hand, innovative approaches are advocated on the basis that the solutions of the world's water problems are not quick technical fixes, while there are loud voices claiming that there is sufficient knowledge and data available to rely on. While water is on the political agenda and humanity's water awareness is on the rise, funding for water-related research and education is on the decline. It is a sad fact for example that hydrological observation networks in crucial parts of the world are generally in a worse shape now than 20 or 30 years ago. Likewise, many signs and statements document that the environment, water quality and security are worse off than they were at the time of the Rio Conference. This is our biased "water world".

The intention of the Ministerial Conference in The Hague in March 2000 was to explore new avenues, to seek the dialogue of political decision-makers with the stakeholders of the "water world". With reference to the above contradictions, it is not surprising that water professionals were not considered originally as stakeholders. Finally, a single representative from science and the water professionals had the possibility to address, in five minutes, the assembly of ministers, next to stakeholder representatives of gender, youth, grassroots NGOs, trade unions and the business community. The dialogue with the ministers remained limited to these "new stakeholder" groups.

The motto of the WWV consultation and the subsequent Forum was: "Water is everybody's business". Yes! But first of all it is, and should be, the business of the professional community. How can we otherwise expect that these international forums would really influence water policy, strategy, planning and operation of water resources systems?

It has to be acknowledged that the WWV and the 2nd WWF revealed a credibility gap. The professional community is challenged to close it. Thus, the analysis of the Vision and the Forum is not that of a critic. It is rather a reflection that the professional and scientific community of the "water world" has to develop strategies on how to recapture the leadership in the debate. Water professionals always carry out a societal mandate. Because many previous water development projects are now regarded as "unsustainable", the professionals who built and managed them are considered responsible. But no one questions the wisdom of former social aspirations. This suspicion towards the professional community still exists. It can be seen as a serious hindrance to translate political recommendations into implementable water policies. Leaving out those from formulating policy principles, whom you expect to implement these, contradicts, next to common sense to the principle of participative decision-making.

What are the prevailing water issues at the turn of the millennium? The Ministerial Declaration of The Hague summarises the main water-related challenges. It is a fair assumption that this negotiated international document, accepted by 120 ministers and delegations from all over the world, reflects well the "public perception" of water problems and thus defines the "political environment" within which professionals, individuals and organisations alike, have to act. The key statement is "Water Security" for the 21st century – an appealing slogan, since no one is against security. This term is interpreted in its multiple dimensions:

- Meeting basic needs (water supply, sanitation, health).
- Securing the food supply (equitable allocation of water for food).
- Protecting the ecosystems (integrity and sustainability).
- Sharing water resource (between different users and between different states).
- Managing risks (floods, droughts, pollution, etc.).
- Valuing water (economic, social, environmental, cultural, including "careful" (socially cushioned) pricing.
- Governing water wisely (good governance and stakeholders involvement).

We may subdivide these seven attributes of "Water Security" into two categories: that of the first three, describing and "politically" prioritising the legitimate water demands, and the remaining four highlighting the "hows" of water resources management. It must not be overlooked that, of these four categories, the three that are underlined: Sharing, Managing, Valuing, Governing, are almost entirely beyond the classical domain of water resources engineering.

The Ministerial Declaration is profuse in defining priority issues and steps to meet the above-outlined challenges.

The Integrated Water Resources Management (IWRM) was adopted as the framework for action to be taken. IWRM is defined as the approach which takes into account social, economic and environmental factors and integrating surface water, groundwater and the ecosystems through which they flow. Quantity and quality aspects are to be considered.

"Water Security" and the sustainable management of water resources are treated as synonymous. Collaborative partnerships across whole society and coherent institutional policies to counteract fragmentation of the sector are called for.

The Ministerial Declaration reiterates the decision of the Commission of Sustainable Development of the UN to call on the UN System, its specialised agencies and programmes, to develop appropriate indicators and periodically reassess the state of freshwater resources as well as to document this in subsequent World Water Development Reports.

The Ministerial Declaration realises the need for a new strong "water culture" to be developed through the cooperation of all stakeholders.

"Best practices" are to be identified through enhanced research and knowledge generation, knowledge dissemination and sharing among individuals, institutions and societies. This includes, not only technological transfer and capacity-building in developing countries, but also strengthening humanity's capability to cope with water-related disasters.

Beyond any doubt, the 2nd WWF set the stage for entering the 21st century of the "water world". The seven challenges identified in The Hague became the basis to structure the WWAP, the ongoing UN-wide effort launched by the Director-General of UNESCO, Mr Koïchiro Matsuura, at the 2nd WWF. This programme aims to create the database, but also to develop the methodologies and concepts of policy-relevant assessment of the resource, its use and protection.

The 2nd WWF, through its success of bringing water to the political conscience of the world, proved the viability of "composite events", having parallel and interactive meetings of various stakeholder constituencies. Ever since, all major water-related events, including the WSSD, have been following this model.

3 TOWARDS THE WORLD SUMMIT ON SUSTAINABLE DEVELOPMENT: THE INTERNATIONAL CONFERENCE ON FRESHWATER

The "water world" and water policies after Johannesburg cannot be characterised appropriately without analysing the outcomes of major events shaping the "water world" and influencing the spirit and outcomes of the summit in Johannesburg. Without doubt, international conferences with a clear focus on water show the prevailing trends and describe the state and mentality of the "water world" more clearly than a global event, such as the WSSD. Therefore, it is warranted to highlight the outcomes of the most recent water conference, that of the International Conference on Freshwater, held in Bonn in December 2001, preceding by a mere 8 months the Johannesburg Summit. In this paper, only two outcomes of the Bonn Conference, the "Recommendations for Action" and the "Bonn Keys" will be analysed. Bonn's motto: "Water – key to Sustainable Development" clearly describes its aim: to prepare "water-wise" the WSSD to provide a "key" to open the "lock" in Johannesburg.

3.1 *Recommendations for action*

The "Recommendations for Action", while politically "lower traded" than the Ministerial Declaration, is still a negotiated text. On 13 pages covering 27 points, it diagnoses the "water world" and prescribes the "medicine". Due to the nature of a negotiated text, the "Recommendations" contains several general statements. Yet careful reading enables the subtle order of priorities to be found. The following citation is an excellent example to prove it:

> "Water is a key to sustainable development, crucial to its social, economic and environmental dimensions. Water is life, essential for human health. Water is an economic and a social good and should be allocated first to satisfy basic human needs. Many people regard access to drinking

water and sanitation to be a human right. There is no substitute for water: without it, humans and other living organisms die, farmers cannot grow food, businesses cannot operate. Providing water security is a key dimension of poverty reduction".

While these are not in contradiction to the Ministerial Declaration of The Hague, the Bonn Conference achieved an additional step, to focus on priority actions grouped under three headings:

– Governance.
– Mobilising financial resources.
– Capacity-building and sharing knowledge.

Furthermore, it identified specific roles for the most crucial stakeholder groups represented in the conference.

The above three main groups as priority actions reveal implicitly that we are very much at the beginning of the long process to secure a sustainable "water world". The framework of this paper does not allow a line-by-line analysis of the text. In fact, this should be reserved for a later scientific review of the evolving "water world" at the change of the millennium. Rather, the attempt is made to highlight the document on the basis of the "macro-level" of 21 recommended actions in the three groups and the list of the 6 stakeholder groups present.

Actions in the Field of Governance:
1. Secure equitable access to water for all people.
2. Ensure that water infrastructure and services deliver to poor people.
3. Promote gender equity.
4. Appropriately allocate water among competing demands.
5. Share benefits.
6. Promote participatory sharing of benefits from large projects.
7. Improve water management.
8. Protect water quality and ecosystems.
9. Manage risks to cope with variability and climate change.
10. Encourage more efficient service provision.
11. Manage water at the lowest appropriate level.
12. Combat corruption effectively.

Actions in the Field of Mobilizing Financial Resources:
13. Ensure significant increase in all types of funding.
14. Strengthen public funding capabilities.
15. Improve economic efficiency to sustain operations and investment.
16. Make water attractive for private investment.
17. Increase development assistance to water.

Actions in the Field of Capacity-Building and Sharing Knowledge:
18. Focus education and training on water wisdom.
19. Focus research and information management on problem solving.
20. Make water institutions more effective.
21. Share knowledge and innovative technologies.

Roles:
22. Governments.
23. Local Communities.
24. Workers and Trade Unions.
25. Non-Governmental Organisations.
26. The Private Sector.
27. The International Community.

The simple list of 21 actions (to be pursued by the 6 stakeholder groups), reveals a lot of important features.

- In the group of governance-relevant actions:
  - Some actions, while obviously important as far as water management is concerned, are much for far-reaching than the "water world" alone. Actions 3 and 12, but also 5 and 6 belong to this sub-group.
  - Some actions target the translation of ethical principles into management practices, like Actions 1 and 2, while Action 11 transposes the principle of subsidiarity into water resources management.
  - The remaining actions: 4, 7, 8, 9 and 10 seem to be the ones focusing on the "how to manage what" issues.
- In the group of finance-relevant actions, the proposed actions indicate the need for activities on all fronts. This reveals a certain lack of strategy and in-depth analysis of real opportunities on how to raise the necessary financial means[1].
- In the group of actions relating to capacity-building, the four proposed actions are well-focused, covering a wide range from ethical imperatives, institutional changes through research reorientation to the human dimension. Being the least politicised of the three groups, the actions proposed reflect a professional concern and readiness to act. It has to be emphasised that the prominent place given to capacity-building and knowledge sharing in the Recommendations for Action underlines the acknowledged interlinkage. Improved financial provisions alone would not solve the problems without creating the broad knowledge base and institutional frameworks. Without knowledge and wisdom, the best governance model would fail.

In the light of this well-balanced list of actions, the more astonishing is to review the six classes of stakeholders and their assigned roles. It is fair to say that representation of the community of water professionals and that of the respective scientific communities would have been warranted, in particular with regard to actions referring to research, knowledge base and education. It is worth noting in this regard that the International Water Associations Liaison Committee is actively rectifying this lack of professional presence as a stakeholder group at the 3rd WWF.

3.2 The Bonn keys

Next to the Recommendations for Action the five "Bonn Keys" represent a very well-condensed message from the International Conference on Freshwater (Bonn, December 2001):

- To meet the water security needs of the poor.
- Decentralization: the local level where national policy meets community needs.
- The key to better water outreach is new partnerships.
- The key to long-term harmony with nature and neighbor is cooperative arrangements at the water basin level, including across water that touches many shores.
- The essential key is stronger, better performing governance arrangements.

From an ethical imperative (Key 1) to the embedding of water resources management into the overall responsibility of governments and society as a whole to create the effective legal and regulatory framework within which better water governance can be developed (Key 5), the five keys cover a wide range of issues. These five keys – as presented by the facilitator of the conference, Margaret Catley-Carlson, Chairperson of the Global Water Partnership (GWP), are an excellent example of a combined conference summary, political manifesto and strategic programme in a nutshell.

Next to its strong ethical content, Key 1 refers to the looming vulnerability of the social fabric and the risk of irreparable environmental damages should the plea of the poor remain unanswered. Key 1 is also the link between the United Nations Millennium Development Goals (MDGs) on access to safe drinking water and the commitment of WSSD to extend it by also halving the number of those without adequate sanitation.

---

[1] Important to note in this respect that, since then, the World Water Council and the GWP convened jointly a high level Finance Panel addressing these issues and preparing a focused strategy for the 3rd WWF in Kyoto.

Key 2 could be called the "battle cry" for a practical application of the principle of subsidiarity. It implies also that no policy (at national or global scale) can be effective unless addressing the "scale issue"; how to translate ideas and ideals to actions at "human scale". Key 2 is thus not only valid for water resources, but also for the management of other resources and for the administration of human society in general.

Key 3 is again a link towards Johannesburg. The "spirit of partnership"; the endorsement of the so-called "Type 2" partnerships at the WSSD has one of its roots in this statement of the Bonn Conference.

Key 4 links the ethical principle of good neighbourly behaviour, the practical necessity of arrangements over shared water resources with IWRM as a practical framework. What is spelled out here for river basins must hold also for shared aquifers. The ideas expressed in Key 4 are germinating. In November 2002, in Thonon-les-Bains (France) international river basin organisations will meet in an assembly to form their own network within the International Network of Basin Organisations (INBO).

Key 5 is a strong political appeal to the governments. It is rightly called the "essential" key. It implicitly admits that the good intentions, principles, approaches and methods specified in Keys 1–4 can only be effective if an enabling "political climate" prevails. In this respect, the WSSD with its broad scope on sustainable development is more relevant than the water-focussed International Conference on Freshwater in Bonn. How far the WSSD improved this "climate" can only be judged once the "political seeds" sown in Johannesburg grow into "plants" of legal framework, regulations, institutions and human capacities enabling the realisation of the substance of Keys 1–4. In fact, this key claims a global cultural adaptation with regard to water.

The Bonn Conference was convened in December 2001 with the claim of its organisers that it would be a link between the 2nd WWF and the WSSD. Legally, it was not recognised as part of the official inter-governmental preparatory process towards Johannesburg. Back in December 2001, it was to be seen whether, and what, would be the impact of "Bonn on Johannesburg". After less than one year the influence of the Bonn Conference on the "water world" and on the outcomes of the WSSD can be well-documented. It is an analytical recognition of a job "well done".

A very subtle prioritisation, which as well as being found in the WSSD declarations and objectives can also be traced back to Bonn. The last lines of the "Bonn Keys" read as follows:

"Water is essential to our health, our spiritual needs, our comfort, our livelihoods and our ecosystems. Yet, everywhere, water quality is declining and the water stress on humanity and our ecosystems increases. More and more people live in very fragile environments. The reality of floods and droughts touches increasing numbers and many live with scarcity. We are convinced that we can act, and we must. We have the keys".

This human-centred view can also be found in the WSSD declarations. It is a clear statement that sustainable human existence requires a sustainable environment without putting the cart before the horse.

4 KEY CONCERN AREAS: WEHAB AND THE PREPARATORY PHASE BEFORE JOHANNESBURG

The preparatory phase of the Johannesburg summit reflected several new tendencies in the international, political and, in particular in the resource and environment related arena. Obvious government reluctance on one side to make binding commitments on deadlines, quantified objectives and financial allocations and the growing importance of pressure groups, NGOs and direct stakeholder involvement on the other side, harangued a new type of summit. Already the evolution of the World Water Forums show that the hitherto separated events of scientific conferences, public rallies and ministerial conferences tend to merge into a new type of articulation of concerns, objectives, commitments and strategies. The WSSD certainly cannot be seen and separated from its side events[2].

---

[2] As far as water is concerned the Water Dome in Johannesburg can be clearly identified as a melting pot of all stakeholders and their ideas.

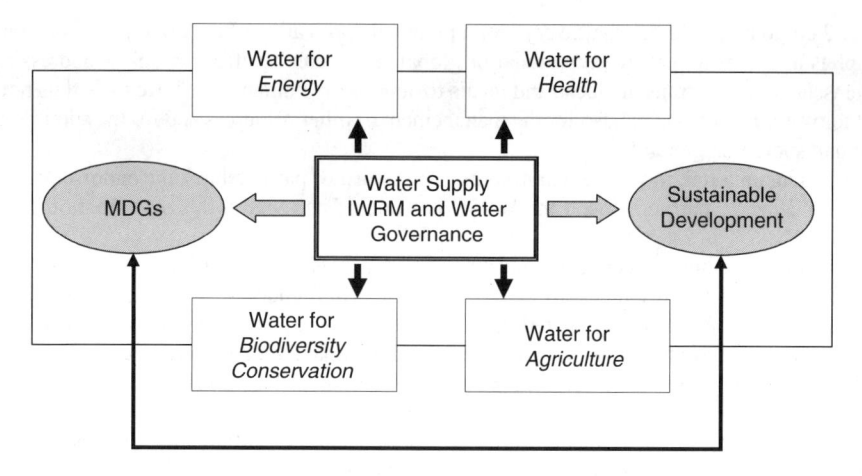

Figure 1. Water – the Link and Key in Sustainable Development.

The "unorthodox" preparatory phase for a new style summit is the background of a document summarising the key concern areas: Water, Energy, Health, Agriculture and Biodiversity (WEHAB). An inter-agency working group of the UN system was entrusted to develop what may be called a background or position paper. The five key areas of WEHAB again reflect a human-centred focus adopted and maintained during the WSSD.

In the subsequent drafts of the water chapter of the UN's WEHAB document the role of water in relation to all other four key concern areas are described and put it into the strategic axis of the MDGs and sustainable development. The human-centred objectives relating to water supply and sanitation, the most urgent need to redefine and to reform water governance and IWRM as the endorsed approach to achieve it, are shown in Figure 1.

It is a fair statement that the chapter on water from the WEHAB document served as the most important basis for reaching the water-related objectives in the WSSD. Many of the priority actions identified during the Bonn Conference and the Bonn Keys are reflected in this document, thus clearly showing the "seepage path" and how the thoughts, concerns and dedication of the subsequent events and preparatory documents channelled thoughts to the WSSD.

5 THE WORLD SUMMIT ON SUSTAINABLE DEVELOPMENT: ANALYSIS OF RESULTS FROM WATER RESOURCES POINT OF VIEW

5.1 *Key outcomes, commitments, targets and initiatives*

UN Department of Economic and Social Affairs summarised the "results" of WSSD in a brief publication on the Summit website immediately after the Summit was over.

As far as "water" is concerned, as key outcome, the explicit acknowledgement is given that sanitation issues become critical and crucial elements of the negotiations and are reflected upon in the final declarations, in more depth than ever before.

Water can be seen as a "winner" of the WSSD. As far as water and sanitation are concerned, the Summit confirmed the water supply-related goal set by the MDGs, i.e., to halve by the year 2015 the proportion of people without access to safe drinking water. WSSD – in the spirit of the above cited outcome – added the quantifiable objective to halve by the year 2015 the proportion of people who do not have access to basic sanitation.

The achievement of these objectives implies that up to the target year 2015, annually, about 50 million people should be added to those having access to safe drinking water and approximately 100 million to be added each year to those having adequate sanitation. Even if the Summit does not come up with the inherent financial commitments, the political aim is well quantified, especially in comparison with the overwhelmingly verbal expression of aspirations for other aspects of sustainability.

Besides this clearly human-centred water and sanitation commitment, water also surfaces as part of the subheading on "Management of the Natural Resource Base". The commitment "to develop IWRM and water efficiency plans by 2005" is, however, very vague. No specification is given as to whether these plans should be national or basin-oriented. Compared with the transitional provisions and implementation phases foreseen in the recent European Union (EU) Water Framework Directive (WFD), it is highly unrealistic that the 2005 deadline can be achieved at global level. The more so that the IWRM still cannot be defined as a universally accepted, standardised methodology to be implemented routinely under whatever institutional framework. The lack of reliable databases is an additional factor making the 2005 target date illusory. While common sense can imagine and "invent" the content of a "Water Efficiency Plan", this terminology is entirely new. Again, no spatial (or temporal) reference is given to guide the scope and extent of these plans.

Among the key initiatives announced during the Summit, mention should be made of the "Water for Life" initiative of the EU with its focus areas in Africa and in Central Asia. This initiative clearly reflects the emerging partnership "spirit" which is likely to be the strongest legacy of Johannesburg. Annex 1 shows the Johannesburg Declaration "Battle for the Planet" of GCI to illustrate through this NGO statement the broad front of dedication to address the sustainability issues.

5.2 The Johannesburg declaration on sustainable development

This is the political statement of the Summit. Among the 37 paragraphs only one, paragraph 18, refers to the afore-mentioned water and sanitation focus, besides a short reference to water pollution in paragraph 13. This relatively modest place in the political statement is "compensated" however by the fact that clean water heads the list of basic human requirements.

Annex 2 compares "waterwise" the final Declaration with its draft published one day before the end of the Summit in a Johannesburg daily newspaper "Star" (Star, 2002). A certain "downgrading" of water during the last minute negotiations can be sensed. This comparison, besides its historical value showing the evolution of a segment of the text through the last night of negotiations, also demonstrates that behind the negotiated text there are frequently well articulated, concrete aspirations capturing the spirit and momentum of a conference. This "untold" message is also a very important, invisible output of a forum, congress or summit.

5.3 The WSSD Plan of Implementation (WSSD/PI)

While looking for the term "water" in a document to assess its importance with regard to other areas of concern may seem a mechanical approach, it is still a very legitimate and astonishingly accurate measure, if applied in the case of negotiated, political texts.

Out of the 170 paragraphs of the (unedited) 4th September version of WSSD/PI, "water" explicitly appears at least once in 24 paragraphs: 3 paragraphs in Chapter II – Poverty Eradication; 11 paragraphs in Chapter IV – Protecting and Managing the Natural Resource Base of Economic and Social Development; 1 paragraph in Chapter VI – Health and Sustainable Development; 2 paragraphs in Chapter VII – Sustainable Development of Small Island Developing States; 2 paragraphs in Chapter VIII – Sustainable Development for Africa; and 5 paragraphs in Chapter VIII bis – Other Regional Initiatives (2 for Latin America and the Caribbean, 2 for the West Asia Region and 1 for Europe). This confirms the importance of water in many facets of sustainability and in particular, its crucial role in developing regions.

The following Table 1 provides a quick reference to locate the water-relevant statements, goals and recommendations in the text of the PI.

Without doubt the most comprehensive focus on freshwater-related issues are in paragraphs 23–28. Therefore, the entire text of these paragraphs is reproduced as Annex 3.

Paragraph 23 calls, in particular, for the integrated management of land, water and living resources.

Table 1. How to trace "water" in the WSSD/PI (list of paragraphs where water is mentioned).

II. Poverty Eradication	6a,c,l,m; 7g; 9e;
IV. Protecting and managing the natural resource base of economic and social development	23; 24a-e; 25a-f; 26; 27; 28; 31c; 34; 35d; 36; 38b,d,I,k; 39d; 40b; 43;
VI. Health and sustainable development	47l;
VII. Sustainable development of small island developing states	52d; 54c;
Sustainable development for Africa	57; 60a,d;
VIII.bis Other regional initiatives	
Latin America and the Caribbean	67; 70;
Sustainable development in the West Asia region	71; 72;
Sustainable development in the Economic Commission for Europe Region	74;

Paragraph 24 calls for an action programme (financial and technical) to achieve the quantitative objectives in providing safe drinking water and basic sanitation. While this paragraph appeals to "mobilise international and domestic financial resources at all levels" it falls short of recalling the estimated increased investment need (World Water Vision 2000) of approximately $100 billion for each year or the respective estimates in the "Recommendations for Action of the Bonn Conference". This paragraph is more elaborate on management options and practices:

- Public information and participation.
- Focus on poor and gender sensitivity.
- Capacity-building and popularisation programmes.
- Setting water as priority in national and international development agendas.
- Intensifying water pollution prevention.
- Mitigating groundwater contamination.
- Exploring innovative technologies.

No doubt, this list is quite exhaustive but does not reveal any new aspect which would not have been mentioned either in the 2nd WWF or in the International Conference on Freshwater (Bonn, 2001).

Paragraph 25 elaborates further the necessary/possible water management options:

- Development and implementation of national/regional strategies, plans and programmes in an integrated context.
- Improved efficiency and loss reduction within the water infrastructure and recycling.
- Full employment of policy instruments: regulation monitoring, voluntary actions, market mechanisms, land use management, cost recovery and integrated basin wide approach.
- Prioritisation of water uses to cover basic human needs and water allocation among competing uses.
- Programmes to mitigate the effects of extreme water-related events.
- Technology transfer and capacity-building including non-conventional methods, sea water desalinization.
- Public-private partnerships, respect of local conditions, accountability, etc.

This reads more like an (unedited) table of contents of a textbook on water resources management rather than a focused political statement. It mentions all the components but gives little guidance or reports commitment on implementation.

Paragraph 26 calls for cooperation on behalf of developing and transitional countries to assess their water resources, create monitoring networks, databases and national indicators. Again, this paragraph addresses an important problem. However, monitoring and data collection is on the decline even in the so-called developed countries. Calling for national indicators in resource assessment seems to contradict the basin-wide integrated approaches advocated in most parts of the WSSD/PI.

Paragraph 27 is a welcome exception in a political text. It calls for science to improve the understanding of how the hydrological cycle functions. It calls for knowledge sharing, for capacity-building and for the application of modern earth observation technologies to create the science-based assessment of resource availability and variability.

Paragraph 28 calls for close co-operation of international and intergovernmental bodies in addressing water-related issues. It refers also to the International Year of Freshwater 2003.

In particular, paragraphs 26–28 are important as they provide the political endorsement and mandate for the WWAP of the UN System and for international science programmes, like the International Hydrological Programme of UNESCO.

The other references to water in Chapter IV. Protecting and managing the natural resource base of economic and social development addresses the role of water to carry pollutants from land-based sources to the seas, it refers to water as a source of risk (floods, droughts) and its vulnerability being affected by climate change.

Paragraph 38 refers to water in agriculture. Given that 70%–80% of the water consumed worldwide is related to irrigation and other agricultural uses, the WSSD/PI is astonishingly short and general on this issue. The professional wisdom that water shortage can and must be dealt with first of all in an agricultural context does not seem to be adequately reflected and worded in the document. Calling for market-based incentives may be a sensible approach but it is also a very sensitive issue/subject. By avoiding to address this key issue adequately, the political document failed in this respect to be policy-relevant.

The role of water to combat desertification and secure mountain ecosystems, including their relationship with erosion, deforestation, land degradation and biodiversity, is, however, well documented.

The following chapters focus on the reciprocity between access to safe water and sanitation and health, as well as highlighting certain water issues in a regional context.

6 HOW DOES THE WSSD INFLUENCE THE "WATER WORLD"?

As already stated, a consolidated assessment of the impact of WSSD on the "Water World" and consequently, on water (resources management) policies will be first possible after the 3rd WWF. Nevertheless, tendencies can be summarised in the following bullet points:

- WSSD did not produce any substantial paradigm shift as far as water is concerned. It confirmed IWRM as the endorsed approach reflecting holistic concerns.
- WSSD, while remaining "holistic" at the level of describing components of sustainability, placed availability of safe water and sanitation, equitable use and integrated management of this resource, as the key factors for combating poverty and securing sustainable development. This role is convincingly visualised in the WEHAB document. Thus, calling "water" the winner of Johannesburg is justified.
- WSSD reconfirmed the MDGs, hence further strengthening the political commitment of providing safe drinking water and adequate sanitation for hundreds of millions of people deprived of these basic human needs.
- By adding sanitation to the MDGs on safe drinking water, a crucial step was made towards seeking comprehensive, remedial action for improving the quality of human life.
- With these human-centred core objectives, WSSD clearly set priorities. It is up to the UN system and governments, as well as a task of NGOs and the private sector, to act along these ethical lines.
- The WSSD/PI calls for IWRM and water efficiency plans by 2005 and for the development of integrated land management and water use plans. However, it does not elaborate in detail on the key linkage between (frequently unsustainable) water use in irrigation and the inherent conflicts it creates, especially between food and environmental security.
- Water resources assessment and cooperative efforts among sovereign states are mentioned, yet the WSSD fell short of explicitly declaring hydrological and meteorological data-sharing as an imperative.

7 WATER POLICIES FOR THE 21ST CENTURY: THE LEGACY OF WSSD

The preceding sections clearly document the evolutionary process of the recent events shaping principles and concepts towards water resources policies of the "water world". The WSSD gave, in fact, a high level endorsement to these ideas. It even added, with the explicit call for "water efficiency plans", a new challenge and task to the professional community. The following Table 2 summarises and comments on the water policy and the planning of relevant WSSD/PI statements. This can serve also as a checklist to identify the areas of need for further research, development and knowledge sharing.

As this table reveals, the summit of Johannesburg can serve as a reference point in shaping water policy and its inherent activities. Its strongest legacy is expected to be its endorsement of concepts and appeals elaborated by preceding water-related conferences and ministerial declarations. Simultaneously, the WSSD passes the mandate to subsequent events, first and foremost to the 3rd WWF to follow up its water legacy. The shift from commitment to action refers also to water policy and planning to move from concepts to implementation. Water did not only win in Johannesburg. We also obtained a water mandate.

Table 2. Statements with relevance to water policy and planning in the WSSD/PI.

Statement of WSSD/PI	Comment	Consequence
1. Integrated Water Resources Management (IWRM)	Endorsement of previous statements by other conferences and declarations	Further development and practical use of the concept needed
2. Integrated sanitation into IWRM	Should be an obvious part of IWRM	Monitor in practical implementation
3. IWRM to optimize upstream/downstream benefits	Political mandate of IWRM to be a tool of water conflicts resolutions	IWRM to be implemented also in international context
4. Effective Integrated management over all uses	Should be an obvious part of IWRM, mandate to resolve sectoral water conflicts	Monitor practical implementation
5. Integrated land management and water use plans (in agricultural context)	Should be a focused subject of IWRM	Should result in land use changes and better use of water irrigation
6. Land use and linked water resources management plans to offset floods & droughts	Should be a focused subject of IWRM	Develop as part of disaster preparedness + links to IWRM
7. Combat desertification: land-water-forest management plans	Explicit linkage of IWRM with creeping environmental degradation	Implement plans and monitor impacts improvements
8. National strategies for disaster management	Identification of a serious need. Lacking in most of the countries	Link plans with IWRM at all scales
9. Watershed planning	Should be an obvious part of IWRM	Assess efficiency both locally and basinwide
10. Freshwater Programmes for small island developing states	Recognizing particular vulnerability. Not spelled out who should do it	Additional focused funding and capacity building needed
11. Transboundary Water Treaties	Explicit acknowledgement of the need of legal framework for water sharing	Prepare concepts, review best practices
12. Promote better land use and water management practices	Awareness raising, outreach mandated	Develop dissemination strategies and material
13. Water Efficiency Plans	New term, not defined contentwise	Develop concept and guides for this plan

8 FROM POTENTIAL CONFLICT TO CO-OPERATION POTENTIAL: WATER FOR PEACE

Table 2 contains two statements (3 and 11) related to the principle of equitable sharing of water-generated benefits along a river. One in context of IWRM in general and one in particular by referring to the possible international implications of this task.

In recent years, there has been an emergence of the fear of "water wars" in the 21st century, after several related statements in the media. UNESCO and GCI took the initiative to launch a comprehensive programme to investigate how far these fears are justified and how could the "water world" be prepared to respond to challenges inherent in the management of shared water resources in order to avoid these gloomy prophecies becoming true. This ongoing programme is to illustrate how science, but also public awareness raising programmes, should be conceived to provide policy relevant advice, background information, change perception and develop methodologies.

8.1 Background

During the 20th century, the world population has tripled while world demand for water has increased seven-fold. The signs of a looming water crisis are evident. Since water is essential to every aspect of life, this crisis affects everything – from health to human rights, the environment to the economy, poverty to politics, culture and civilization to cooperation and conflict. Just as water defies political boundaries, the crisis is also well beyond the scope of any individual country or sector and cannot be dealt with in isolation. The need for integrated, cooperative solutions is particularly urgent in the 261 river basins which are shared by two or more sovereign states. These basins cover nearly half of the world territory and provide home for almost 50% of the world population.

The Hague Ministerial Declaration, signed at the 2nd WWF in March 2000, identified seven key challenges for achieving water security. One of these challenges "Water sharing" gives the proper context for the project "From Potential Conflict to Co-operation Potential: Water for Peace" (PC→CP:WfP).

UNESCO and GCI are contributing to this international initiative by jointly examining the potential for shared water resources and how to become a catalyst for regional peace and development through dialogue, cooperation and participative management of river basins. An increasing number of states are experiencing permanent water stress, yet in most cases, mechanisms and institutions to manage conflicts over water resources are either absent or inadequate. Competition over this precious resource could increasingly become a source of tension – and even conflict – between states and water use sectors. But history has often shown that the vital nature of freshwater can also be a powerful incentive for cooperation; it can impel stakeholders to reconcile their diverging views, rather than allow opposing interests to escalate into harmful confrontations, which could jeopardize water supplies, for all parties involved in a dispute.

A team of scientific researchers from the Department of Geosciences at Oregon State University, led by Prof. Aaron T. Wolf (Wolf, 2002) has been conducting an interesting work confirming this aspect of the resource as a unifier. They have compiled a systematic database for water conflict/cooperation reported during the last 50 years in the international media. Of the 1,831 events – 507 conflictive, 1,228 cooperative, and 96 neutral or non-significant have been identified. To define the intensity of the events a scale going from -7, the most conflictive (war), through 0 (neutral events), and up to $+7$, the most cooperative (voluntary merging of countries) were used. The events, measured on this scale revealed the following:

– No events were observed on the extremes.
– Most interactions are cooperative.
– Most interactions are mild.
– Water acts as an irritant.
– Water acts as unifier.

- Overall, the major water-related issues likely to lead towards conflicts are quantity and infrastructure related.
- Nations cooperate over a wide variety of issues.

UNESCO has launched the project PC→CP to assess the available material for the prevention and the resolution of water conflicts, as well as to develop decision-making and conflict prevention tools for the future. The "Water for Peace" project initiated by GCI – developed with the input of civil society in several international basins – aims to enhance the awareness and participation of local authorities and the public in water conflict resolution and integrated management, through facilitating more effective dialogue between all stakeholders.

The joint PC→CP:WfP project will address the obstacles, identify the incentives and promote the means to achieving the integrated, equitable and sustainable management needed to make international watercourses natural thoroughfares for stability and sustainable development across the world. The following summary refers to the PC→CP component.

8.2 Scope of PC→CP

PC→CP addresses specifically the challenge of shared water resources primarily from the point of view of Governments and intergovernmental organization. All PC→CP efforts were conceived with the idea that, although shared water resources can be a source of conflict, their joint management should be strengthened and facilitated as a means of cooperation between various water users. Thus PC→CP aims to demonstrate that a situation with undeniable potential for conflict can be transformed into a situation where cooperation potential can emerge. PC→CP's thematic focus is on this very transition – from PC to CP.

The goal of PC→CP, in accordance with the mandate of WWAP, is to render services to the Member States and to foster cooperation between Nations. It is also guided by UNESCO's paramount mandate: to nurture the idea of peace in human minds. In its first phase (2001–2003), PC→CP gives priority to water conflicts, which are international in nature and may cause tension or even open conflict between sovereign states.

8.3 Target groups

PC→CP's role is to help water resources management authorities to tip the balance in favor of cooperation potential away from potential conflict.

The priority target groups of PC→CP are therefore institutions and individuals that manage shared water resources.

These include Governments, to which the WWAP is essentially addressed, then donor and funding agencies, which need information on actual or potential water conflicts.

8.4 Objectives of PC→CP

The purpose of PC→CP is to promote water security through cooperative management of shared water resources.

PC→CP aims to foster cooperation between stakeholders in the management of shared water resources and mitigate the risk that potential conflicts turn into real ones. It will help the parties involved in potential water conflicts to negotiate the way towards cooperation.

PC→CP's overall purpose can be reached through the achievement of these following five operational objectives:

- Defining and surveying conflicts in water resources management.
- Developing indicators to monitor cooperation in large basins.
- Developing educational material targeting all respective levels.
- Providing decision-support tools, by indicating how best to transform PC into CP.
- Disseminating results and good practices.

PC→CP activities will be guided by these operational objectives and will develop along three major tracks:

- A *disciplinary track*, which will investigate the professional approaches as well as the scientific background to conflict management, water-related negotiations and cooperation building techniques and methods.
- A *case study track*, which will survey and consider a selection of real-world cases of water conflicts and cooperation, in order to draw lessons on both the root causes of such conflicts as well as the successful cooperation in shared water resources management.
- An *educational track*, which will concentrate on how to develop skills for successful management of shared resources, at all levels – from professionals to decision makers. This track will also focus on public information needs.

8.4.1 Disciplinary track

The disciplinary analysis will develop along the following four main axes:

- History and future.
- Law.
- Negotiations, facilitation, mediation.
- Systems analysis techniques.

Outputs:

- Think pieces or essays on the history and future of shared water resources.
- A state of the art report on the role of law and institutions in the transition from PC to CP; and another one on the legal protection of water facilities during time of war.
- A report on the negotiations techniques.
- A report on the role of NGOs and the civil society in negotiations.
- State-of-the-art report on the negotiation process in the international context.
- A report on systems analytical techniques and their role in cooperative water resources management.
- Critical analysis of management models of shared water resources.

8.4.2 Case study track

The case study track analyses "real-world" cases in an interdisciplinary context.

In line with the philosophy of PC→CP, the case studies selected are examples of good practices, and rely on existing and evolving institutional mechanisms which facilitate cooperation. The project's objective is not to achieve a full geographical coverage but to prove as much as possible that the transition from PC to CP is possible.

The following in-depth case studies are envisaged for the 2001–2003 period (outputs, in addition to an overview of several other cases):

- Rhine River basin.
- Aral Sea basin.
- Incomati River basin.
- Mekong River basin.
- Jordan River basin.
- Danube River basin.
- Columbia River basin.
- Nile River basin.

8.4.3 Educational track

The most valuable contribution to future education in water conflict management will come from the other two tracks of PC→CP: the generic research through the disciplinary track and the empirical material elaborated through the case study track. Therefore, the main educational use of PC→CP outputs will follow after the current phase of the project.

The educational track will develop an interface with the present and future users of PC→CP outputs, generate a vehicle for self-instruction of actual and future decision makers, experts, and trainers. It will produce educational tools and the experience gained will be put at the disposal of educational institutions with an interest in water management.

The outputs of the educational track in the First Phase are the following:

– Design one postgraduate course module on Conflict Prevention, Diplomacy and Cooperation in International Water Resources.
– On the basis of the above-mentioned module, a short (5 days) course specifically targeting highly placed decision-makers and diplomats will be developed and tested: workshops, role-plays, simulation games, meta-plan sessions, lectures, small group works and exercises.

International symposia and conferences are among the most efficient means to disseminate the results and ideas of a project. In November 2002 an international conference on issues, concepts and challenges in the area of international water resources management will take place in Delft in order to prepare the presentation of the results of the first phase of PC→CP in Kyoto at the 3rd WWF. In the spirit of the partnership of Johannesburg, UNESCO suggests a partnership with the new International Basin Network of INBO to develop regular water cooperation assessment of river basins thus contributing to implementation and monitoring of shared water resources in response to the water policy imperative identified at the 2nd WWF.

9 CONCLUSION

The preceding analysis of three major international events clearly reveals the converging trend towards broadly shared principles and concepts to serve as the basis for national water policies and subsequent strategic and operational actions in national, basin and local levels. The recent WSSD confirmed and set the recommendations and aspirations of specific water events in a broad sustainable development focused framework. The water world is now encouraged to take a further step towards implementation. The forthcoming 3rd WWF has a clear mandate: to guide the process from principles through policies to actions.

REFERENCES

2nd WWF. 2000. *Final Report.* 2nd World Water Forum. The Hague, From Vision to Action 2000. Water Management Unit, C/o Ministry of Foreign Affairs, The Hague. http://www.worldwaterforum.net
Bonn. 2001. *Bonn International Conference on Freshwater.* http://www.water-2001.de
Cosgrove, W.J. & Rijsberman, F.R., 2000. *Making Water Everybody's Business.* World Water Vision. EARTHSCAN Publishing.
MDGs. List of the Millennium Development Goals: http://www.un.org/millenniumgoals/index.html
Star. 2002. *Star* Newspaper. Issue of 03.09.2002
UNESCO Division of Water Sciences: http://www.unesco.org/water/
Wolf, A., 2002. Oral communication
WSSD. 2002. Johannesburg Summit on Sustainable Development: http://www.johannesburgsummit.org

ANNEXES

Annex 1:   List of abbreviations.
Annex 2:   GCI Statement: Battle for the Planet.
Annex 3:   Reflection on Water in the Johannesburg Declaration on Sustainable Development ("waterwise" comparison of the draft and the final declarations).
Annex 4:   Excerpts from the World Summit on Sustainable Development Plan of Implementation (key paragraphs concerning water).

ANNEX 1 – LIST OF ABBREVIATIONS

GCI:	Green Cross International.
GWP:	Global Water Partnership.
EU:	European Union.
IGOs:	Intergovernmental Organisation (like UN Agencies).
INBO:	International Network of Basin Organisations.
IWRM:	Integrated Water Resources Management.
MDGs:	Millennium Development Goals.
NGOs:	Non-Governmental Organisations.
PC→CP:	From Potential Conflict to Co-operation Potential.
PI:	Plan of Implementation (Output of the WSSD: WSSD/PI).
UN:	United Nations.
WEHAB:	Water-Energy-Health-Agriculture-Biodiversity (key areas by UN before WSSD).
WFD:	Water Framework Directive.
WfP:	Water for Peace.
WSSD:	World Summit on Sustainable Development (Johannesburg Conference).
WWAP:	World Water Assessment Programme.
WWF:	World Water Forum.
WWV:	World Water Vision.

ANNEX 2

Battle for the Planet
Johannesburg Declaration

We, stand at a critical moment in Earth's history: a time when humanity must choose its future. As the world becomes increasingly interdependent and fragile, the future holds great peril, and we, the People, have great responsibilities. Ten years ago, the world was swept up in a wave of optimism with the end of the Cold War and the adoption of Agenda 21 at the Rio Earth Summit. Today, however, the report on the state of the planet is overwhelming: the environment continues to deteriorate, poverty is increasing and the number of armed conflicts is on the rise. In addition, the globalizing economy appears to thwart all possibility for change. The main causes of unsustainable future – besides the patterns of production and consumption and the unjust distribution of wealth – include: growing populations, which need more energy and resources; affluence, which increases material consumption and waste; poverty, which limits the access of poor to resources and choices for environmentally sound options; technologies, which use energy and dispose of waste inefficiently; insecurity, which leads to massive spending on military budgets and the construction of nuclear, chemical and biological weapons of mass destruction; and financial institutions and policies, which avoid addressing the most pressing problems and exclude stakeholders and key social actors, especially women, indigenous peoples and the poor.
  We, know that today:

- 1.2 billion people live on less than $1 a day.
- 800 million people are suffering from hunger.
- 1.5 billion human beings do not have access to safe drinking water.
- 2.5 billion people lack adequate sanitation services.
- 5 million individuals, predominantly women and children, die every year from diseases related to water quality.
- 2 billion people do not have access to electricity.
- 25 million refugees have fled their homes for ecological reasons.
- The standard of living of the average African family has decreased by 20% in the past 10 years.

- 36 million human beings are infected with the HIV virus; 23 million of whom are living in Africa without access to any treatment.
- The average level of Overseas Development Aid is at 0.22% of GNP in OECD countries; a figure to be compared to the 0.7% which nations committed to during the Rio World Summit in 1992.
- The urban population of 2.5 billion will increase to 5 billion in the next 25 years.
- 12% of the 1.7 million known species are threatened with extinction.
- Average world temperatures are projected to increase further, by 1.2° to 3.5°C (2° to 6°F) over the course of the 21st century, which could exacerbate flooding, fires and other natural disasters across the world, melt glaciers and the polar ice caps, raise sea levels and pose threats to hundreds of millions of coastal and island dwellers.
- The stabilization of levels of carbon dioxide in the atmosphere to a range that is considered safe will require overall reductions in the order of 60% or more in the emission of the "greenhouse gases" that are responsible for global warming.
- The known reserves of petrol and natural gas could become exhausted in the next half century.
- Developing countries loose $100 billion a year due to imbalances and unjust trade tariffs imposed by developed countries.

We, are declaring our *Planet in Danger*, and accuse the self-interested politics of "business as usual" pursued by governments of creating a social, economic and ecological impasse for the six billion inhabitants of the planet today, and of compromising the survival of the 11 to 12 billion people who will likely inhabit the Earth at the end of the century.

We, declare our responsibility to one another, to the greater community of life, and to future generations, and promise to bring forth a sustainable global society founded on respect for nature, universal human rights, economic justice, and a culture of peace.

We, Signatories of this Declaration, demand to Heads of State and Governments to:

1. Acknowledge and act on their responsibility to turn rhetoric into real action and achieve sustainable development.
2. Respect the principles, commit the necessary resources, and create adequate instruments to achieve the Millennium Development Goals (MDGs), in particular:
   - To halve, by 2015, the proportion of the world's people whose income is less than one dollar a day.
   - To halve, by 2015, the proportion of people who do not have access to safe drinking water.
   - To ensure, by 2015, all children complete a full course of primary education.
   - To halt, and begin to reverse, by 2015, the spread of HIV/AIDS.
   - To grant the goods produced in poor countries free access to the markets of developed countries.
3. Adopt a common ethical framework in order to achieve sustainability and to reinforce the goals established in the Millennium Declaration. To recognize the Earth Charter as a valuable contribution to the development of a shared vision of fundamental values and the creation of strong, equitable global partnerships for sustainable development.
4. Implement all principles contained in the Rio Declaration and Agenda 21, and enforce the principles of democracy and good governance.
5. Curb currently unsustainable patterns of consumption and production, and support greatly increased research, development and implementation of renewable energy sources and other eco-efficient alternatives.
6. Create a legal basis motivating the business community to become committed agents of Sustainable Development.
7. Reform the United Nations system in order to give more power for actions, for enforcement of UN decisions, for peace and stability. A more active role of civil society and terms of international accountability must be key components of such a reform.
8. Ratify all International Conventions and Protocols without delay, and implement their terms with courage and determination: including those related to Climate Change, Biodiversity, Desertification, Wetlands, International Watercourses, and others.

9. Reverse the tendency of the last ten years by increasing Overseas Development Aid spending and abolishing environmentally harmful and trade-distorting subsidies, in order to allow developing countries to eliminate their crippling debts, cover their basic human and ecological needs, and have access to modern technologies that use materials and energy efficiently and with a minimum of waste. The objective of 0.7% of GDP should be reached by 2012.

Appeal signed by:

- *Laureates Peace Nobel Prize*: Rigoberta Menchu Tum, Oscar Arias, Betty Williams, Jody Williams, Mikhail Gorbachev, Desmond Tutu.
- *Mayors of large Cities*: Lyon, Curitiba, Rome, Durban, Auckland, Ouagadougou.

ANNEX 3

Reflection on Water in the Johannesburg Declaration on Sustainable Development ("waterwise" comparison of the draft and final declarations)

Draft: 3rd September, 2002 (published in a Johannesburg newspaper "Star")	*Final Version: 16th October, 2002*
Fragments from the draft being reflected in paragraph 13 of the final version	*Part of "the Challenges we face", paragraph 13*
"The most pressing challenges of our time remain…, environmental degradation…" "We reaffirm the Rio principle that human beings are entitled to healthy and productive lives in harmony with nature" "We agree to protect and restore the integrity of our planet's ecological system, with special emphasis on preserving biological diversity, the natural processes that sustain all life on earth as well as addressing the process of desertification" "We are committed to the reduction of the economic, social and environmental impact of natural disasters…"[3]	13. The global environment continues to suffer. Loss of biodiversity continues, fish stocks continue to be depleted, desertification claims more and more fertile land, the adverse effects of climate change are already evident, natural disasters are more frequent and more devastating and developing countries more vulnerable and air, water and marine pollution continue to rob millions of a decent life.
Part of paragraph 18 in the draft.	*Part of "Our Commitment to Sustainable Development", paragraph 18.*
"We welcome the focus of the Joburg commitment on the basic requirements of human dignity, access to clean water and sanitation, energy, health care, food security and bio-diversity. At the same time, we acknowledge the central importance of technology, education and training and employment creation"	18. We welcome the Johannesburg Summit focus on the indivisibility of human dignity and are resolved through decisions on targets, timetables and partnerships to speedily increase access to basic requirements such as clean water, sanitation, adequate shelter, energy, health care, food security and the protection of bio-diversity. At the same time, we will work together to assist one another to have access to financial resources, benefit from the opening of markets, ensure capacity building, use modern technology to bring about development, and make sure that there is technology transfer, human resource development, education and training to banish forever underdevelopment. *(continued)*

Annex 3 (*continued*)

Core statement on water in the draft:	*No reflection at all in the final version.*
"We agree that water is essential for life. It is the key resource for good health, for irrigating crops, for providing hydropower, for protecting ecosystems. Adequate supplies of water and sanitation, given rapid rates of urbanisation and the needs of the rural poor, are therefore central to the achievement of the objective of sustainable development"	

[3] Note that most of the commitment-type statements of the draft have been "weakened" to a challenge statement.

ANNEX 4

Excerpts from the World Summit on Sustainable Development Plan of Implementation (key paragraphs concerning water)

IV. Protecting and managing the natural resource base of economic and social development

23. Human activities are having an increasing impact on the integrity of ecosystems that provide essential resources and services for human well-being and economic activities. Managing the natural resources base in a sustainable and integrated manner is essential for sustainable development. In this regard, to reverse the current trend in natural resource degradation as soon as possible, it is necessary to implement strategies which should include targets adopted at the national and, where appropriate, regional levels to protect ecosystems and to achieve integrated management of land, water and living resources, while strengthening regional, national and local capacities. This would include actions at all levels.
24. Launch a programme of actions, with financial and technical assistance, to achieve the millennium development goal on safe drinking water. In this respect, we agree to halve, by the year 2015, the proportion of people who are unable to reach or to afford safe drinking water as outlined in the Millennium Declaration and the proportion of people without access to basic sanitation, which would include actions at all levels to:
    a) Mobilize international and domestic financial resources at all levels, transfer technology, promote best practice and support capacity-building for water and sanitation infrastructure and services development, ensuring that such infrastructure and services meet the needs of the poor and are gender-sensitive.
    b) Facilitate access to public information and participation, including by women, at all levels, in support of policy and decision-making related to water resources management and project implementation.
    c) Promote priority action by Governments, with the support of all stakeholders, in water management and capacity-building at the national level and, where appropriate, at the regional level, and promote and provide new and additional financial resources and innovative technologies to implement chapter 18 of Agenda 21.
    d) Intensify water pollution prevention to reduce health hazards and protect ecosystems by introducing technologies for affordable sanitation and industrial and domestic wastewater treatment, by mitigating the effects of groundwater contamination, and by establishing, at the national level, monitoring systems and effective legal frameworks.
    e) Adopt prevention and protection measures to promote sustainable water use and to address water shortages.

25. Develop Integrated Water Resources Management (IWRM) and water efficiency plans by 2005, with support to developing countries, through actions at all levels to:
    a) Develop and implement national/regional strategies, plans and programmes with regard to integrated river basin, watershed and groundwater management, and introduce measures to improve the efficiency of water infrastructure to reduce losses and increase recycling of water.
    b) Employ the full range of policy instruments, including regulation, monitoring, voluntary measures, market and information-based tools, land-use management and cost recovery of water services, without cost recovery objectives becoming a barrier to access to safe water by poor people, and adopt an integrated water basin approach.
    c) Improve the efficient use of water resources and promote their allocation among competing uses in a way that gives priority to the satisfaction of basic human needs and balances the requirement of preserving or restoring ecosystems and their functions, in particular in fragile environments, with human domestic, industrial and agriculture needs, including safeguarding drinking water quality.
    d) Develop programmes for mitigating the effects of extreme water-related events.
    e) Support the diffusion of technology and capacity-building for non-conventional water resources and conservation technologies, to developing countries and regions facing water scarcity conditions or subject to drought and desertification, through technical and financial support and capacity-building.
    f) Support, where appropriate, efforts and programmes for energy-efficient, sustainable and cost-effective desalination of seawater, water recycling and water harvesting from coastal fogs in developing countries, through such measures as technological, technical and financial assistance and other modalities.
    g) Facilitate the establishment of public-private partnerships and other forms of partnership that give priority to the needs of the poor, within stable and transparent national regulatory frameworks provided by Governments, while respecting local conditions, involving all concerned stakeholders, and monitoring the performance and improving accountability of public institutions and private companies.
26. Support developing countries and countries with economies in transition in their efforts to monitor and assess the quantity and quality of water resources, including through the establishment and/or further development of national monitoring networks and water resources databases and the development of relevant national indicators.
27. Improve water resource management and scientific understanding of the water cycle through cooperation in joint observation and research, and for this purpose encourage and promote knowledge-sharing and provide capacity-building and the transfer of technology, as mutually agreed, including remote-sensing and satellite technologies, particularly to developing countries and countries with economies in transition.
28. Promote effective coordination among the various international and intergovernmental bodies and processes working on water-related issues, both within the United Nations system and between the United Nations and international financial institutions, drawing on the contributions of other international institutions and civil society to inform intergovernmental decision-making; closer coordination should also be promoted to elaborate and support proposals and undertake activities related to the International Year of Freshwater 2003 and beyond.

Water policies in Europe after the Water Framework Directive

B. Barraqué
Laboratoire Techniques, Territoires et Sociétés
Unité de Recherche Associée au CNRS, France

ABSTRACT: Adopted in October 2000, and published in December of the same year, the Water Framework Directive (WFD, 2000/60 EC) is a very ambitious piece of legislation indeed: not only is it the first time that the European Union takes the improvement of the aquatic environment as a direct goal, but, beyond this, the real goal is a sustainable development in the field of water. In this paper, we would like to give a more precise definition of the very notion of a sustainable water policy, then apply it to the particular case of water services and hydraulic infrastructures. In a second part of the paper, we'll try to show how the WFD pushes European member States, and in particular the Mediterranean ones, toward demand side management, which might be called the "third age" of water industry.

1 INTRODUCTION: THE WFD AND THE *3ES* DEFINITION OF SUSTAINABILITY

Environment, economics and ethics are three key words (Figure 1) which help make the definition of sustainability a little more precise than the notion of intergenerational equity of the Brundtland report: these three dimensions of sustainability are retained in the United Nations' definition, which have been adopted and applied to water policy by our research partnership for DG XII of the European Commission, Eurowater. In the Water 21 project[1], we have tested these three dimensions of sustainability in two major areas of water policy: integrated river management, and water services provision. We have also discussed the appropriate "subsidiary" levels of water management and their

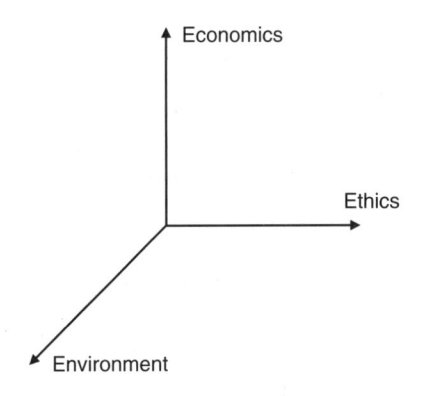

Figure 1. "Dimensions" to define "intergenerational equity"

---

[1] Eurowater and Water 21 are research projects funded by DG Research of the EU and by other institutions at national level. The coordinator is Francisco Nunes Correia, Instituto Superior Tecnico, Lisbon, and the teams include R. Andreas Kraemer (Ecologic, Berlin), Erik Mostert (RBA Centre, Delft University), Thomas Zabel (Water research centre, Medmenham, UK), and myself from LATTS.

articulation in the 5 member States under study: France, Germany, The Netherlands, Portugal and United Kingdom. Doing this, we were in tune with the issues in the preparation of the WFD. Apparently issues are very complex, as indicated by the unusual length and reading difficulty of the written document (EC 2000/60). Yet we can propose this 3-dimensional presentation to simplify the reader's comprehension. Basically, the WFD tries to reach 3 key goals (Figure 2): a good ecological status of all aquatic environments in Europe, both surface and ground water; a recovery of the costs of the policy by the water users, in conformity with the economic theory's conception of efficiency; and public information and participation, on the ethical side.

In fact, the first requirement of the WFD is to identify appropriate territories for water management, which should be hydrographic districts, i.e., groups of river basins, including whole international ones. The second task is to develop the knowledge of the quality (and quantity) of the aquatic environment, and to do a regular reporting compared to a certain state of reference. Then member States must devise 6-year programmes to get closer to the state of reference. This will imply costs, which will have to be reflected in water tariffs. Lastly, the information on the status of the aquatic environment, the programmes, and the impact on water tariffs should be made available to the public. Now, the implementation of these apparently simple goals raise a lot of scientific and policy issues. We have identified six major ones, which we present now.

2 SIX MAJOR ISSUES IN THE WFD

Six issues can be raised as of now; the two first ones are involving primarily natural sciences, but not only: how will hydrographic districts be set up, and what is the meaning of "good ecological status"; two other ones deal primarily with social sciences: the economic approach to costs and prices, and the public participation; lastly there are two unsettled issues: the list of dangerous substances to be phased out, and the groundwater issue.

2.1 Hydrographic districts

In most member States, the appropriate territory being a group of catchments, is at odds with the traditional administrative boundaries of water management. So, not only is there a debate about the optimal size, i.e., groups of catchments, which should be taken together (depending on similar/different geographic conditions on the one hand, policy and politics conditions on the other), but there is also another debate about the type of authority which should govern the hydrographic districts. In France, the 30 year old *agences de l'eau* are only the managers of compelled mutual savings of the water users, and providers of cheap money for environmentally friendly investments. Police powers largely remain in the hands of the prefects at the level of each *département*, while

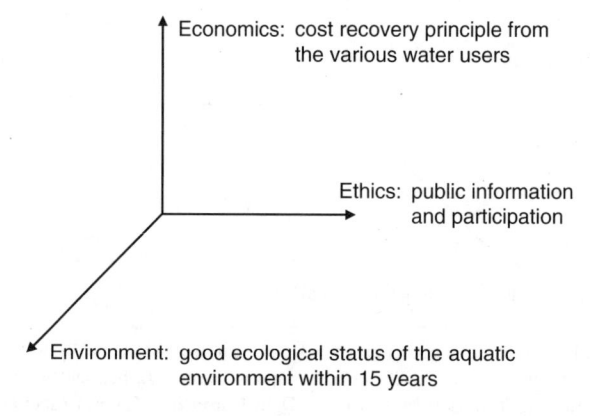

Figure 2. WFD goals for every dimension.

planning is rather in the hands of the regional prefects. There is a high probability that France will transform the existing "bassins" into the hydrographic districts[2]. Will the WFD approach develop the contractual and co-ordination approach led by the *agences de l'eau* and the River basin commissions, or give more importance to command-and-control policies led by prefects; and which level of prefects? Will the action be based on efficiency or on sovereignty as a prime legitimization? If even in France, the catchment is not always felt as the locus of the best approach to integrated river management (meaning that most people prefer to keep the pre-existing territorial organizations and institutions, and to co-ordinate between them), what is going to happen in other member States? And concerning international rivers in the European Union (EU), how will concerned member States co-ordinate their action? Are we heading toward supra-national basin authorities, or toward more flexible arrangements like the Rhine Commission?

2.2 *Reaching the state of reference*

In a continent where humankind has interacted with water since the Neolithic era, what can the state of reference be? What does the notion of good ecological status mean? Certainly not the virgin and climax status, but which alteration can be considered as positive? Where will be the limit of the strongly modified, artificial, body of water? Second, what set of indicators is needed to assess the reference state and the present discrepancy from it, and how much will it cost to monitor them? How can member States develop an information system suited to the hottest issues they have to treat first? To which extent will the specificity of member States be respected by the Commission, given the fact that on the one hand there can be passes through to the 15 years deadline (provided motivated), but that on the other hand there are now possibilities of financial penalties?

2.3 *Economic analysis and cost recovery*

In most member States, the water industry is particularly concerned by a key issue in the proposed project: following a widespread opinion among economists and some water operators, the DGXI supports the full-cost pricing for water services. But this has almost never been the case in the past, and if the leading countries are getting close to it now, one should not forget how the systems were first installed: health and welfare considerations led to massive State subsidies for the initial equipment period, earlier in this century. This has considerably lowered the prices of the services, and it probably still has a long lasting influence now. This should be checked by economic or science and technology historians. Anyway, it may seem unreasonable, if not unfair, to condemn the subsidies practiced in the Southern member States and Ireland, which are still in the process of completing their initial or basic equipment. This point is raised in the Institute for Prospective Technological Studies (IPTS) report on the Mediterranean action plan prepared for the DG XVI (EU, 2000).

In Northern member States, water prices have been rising sharply in recent years, with a good reason: once the initial investment phase is completed, national governments tend to stop subsidizing the water services sector. Then water supplies have to face the long term reproduction of the very important capital that has been built up, at the very moment when new environmental constraints are imposed; but this moment is unfortunately also one of economic crisis, and the social consequences of the economic crisis can be felt in the water sector. So the issue is: can European water bills afford environmental upgrading on top of long term reproduction of present-day water technologies and systems? And, even if this policy is affordable, will it be politically acceptable? Sustainability might then even mean to reconsider the supply side model that has prevailed in the last century. Can we still dream to connect practically all water uses to a single drinking water system, including lawn sprinkling and toilet flushing, and to a generalized sewage collection and treatment system which people

---

[2] There will have to be adaptations: e.g., the Sambre is in Artois Picardie basin, but is a tributary of the Meuse, which it reaches in Belgium. It will be part of the Meuse international district. Between the Ebro Confederation and the Adour-Garonne Basin, there will have to be at least symbolic exchanges of representatives, since several rivers cross the border.

are totally unconscious of? Some member States should rather develop public services to manage decentralized sewerage in low density areas, organize recycling of water to serve lower quality uses, and all of them should consider supplementing the traditional public works and civil engineering approach with land-use control policies. The Commission itself will end up in contradiction with its own full cost recovery criterion, if it gives subsidies to fancy long distance water transfers in Spain, and to the Alqueva reservoir in Portugal (Vergés, 2001). These projects are still supporting in practice an outdated idea under which water is life, so it should be free, in particular to develop irrigation cash crops at the expense of drinking water for cities, and of course of the ecosystem needs. One very important issue is whether the level of cost recovery will be calculated only sector by sector at national level, or if it will also be done in a cross-sectorial manner at catchment or hydrographic district level. As the example of Northern Europe shows concerning water quality contamination by diffuse pollution from agriculture, the most efficient solution is for water supply undertakings to purchase to the farmers the loss of productivity incurred from bans on nitrates and pesticides.

2.4 *Public information and participation*

There have been a lot of debates about this point: in particular, at what time in the knowledge and planning process should the public be involved? Do the hydrographic districts commissions make the assessment of the state of reference and the programmes to get there first, and then present them with maps to the public, which in turn can complain if he isn't satisfied? Won't it push the public into a judiciarization process has is witnessed already for some Directives, and also in the US? or is public participation going to take place upstream, in the making of the reference state and programme of action, but then in a more qualitative manner (i.e., through NGOs)? There is an obvious need for both types of participation, but, in some member States, authorities might be tempted to interpret the WFD so as to "drown" the qualitative participation of water users and ecosystems supporters under the anonymous user giving his say at the end of the planning process.

2.5 *Dangerous substances*

It was difficult to get a compromise on the list of substances, in particular on the black listed ones to be phased out within 20 years, in a process of listing to be reproduced every 4 years, because we don't know what the second degree impact will be on industry and agriculture policies; we also don't know how to suppress these substances in the environment once they have been diffused. So there is a wide debate going on about the number of substances: the Commission has listed 32, already, of which 21 as hazardous, but some ecologists think that more than 400 should be phased out. The Directive also does not clearly compel manufacturers to provide data on all chemical substances they produce or they use. Yet regulation on hazardous ones is then impossible.

2.6 *Groundwater*

Groundwater is a serious issue, because the consequences of their over-exploitation and of their contamination are usually lasting much longer than for surface water. Besides, the European policy had been quite limited, since the only Directive available addressed the contamination by toxic substance from industry. Since then, contamination by intensive agriculture has become the most important issue. When they joined the EU, the Swedes were very active in preparing a revision of the groundwater Directive (Hilding & Johansson, 1998). Any action plan in this sector has to deal with the fact that it is virtually impossible to exert a tight control on farmers scattered on the countryside. In Latin countries, this problem is doubled with the fact that under Roman law, groundwater is appropriated by landowners, which makes it time consuming to get farmers to recognize the common property character of aquifers, particularly when they are over-exploited. Protecting groundwater for the sake of having cleaner resources to make drinking water implies to supplement water policies with land-use control. However, observers in Northern member States contend that the WFD is potentially making things worse for groundwater, since the former zero-emission on some substances (1980 Directive) is replaced by a "no-deterioration" principle which will be very difficult to enforce and even to monitor. Conversely, in some Southern member States, traditionally water engineers are only interested with surface water, ignoring not only the potential of groundwater, but also the interconnection between aquifers and rivers. This is what Prof. Llamas calls hydroschizophrenia (Llamas et al., 2001). More generally, several Spanish authors have pointed the discrepancy between their national water policy, still oriented toward supply sided big hydraulic projects on just surface water, and the WFD, pushing toward demand side management. But should bad grades be given only to Spain? In France, for instance, while some observers superficially argue that "WFD is easy for us, since we gave our model to the EU", others contend that the implementation will be very costly, in particular in terms of all the monitoring requested by the Commission. Also, economic and financial calculations might well lead to surprises in terms of present cost recovery.

3 SUSTAINABILITY OF WATER SERVICES

If we take the public water services sustainability, implementing the *3Es* approach would mean providing simultaneous answers to three questions:

a) *Economics*: how is the enormous capital accumulated in water services technologies since 150 years maintained and reproduced on the long run? Are depreciation schedules correct, how do they reflect in the water bills or charges, in particular since subsidies are usually being phased out by governments after initial investments? What compromise is found between the risk of overmanning, for the sake of very good maintenance, and the inverse life time reduction of infrastructure due to poor maintenance? Can one separate normal maintenance and replacement of old infrastructure? Tracking water losses is a good example of this search for compromise: there obviously is an optimum of maintenance which is away from a zero leak objective, and relining leaky pipes is often cheaper than changing them. But what will be the impact of all this on water bills?
b) *Environment*: in order to improve the environmental performance of existing systems, what kind of new investments are needed to comply with European Directives and with national standards? How will operation costs increase (e.g., with increased volumes of sewage sludge)? What would be the extra burden on water consumers, of extra investment and operation costs, in particular if the loans would have to bear normal commercial rates? How far are national policies from the formally accepted polluter-pays principle and user-pays principle?
c) *Ethics*: if all the "sustainability costs" (long term reproduction and environment) are passed on to the consumers, can they afford it, and is it politically acceptable? If not, what kind of technological, financial or planning adaptations are necessary? What impact of poor reproduction today on future generations' either prices or health and environmental risks? Is it at all possible to do without subsidies, i.e., without some funding from taxes and general budgets at local or supra-local level?

There are also several specific and actual issues to be addressed: concerning drinking water, one is the cost of the potentially increased number of criteria and of their reinforcement (e.g., lead pipes) by treatment technologies, another is choosing alternative policies: e.g., should we go on removing biocides and nutrients from agriculture in the drinking water production processes, or devise planning and economic incentive policies aiming at reducing their use in agriculture? And who should bear the associated cost, water consumers (like in the case of the German Wasserpfennig) or farmers? Another issue concerns metering: in the case of England and Wales, putting individual meters in all homes, as foreseen by the privatization law, would be very expensive, and might not be economically efficient, given the relatively low elasticity of water demand to prices. But with very few meters, it is difficult to track water losses, and there is no incentive to consumers to fix leaky facilities. There probably is an optimum to find. Yet there is pressure from consumer NGOs, experts and politicians to generalize metering for all consumers, for equity reasons.

Concerning waste water collection and treatment, at least four issues are at stake: are industrial effluents acceptable in public sewers, and how much should they be charged? Or which pre-treatment should be requested? What kind of technologies are appropriate in low density areas and in small local authorities, if their long term operation cost is considered? What would be the cost and the environmental benefit of a stronger policy on stormwater pollution and run-off control, and how could it be funded? What is going to be the impact of increased volumes of sewage sludge to process, at a time when farmers are increasingly reluctant to spread it on their fields?

Concerning both types of services, one important issue is personnel redundancies, which are allegedly more frequently found in services operated under direct labor. Are the services overmanned? Does this eventual overmanning allow to reduce the long term rehabilitation (replacement) costs? Is the integration of various parts of the industry into large companies able to do the engineering, the building and the operation (like in the French, and increasingly in the British, models) more efficient than the traditional split linked to public operation and the "shopping list" in the calls for tenders? And above all, what would be the best "subsidiary" size of services? Is local operation preferable to regionalization? What about intermediary formulas, like local operation with regional financing schemes and pooling of charges?

Detailed as it may appear, this is only a partial list of the intermingled questions which arise in an evaluation of sustainability. The same kind of analysis can be proposed for water use by industry or electricity generation, agriculture hydraulics, etc. But now, we need to address the issue of the needed change in water engineering culture to face the WFD. In a conference in Rome on demand-side management approaches in the Mediterranean in October 2002, the representative of the European Investment Bank declared that his institution was not inclined to support water projects, since they usually forgot to develop simple leak control and other conservation measures for the sake of justifying "sexy desalination plants". He didn't even mention long distance water transfers. Now we have to understand why some techniques are considered as part of demand side while others are more supply side (this issue was debated of course in the seminar), but we can do it through a historical overview of water technologies in the Contemporanean period.

4 THE THREE AGES OF WATER INDUSTRY

Let us try to sketch three successive engineering approaches to water services: pipes and reservoirs go with civil engineering; drinking water and sewage treatment plants go with chemical engineering, as well as membranes (bio-chemical); while the most recent outcomes of demand management are related to environmental engineering. To explain their successive development and partial crisis, we can mobilize Tom Hughes' concept of *reverse salient* in infrastructure development, initially proposed for electricity networks. In the case of water, we can hypothesize the following: in the 19th Century, or rather until the Koch and Pasteur discoveries were popularized, Public Water Supply (PWS) developed on the assumption that water should be drawn form natural environments far from cities. Large cities in particular would have to get water from further and further. This was made possible by the possibilities of municipalities to obtain "cheap money", in particular from the early popular savings banks which they controlled. Their bonds were found attractive by the public, and on top of all this Governments would subsidize projects. Take the example of Glasgow: "direct municipal provision seemed to offer several advantages to the city. The existing private company had […] outdated infrastructure consequently was unable to cope with the demands of the rapidly growing population […] Moreover, the company was not in a position to raise the necessary capital for improvements, unlike the Town Council, whose extensive community assets made it eminently creditworthy. Public accountability meant that unpredictable market forces could be over-ridden, and a stable service provided […] Loch Katrine was located in the Perthshire Highlands, some 55 km from Glasgow, and thus well away from the polluted city […] The official opening by Queen Victoria on an appropriately wet autumn day in 1859 was an event of enormous significance for Glasgow […] Loch Katrine was unquestionably the prime municipal showpiece for the city, combining the wonders of Victorian technology with the nurturing quality of pure Highland water"(Maver, 2000). Tarr (1996) has illustrated this broad approach in the US: getting cleaner water from further on the one hand, and using the rivers as sewers on the other, the latter decision relying on the assumption of natural dilution and self purification of rivers.

This overall *problematique* has remained dominant in the New World, and was also extended in the rest of the world after the Second World War, due to the co-occurrence of International Financing Institutions offering cheap money, and of various forms of support for Government intervention in infrastructure provision (Keynesian or socialist). However, large hydraulic projects of the 50s and 60s were increasingly devoted not to cities, but to irrigated agriculture for exportation. Today, still, many States in developing countries base their water policy on large water transfers, so as to indirectly subsidize the production of irrigated cash crops to integrate in the world market. A small part of the water is diverted to cities, which help reduce deficits of the projects, but overall, water transfers offer a subsidized water which is usually managed inefficiently (wastage and low price make a vicious circle). In some cases the ceiling of extractable water resources has been reached, and the present crisis offers the possibility to check the unsustainability of these past policies.

5   A TECHNICAL *REVERSE SALIENT*: FROM FURTHER QUANTITIES TO NEARER QUALITIES

Interestingly enough, a similar crisis occurred a long time ago in (Northern) Europe: growing population densities and smaller natural resources increased competition for pure water resources, while the development of bio-chemical analyses showed the growing extent of contamination. Irrigation being unnecessary in Northern Europe, there wasn't so much a resource quantity problem than a quality problem. In the end, it was decided that water should be filtered (end of 19th Century), and later chlorinated, ozonized or disinfected through GAC beds (around First World War), which ever source it would come from. But then, taking water from the river just upstream the cities would not change anything in term of public health, while it would save a lot of investments. So in that still early period, large European cities changed strategy from investments to increase quantities to investments to improve quality, with of course a serious rise in operation costs. But delivery of pressure water within the homes changed status, from a luxury good to a normal comfort, and made it possible to have customers pay water bills.

This is exactly what happened in the city of Paris a century ago. From Napoleon the idea had prevailed that Paris should get water from distant sources, and indeed, the work of engineer Belgrand under Baron Haussman's general approach of the public service as a monopoly, was turned towards securing longer distance sources of water (around 100 km). It was even felt that one day the capital city would have to get water from the Loire, except that this river has very low flows in the summer, precisely when water demands would request it. Then, in 1890, an engineer named Duvillard came up with a project to draw water from lake Leman, i.e., 440 km away! It sounds as a fancy project, because of the Alps and of the international character of the Rhone (would the Swiss accept?), yet in fact it was technically quite simple, even at that time. Proponents of the project soon came up with all sorts of arguments to convince Paris Council and the State: a Capital of the world would need at least 1000 l/(capita day) to have more luxurious fountains, more street cleaning[3], better domestic comfort and hygiene; besides, such a quantity of water would alleviate navigation problems in drought periods, help flush waste water from the new sewer system away to the sea, and let other water resources available for local economic development. In the end they said, a huge transfer

---

[3] Paris is one of the few cities in the world where streets are washed clean: for political and business reasons linked with Haussmann's decision to merge suburban communes and extend Paris from 12 to 20 arrondissements, it was decided that water would be produced by the city and delivered for free through a public network, while a second network would serve domestic and other private homes under pressure the payment of water bills. This is why Paris still has 2 PWS, one potable and the other non-potable by today's standards. The non potable network produces hardly filtered Seine water to flush the sewers, to supply the lakes in the Bois, and to clean the streets. Other public uses like fire hydrants have quit because of unreliability, lower pressure, sprinkler clogging, etc.

would assure Paris PWS forever, and the bigger it would be, the cheaper each cubic meter would be! It's indeed very amusing to see that it's the same type of arguments which have recently been raised by the Franco-Spanish partnership to transfer water from the Rhone to Barcelona (350 km)! This also shows how irrigation makes Southern member States of the EU change the whole picture: cities have to get water from ever further because all the local surface and ground water is given away for quasi-free to farmers. Back in Paris a century ago, while proponents were finalizing the studies, some epidemic disease broke out, and it was found that one of the distant natural intake points (the Loing springs) was to blame; even distant pure water could be contaminated. So that water should be filtered and treated even in that case! In 1902 Paul Brousse, one of the father founders of French so called "municipal socialism" (equivalent to what was derided in England as "water-and-gas socialism"), inaugurated the new water filtration plant in Ivry, just upstream Paris (the one which has been redesigned a few years ago to serve as a showcase of French water technological know-how). A long lasting choice was being made. And chlorination was decided after the First World War. Water demands were growing incrementally and the big jump then appeared as too risky. Lastly, the project was discarded by Paris Council for national defense reasons: what would happen if next war the Germans would rush on the aqueduct and cut it off?!! Even though this decision didn't help much for the next war (are there ever real water wars?), it's clear that quality investments replaced quantity ones just as chemical engineering replaced civil engineering.

After the Second World War, the Prefect of the Seine took advantage of a severe flood to obtain the construction of three large upstream reservoirs on the Seine, the Marne and the Aube, in fact to increase summer flows and meet Paris water demands even in very serious droughts like happened in 1976. Interestingly enough, a fourth reservoir was planned by mayor J. Chirac's councilors, the water being then transferred directly to Paris by a pipe, but it was abandoned for the same reasons (we must and can purify the water anyway, said the giant water supply companies) plus the fact that Paris water demand went down by 13% between 1990 and 1996 (Cambon, 2000).

Now on the whole, the development of chemical and bio-chemical engineering allowed water services to develop local solutions to their problems, which helped greatly to maintain local authorities as providers of the service, eventually through joint boards. This fact is too often ignored in the debate over the world about public versus private service provision in the world, since over the centuries, Europeans found a good solution to connect all their population to water services: the municipality is at the core, but partnerships with the private sector and other levels of government are possible, very diverse, and welcome. The next phase of the water services history illustrates this.

6 THE CRISIS OF MUNICIPAL WATER SERVICES: A NEW *REVERSE SALIENT*?

In France like in other Northern European countries, important efforts were made on city sewerage from the 50s onwards, and on sewage works from the 70s. To make the investment funding easier, it was decided to change the status of Public Sewage Collection and Treatment (PSCT) from an imposed system to a service, and then to have it paid in the water bills. The indirect effect was to internalize the external costs generated by water use in cities on the environment, which is good, and led some water uses which did not really need potable water to quit the system or to recycle their water. But in the same period, PWS itself became a mature business, i.e., it had to face the issue of renewing aging infrastructure without any more subsidies. This is an essential reason why municipalism adapted in various ways towards legal private status: traditional public accounting could not depreciate the assets nor make renewal provisions, while private accounting could. PWS, and later PSCT, slowly turned towards depreciation and provision practices, and this of course meant a rise in water bills. Besides governments are under the influence of economists arguing in favor of full or at least fair cost pricing, and they phase subsidies out. In turn water bills rise dramatically, and an increasing number of large users (industry, services) either change their processes or fight their leaks. This explains the recent stagnation of volumes sold. In some countries even domestic consumers have reduced their demand for PWS, through changing fixtures and domestic machines, different garden design, and also with rainfall storage or other alternative sources of water for non drinking uses. If this movement would develop and quicken, PWS and PSCT could eventually go bankrupt.

At the same moment, water suppliers discover that it's going to be ever harder to produce a water complying with the Drinking Water Standards (DWS) all the time at reasonable costs. Ecotoxicologists' control over drinking water standards' production tends to privilege a traditional "no-risk" strategy without taking the implied costs into account: in Europe, the lowering of the lead content from 50 to 10 $\mu g/l$ will cost up to 35 bn Euros while there is no evidence that the former level provokes lead poisoning. The multiplication of criteria is slowly bringing the situation into over-complexification: chlorination byproducts give cancer. There are many other examples, and anyway, year after year, the media can report a growing proportion of people receiving non-complying water, even though the treatment is improving on the long run. To lower the risk of being unable to make it, along with local, national and European authorities, water supplies turn to a new strategy: land-use control on water resources protection areas, usually implying agriculture re-extensification, and farmers compensation programs (for a history of drinking water criteria and the present result, see e.g., Okun, 1996). This policy turns out to be cheaper than water treatment sophistication, and to be a positive sum-game with farmers. Yet it seems to be a long way to a new and more sustainable equilibrium. And worse, in the meantime, the drinking water criteria are being reinforced regularly as new risks are discovered, and the spiral of treatment costs and its negative effects is going on. This crisis of the "age of chemical engineering" opens the way to a third age, of environmental engineering, where all these issues need to be tackled simultaneously with a systems approach.

In any case, the present crisis is not appropriate for big long water transfers (Barraqué, 2000b). What happened in California starting in the 70s is now general in Europe: it is getting ever harder to build dams, because environmental movements have been joined by economists and liberals who advocate full cost payment of water infrastructures by their beneficiaries. Then the new slogan seems to be "save first and manage the demand, there is no cheap money in sight for subsidized water transfers". In our review of water transfers, we found many more than in Spain: Copenhagen is not going to buy water from Sweden, Palermo isn't either going to have Albanian water, London will have to cut the leaks down before it eventually can purchase water from Scotland. And many of the other fancy projects in other continents would be dying like hydrodinosaurs if national governments opened their eyes on the financial inefficiency, and on the fact that food exchanges vehiculate far greater quantities of "virtual water" than there is in their projects (Allan, 2000)…

New York City, like many US and Canadian cities, could follow a different path from European ones, because of the abundance of clean water. The tradition grew to use a lot of water and to take from further and further, while protecting the water intake points through extensive land-use control. Yet, the metropolis might have to catch up with the European story, because clean natural resources are not immune from cryptosporidium and other new (lethal) diseases. A US Environment Protection Agency experts' panel concluded that New York should not be given any further derogation on the need to filter and treat water extensively, while City engineers were arguing that increased land use control would suffice. The new requested treatments are likely to seriously increase water prices[4], which might lead to a water demand collapse, and to raise a further question: in that case, why not just pump the water in the Hudson river, and forget about the damned Quebecois!

If, in Northern Europe, PWS are going to purchase the quality of water resource to farmers, by compensation of the losses in productivity when going to biological agriculture, a similar evolution is possible, but this time in terms of quantity, in Mediterranean Europe: the 80% of water used in agriculture generate much less economic value than water for cities and for tourism, so that contracts are possible in case of droughts, without any need to invoke formal water markets. It is obvious that Barcelona has cheaper water in the Rialp reservoir in Western Catalonia than across the Pyrenees, if only a flexibilization of water uses can be developed. And it is future water uses, not present ones: Rialp was initially designed to increase irrigated areas in Western Catalonia by around half, which is highly improbable anyway…

---

[4] In particular, if metering is introduced to replace the outdated frontage rates system, negative redistributive effects could be serious.

The example of California is very interesting here, since water markets have been developed, as well as the "wheeling" policy, due to the shortage of Colorado water. However, PWS were quick to realize that the cheapest water by far was not the one purchased to farmers, but the water conserved by existing users and thus liberated for new users. This is why they have developed the California Urban Water Conservation Council (Dickinson, 1998). And this fits perfectly with the WFD.

7 CONCLUSION

In this paper, we didn't discuss at length the issue of subsidiarity in European water policy. Yet it is a critical issue. Because a subsidiary approach means that there would be a development of communities of water users at appropriate territorial level. These users would develop better management, primarily through a collective learning process which would lower the transaction costs between them. This is what is called water governance, or bottom up water planning. Today, we know that this approach is more promising than top down, or command-and-control policies. Now how will the WFD be interpreted in member States? and in Brussels? Given the dominance of the liberal anglo-saxon approach, there is a possibility that the favored option be a traditional government planning approach, with public participation at the end, and a liberal option concerning cost recovery: hydrographic districts would develop taxation mechanisms like the pure polluter pays principle, and create centralizing regulators to check and confront the "selfish interests" of rational actors. This liberal-statist approach, we contend here, is not suited to water resources, which are neither a public good nor a market good, but a common property to be shared equitably between its users; and is not suited either to water services, given the weight of the investments in the costs, and given the very long periods of time of depreciation of these investments. So that in most member States, various cross-subsidy schemes have been developed to lessen the impact of investments on water bills (Barraqué, 2000a). It would be ironical that the new approach to water policy would make these cross subsidy schemes (which are positive sum games) more difficult, while it would pursue subsidizing huge hydraulic projects, which are not always useful and anyway far from the principles of cost recovery.

Hopefully, the formulation of the WFD remains sufficiently prudent to leave all options possible. It would anyway be very fruitful to develop a more systematic comparison between the European and the American water policies: after all, the United States congress launched a very ambitious water policy back in 1972…

REFERENCES

Allan, T. 2000. *The middle East water question, hydropolitics and the global economy.* London, Tauris, 2000.
Barraqué, B. 1995. Les politiques de l'eau en Europe, La Découverte. Coll. Recherches, 1995.
Barraqué, B. 1998. 'Europäisches Antwort auf John Briscoes Bewertung der Deutschen Wasserwirtschaft'. GWF Wasser-Abwasser, 139(6), 1998.
Barraqué, B. 2000a. Assessing the efficiency of economic instruments: reforming the French Agences de l'Eau. *Market based instruments for environmental management.* Andersen M.S. & Sprenger R.U. Edward Elgar, 2000.
Barraqué, B. 2000b. Are hydrodinosaurs sustainable? the case of the Rhone-to-Barcelona water transfer', in Vlachos Evan, and Correia F.N., *Shared water systems and transboundary issues, with special emphasis on the iberian peninsula.* Luso-American Foundation and Colorado State University, Lisbon, 2000.
Cambon. 2000. A paper by Sophie Cambon Grau on water savings by the large consumers in Paris in Barbier, J.M. et al. 2000. Evolution des consommations d'eau. Dossier in *TSM – Génie Urbain – Génie Rural,* periodical of the AGHTM, n° 2, 2000. Incl.
Correia, F.N. 1998. *Eurowater, Institutions for / Selected issues in / Water resources management in Europe.* Balkema, Vol. I and II, 1998.
Dickinson, M.A. 2000. Water conservation in the United States: a decade of progress. In Antonio Estevan & Victor Viñuales, *La eficiencia del agua en las ciudades,* Bakeaz y Fundacion Ecologia y Desarrollo, 2000.

EU. 2000. *Toward a strategic and sustainable management of water resources, study n° 31*. European Union Commission, Regional Policy Directorate. Publications office of the Commission, Luxemburg, 2000.

Hilding, R.T. & Johanssson, I. *How to cope with degrading groundwater quality in Europe*, proceedings of the international seminar in Johannesberg Sweden, FRN report, n° 98:4, Stockholm, 1998.

Llamas, R.M.; Fornés, J.M.; Hernandez, N.; Martinez, L. 2001. *Aguas subterraneas: retos y oportunidades.* Madrid, Fundacion Marcelino Botin & Mundi Prensa, 2001.

Maver, I. 2000. *Glasgow, Town and city histories*, Edinburgh University Press, 2000.

Okun, D.A. 1996. From cholera to cancer to cryptosporidiosis. *Journal of Environmental engineering*, June 1996.

Okun, D.A.; Craun, G.; Edzwald, J.; Gilbert, J.; Rose, J.B. 1997. New-York city: to filter or not to filter? *Journal of the American water works association*, 89(3) March 1997.

Tarr, J. 1996. *The search for the ultimate sink, urban pollution in historical perspective*, Akron University Press, Ohio, 1996.

Vergés, J.C. 2001. *El saqueo del agua en España, Barcelona*. La Tempestad, 2001.

The risks and benefits of globalization and privatization of fresh water[1]

G. Wolff
Pacific Institute for Studies in Development, Environment and Security, United States

ABSTRACT: This paper discusses the globalization, privatization, and commodification of water; defines terms; reviews cases and examples; and offer principles and standards to guide policymakers in the future. The paper neither opposes nor supports the current trends toward globalization and privatization of fresh water. In some places and in some circumstances, letting private companies take responsibility for some aspects of water provision or management may help millions of people receive access to basic water services. In other cases and circumstances, private provision or management of water or water services is inadvisable, because the risks to human and environmental well-being from such management are simply too high.

1 WHY INCREASE TRADE IN WATER? WHY PRIVATIZE WATER SERVICES?

Humans have manipulated water resources since the beginning of civilization. Early agricultural communities developed where crops could be grown with dependable rainfall and perennial rivers. Simple irrigation canals improved crop production and reliability. Growing villages required increasingly sophisticated engineering efforts to bring water from remote sources and to dispose of human wastes.

During the industrial revolution and population explosion of the 19th and 20th Centuries, water requirements rose exponentially. To meet these needs, engineers built tens of thousands of massive projects designed to control floods, protect clean water supplies, and provide water for irrigation or hydropower. Improved sewer systems helped stamp out cholera, typhoid, and other water-related diseases in the richer, industrialized nations. Vast cities, incapable of surviving on limited local water local resources, have bloomed as engineers brought in water from hundreds and even thousands of kilometers away. Food production has kept pace with growing populations because of the expansion of artificial irrigation systems. Nearly 1/5 of all of the electricity generated worldwide is produced by hydroelectric turbines (Gleick, 1998 and 2000).

Yet despite our progress, more than one billion people lack access to clean drinking water. Nearly two and a half billion people do not have adequate sanitation services. Preventable water-related diseases kill an estimated ten to twenty thousand children each day and the latest evidence suggests that we are falling behind in efforts to solve these problems (WHO, 2000). New massive outbreaks of cholera appeared in the mid-90s in Latin America, Africa, and Asia. The number of cases of dengue fever – a mosquito-borne disease – doubled in Latin America between 1997 and 1999. Millions of people in Bangladesh and India are drinking water contaminated with unsafe levels of arsenic.

The effects of our water policies extend beyond jeopardizing human health. Tens of millions of people have been forced to move from their homes – often with little warning or compensation – to

---
[1] Adapted from Gleick et al. (2002)

make way for the reservoirs behind new dams. More than 20% of all freshwater fish species are now threatened or endangered because dams and water withdrawals have destroyed the free-flowing river ecosystems where they thrive. Certain irrigation practices degrade soil quality and reduce agricultural productivity, threatening to bring an end to the Green Revolution. Groundwater aquifers are being pumped down faster than they are naturally replenished in parts of India, China, the United States (US), and elsewhere.

Population growth throughout the developing world is increasing pressures on limited water supplies. Most organizations that work on water problems – from the United Nations (UN) to NGOs to local water agencies – project that the number of people facing water scarcity and shortages in the future will grow in the future, despite efforts to meet human needs.

In the last few years, however, the way we think about water has begun to change. Around the world, large water infrastructure built and operated by governments was considered vital for national security, economic prosperity, and agricultural survival. Until very recently, governments and international financial organizations like the World Bank and multilateral aid agencies subsidized or paid in full for dams or other water-related civil engineering projects, which often have price tags in the billions of dollars. Having seen large amounts of ineffective development in the past, having borne the associated costs (both monetary and otherwise) of that development, and now faced with many competing demands for limited capital, many governments are increasingly reluctant or unable to pay for new water infrastructure.

The focus is slowly shifting back to the provision of basic human and environmental needs as the top priority. To accomplish these goals and meet the demands of growing populations, previous publications from the Pacific Institute have called for smarter use of existing infrastructure rather than building new facilities, which are increasingly considered options of last, not first, resort. The challenge is to use the water we have more efficiently, to rethink our needs and wants, and to identify alternative supplies of this vital resource (Gleick, 1998 and 2001).

Some see the private sector and market mechanisms as a basic part of this changing paradigm. As a result, they are turning to the private sector in a variety of complex and controversial ways. In the following sections, we evaluate these approaches, review experiences in treating water as an economic good, and explore options for policymakers and the public.

2 DEFINITIONS

Because some often-used words and phrases – *globalization, privatization, commodification, water as a social good, water as an economic good* – are critical to the discussion and analysis in this report, we explicitly define these five terms and discuss them in the context of current water management questions.

2.1 *Globalization*

National economies were formed in the last several centuries by intensification of trading within national boundaries, leading to economic interdependence among local economies within each nation, and increasingly among nations in a region. In a similar way, a "global" economy is now being created by the intensification of trading across national boundaries and the transnational character of large corporations, creating dependencies among the national economies of each region and of the world. Over the past few years, new rules and processes governing trade in goods and services have been developed, leading to an expanding influence of multinational corporations and to a series of international agreements with broad implications for consumers, governments, and the natural environment.

These rules and processes have come to be known as "*globalization*" – defined here as the process of integrating and opening markets across national borders. The entire process of globalization is highly controversial, raising great concern about national sovereignty, corporate responsibility, equity

> **Box 1. Private and public goods.**
>
> Economists define private goods as those for which consumption (or use) by one person prevents consumption (or use) by another. Water for consumptive use is a private good. Public goods are those that can be used by one person without diminishing the opportunity for use by others. For example, piped water-supply systems are public goods because, in most circumstances, delivery of water to one household does not prevent delivery of water to another household. The economic definitions of private and public goods should not be confused with public or private ownership of goods. A private good can be publicly owned.

for the world's poorest people, and the protection of the environment. The controversy extends to proposals to encourage large-scale trading of freshwater across borders. Indeed, among the most controversial water issues today are questions about how to implement – indeed, whether to implement – international water trading and sales.

2.2 *Privatization*

"*Privatization*" in the water sector involves transferring some or all of the assets or operations of public water systems into private hands. There are numerous ways to privatize water, such as the transfer of the responsibility to operate a water delivery or treatment system, a more complete transfer of system ownership and operation responsibilities, or even the sale of publicly owned water rights to private companies. Many combinations of public and private functions are possible, such as private investment in the development of new facilities, with transfer of those facilities to public ownership after investors have been repaid. Increasingly, offers to privatize water services are coming from large, multinational corporations. As these efforts intensify, so does opposition at local, regional, and international levels.

When the service being privatized has "public good" characteristics, government regulation or oversight has traditionally been applied (Box 1). For example, water quality, piped water distribution networks, and water-use efficiency improvement programs, are public good characteristics of water management systems.

2.3 *Commodification*

"*Commodification*" is the process of converting a good or service formerly subject to many non-market social rules into one that is primarily subject to market rules. Even with today's sophisticated economies, many goods and services are still traded or exchanged outside of markets. For example, irrigation water users associations often have complex social arrangements that govern access to common water supplies, rationing of these supplies during drought, and so forth. Water exchanges within the community of water users may require commitments to return water in the future or other social commitments, rather than or in addition to payment in currency. Similarly, some types of water rights (e.g., riparian rights) are not transferable. Hence, water possessed via riparian rights is not a commodity because it cannot be marketed.

The processes of globalization and privatization tend to require that water (and water services) be treated as commodities, subject to the rules of marketplaces and free of non-market rules. Of course, water in some forms is already considered a commodity, particularly bottled water of various types. In recent years, however, the sales of different forms of water have boomed, including flavored waters, glacier water, distilled and partially distilled waters, and other "designer" waters. This has led entrepreneurs to begin to explore the possibility of large-scale movements of waters for commercial, rather than purely community purposes. As the International Union for the Conservation of Nature notes (IUCN, 2000): "Water, once revered for its life-giving properties, has become a commodity".

2.4 Water is a social good

There is no single, universally accepted definition of social goods and services.[2] One widely used definition is that social goods are those that have significant "spillover" benefits or costs. Literacy is a social good, for example, because it benefits not just literate individuals but also makes possible a higher level of civilization for all members of a society. Widespread availability of clean and affordable water is a social good under this definition because such availability improves both individual and social well-being.

Access to clean water is fundamental to survival and critical for reducing the prevalence of many water-related diseases (UN, 1997). Indeed, piped water is typically one of the first community services people seek as communities develop, even before electricity, sanitation, or other basic services. Ensuring that the public receives an adequate supply of social goods requires some level of governmental action since purely private markets often do not find it profitable to provide social goods. For example, as noted above, water quality affects public health, both in the short-term and the long-term. However, private water sellers have little or no incentive to mitigate long-term water-quality issues that do not affect the salability of the water (e.g., carcinogens that do not affect the taste, odor, or appearance of water). Similarly, improvements in water-use efficiency and productivity are often economically beneficial to society as a whole, but may reduce revenues to water sellers. Completely "free" markets would not encourage private sellers to improve either water quality or water-use efficiency.

Other dimensions of water supply also have a social good character and therefore require governmental action, oversight, or regulation. Collection, storage, treatment, and distribution of water often require large capital facilities that exhibit economies of scale. When economies of scale exist, there is a spillover benefit (lower average cost) to having a single large reservoir, for example, rather than multiple smaller reservoirs. Since privately owned and operated monopolies will maximize their profits by providing less of their product than is efficient (and thus artificially raising its price), government review and control of capital investments is often appropriate and desirable.

Modern societies usually recognize that markets will be more efficient and effective if social goods are regulated to some degree by government, and in some instances provided directly by government (e.g., energy, communications, transportation, education, criminal and civil courts, police, and military forces). Furthermore, many development economists and theorists urge widespread provision of at least some social goods as a prerequisite for the transformation of poorer economies into highly productive, modern economies.

Because water is important to the process of economic development, essential for life and health, and has cultural or religious significance, it has often been provided at subsidized prices or for free in many situations. In theory, this makes water available to all segments of society, even the poorest. This is politically popular but brings with it a financial burden because society must pay for the subsidy. It can also encourage wasteful use of water, and the perverse result that many of the poor do not have access to clean water at reasonable prices because those who have access use more water than they need. Balancing these public and private benefits is the challenge discussed below.

2.5 Water is an economic good

An *economic good* is any good or service that has an actual or potential value as an item of exchange. A good that is not "economic" is either without value entirely or has value to no one but its owner. Economists would say, in the first case, that the good in question (for example, a small quantity of sand in a desert environment) has neither exchange value nor intrinsic value. In the second case, economists would say that the good in question does not have exchange value, but does have intrinsic value.

Frustration over the failure to meet basic needs for water has been growing over the past decade after the massive effort of the International Drinking Water Supply and Sanitation Decade

---

[2] Economists often mean "goods and services" when they say only "goods". We also use this convention.

(1981–1990). Despite an impressive increase in the number of people with access to clean water, the number without access remains unacceptably high. During the 90s, mobilization of the financial, engineering, and physical resources required to supply clean water to those without it was recognized to be infeasible without more efficient use of water and a rethinking of national and international water priorities and policies. Among these was the potential value of applying economic tools and principles. Consequently, the International Conference on Water and Environment, held in Dublin, Ireland in January 1992, included the following principle among the four so-called "Dublin Principles":

> "Water has an economic value in all its competing uses and should be recognized as an economic good" (ICWE, 1992).

Of the four "principles" enunciated in Dublin, this one has stirred the most debate and confusion. Water is essential for human life. Treating it solely as a commodity governed by the rules of the market implies that those who cannot afford clean water must suffer the many ills associated with its absence. However, making it available at subsidized prices can lead to inefficient use and short supply. The "needle to be threaded" in water management is how to get the most value from water that is available, while not depriving people of sufficient clean water to meet their basic needs.[3] The complete commodification of water, however, is not a necessary consequence of the movement toward management of water as an economic good.

What does recognition of water as an economic good mean?[4] Among other things, it means that water has value in competing uses. Managing water as an economic good, broadly defined, means that water will be allocated across competing uses in a way that maximizes the net benefit from that amount of water. Allocation of water can take place through markets, through other means (e.g., democratic or bureaucratic allocations), or through combinations of market and non-market processes.

A broad economic approach to water management does not inevitably lead to management of water as a commodity in all aspects. For example, water pricing that subsidizes the fixed charge portion (for the physical water connection) of water rates, but imposes a volumetric charge (for actual water used) that reflects the highest value use of water treats each unit of water consumed as a commodity, but treats the piped connection itself as a social good. This pricing scheme could allow the poor to satisfy their basic water needs but also reduce wasteful use of water.

What does "water will be allocated across competing uses in a way that maximizes its value to society" mean, in practice? This is where differing interpretations and vocabulary have caused, and continue to cause, considerable confusion and debate in the international water community. An illustration helps to see the meaning of this phrase:

Suppose that a group of fishermen can leave an additional volume of water in a river to enhance fisheries or sell that water to a nearby factory. If the factory in our example were willing to pay more than the fishermen will benefit by leaving the water in the river, the fishermen can make money by selling the water. The factory makes money after buying the water, or they wouldn't purchase it. This means that a water trade would increase the combined net benefits of water use to the fishermen and the factory. Unless there are adverse effects from the water trade on third parties, or "external costs" that haven't been accounted for (and we note that there usually *are* such costs, such as the oft-ignored ecological values of leaving the water in the river), the fishermen and factory lose, economically, if the water is not sold (allocated) to the factory. Similar logic applies

---

[3] Gleick (1996) discusses the concept of "basic needs" for water in the context of international statements and fundamental human requirements. He estimates a "basic water requirement" for domestic uses and argues that these uses should be considered essential social goods. He also notes that most people can afford to pay for basic water needs, but that when they cannot, governments should subsidize the small amounts of water involved.

[4] Much has been written on this subject (for example, Perry et al., 1997; Rogers et al., 1998; McNeill, 1998; Briscoe, 1996 and 1997; Garn, 1998). Rogers et al. (1998) discuss this issue at length, including examples from Thailand and India. Their discussion emphasizes estimation of the costs and benefits of ecological, cultural, and social factors under current conditions. Costs and benefits can change, perhaps significantly, if property rights and rules, social preferences, technology, or institutions change.

if there is another party who would pay more than the factory for the water. Then the water should be sold (allocated) to that party. When all opportunities to increase net benefits by re-allocating water have been captured, water will have been "allocated across competing uses in a way that maximizes its value to society".

While these principles are clear in theory, the real world is far more complex. Such transactions often entail third-party impacts, and there are many benefits of water that can never be adequately measured in economic terms. Thus, in our example above, the fishermen may never be able to assess the true value of leaving the water in the river, affected downstream users may not be consulted about the negotiations or may be unable to voice their concerns politically, and unexpected chains of ecological impacts may result.

3 SOME SPECIAL CHARACTERISTICS OF WATER

3.1 *Water can be a renewable or a non-renewable resource*

In most forms, water is a renewable resource, made available by the natural hydrologic cycle of the coupled atmospheric-oceanic-terrestrial system. In this sense, continued flows of water are not affected by withdrawals and use. Unlike non-renewable resources such as coal or oil, the amount of water available for use in a basin in the future is not necessarily altered by past withdrawals of water in that basin.

Not all natural waters are renewable, however, and some that are renewable can be made non-renewable through human actions. Some groundwater basins and lakes, for example, have extremely slow rates of recharge and inflow. Water extracted from these basins or bodies of water in excess of the natural recharge or inflow rate is, therefore, equivalent to pumping oil – it reduces the total stock available for later use – and hence, is non-renewable and exhaustible. Contamination of a groundwater stock, similarly, can make a renewable resource into a non-renewable resource. Finally, human actions to modify watersheds, such as cutting forests or paving land, can affect the overall hydrologic balance, reducing recharge or flow characteristics and altering timing, availability, and renewability of water. In extreme cases, this can exhaust a formerly renewable resource.

Whether a particular water resource is renewable or exhaustible is important for international trade discussions. All natural waters, therefore, cannot be treated alike. In fact, how the World Trade Organization (WTO) treats them will depend on their classification as exhaustible or renewable resources; and if renewable, on the minimum flows required to sustain animal and plant life or human health. All exhaustible stocks of water may qualify as non-renewable mineral resources for exemption under the General Agreement on Tariffs and Trade (GATT) Article XX(g). Some renewable flows of water may qualify for an exemption under GATT Article XX (b): specifically those that are "necessary to protect human, animal or plant life or health". These issues are discussed at greater length later in this paper.

3.2 In-situ *water provides ecological benefits*

Unlike mineral deposits, fresh water *in-situ* is often vital to human, animal, and plant health, yet these benefits are rarely protected or even considered by private financial markets or trading systems. Water bodies provide habitat for aquatic life. Riparian systems provide moisture for vegetation and terrestrial biota, nutrient transport between one ecosystem and another, recreational and transportation opportunities, and aesthetic benefits. Larger systems such as the Great Lakes provide broad regional climate and weather services. Reducing water quantity or the quality of a water body by means of large-scale withdrawals or transfers may significantly alter these in situ benefits. Changing the timing of flows in a river, even when quality and total quantity remain unchanged, may also alter ecological conditions.

Diversions or transfers of water from watersheds to other regions have led to many ecological and human health disasters. The diversion of water from the Amu Darya and Syr Darya rivers in Central Asia has caused the destruction of the Aral Sea ecosystem, the extinction of the Sea's endemic fish populations, the dramatic shrinking of the Sea itself, and widespread local health problems associated with the exposure and atmospheric transport of salts. Withdrawals of water from many rivers and streams in North America and Europe have led to reductions and extinctions in many fish populations, particularly anadromous fish, which are born in freshwater rivers, migrate out to the open ocean, and then return to freshwater to spawn. Depletion of river flows has severely damaged river deltas and local communities, such as in the Sacramento/San Joaquin delta in California, the Nile River delta in Egypt, and the Colorado River delta in Mexico.

The transfer of water from one ecosystem or eco-region to another may support economic development, but it also runs the risk of contributing to or accelerating the loss of ecosystem integrity (Linton, 1993) or causing adverse economic effects in the area of origin. Measuring or quantifying these benefits of water in economic terms would be necessary to incorporate them into decisions to trade, market, or manage water, yet as discussed above such measurements are complex, often incomplete and inaccurate, and hence rarely attempted. Until consistent, standard approaches to valuing in situ benefits of water are developed, it will be difficult to develop equitable and consistent rules governing out-of-basin transfers of water.

3.3 *Water has moral, cultural and religious dimensions*

Water has more than economic and ecological importance; it has cultural or symbolic importance as well. It figures prominently in religious rituals such as baptism and ritual bathing, and in the national identities of many native peoples (Graz, 1998). Since water is so fundamental to life in all forms, deep-seated feelings may be relevant to water-management decisions. For example, strong concerns about what is fair or just may arise when water supplied to urban dwellers decreases water availability in rural areas, or when water supply to urban residents whose basic need for water is not being met is blocked to protect rural economies or natural systems.

Moral dimensions of water management intersect with the property rights issues that underlie economically efficient allocation of water. If local people "own" or have a right to water in its natural place, they must be persuaded to voluntarily accept removal of water from its natural place for the reallocation of water to be efficient. Even when outsiders are willing to pay very large sums for water – perhaps enough to make locals extremely rich, in money terms – locals may be unwilling to voluntarily accept such trade. That is, when water in situ or from a particular source has cultural or symbolic significance, as well as its usual uses, it may have very different value to people of different cultures. This is perfectly rational and understandable within economic theory, as discussed further in Box 2.

Box 2. Willingness-To-Pay and Willingness-To-Accept.

Economists define Willingness-To-Pay (WTP) as the value of something to those who don't own it at present. Alternatively, Willingness-To-Accept (WTA) reflects the value of something to those who own it already. WTA is the amount one needs to be paid to voluntarily accept the loss of that thing. In practice, the WTP of any traded good is greater than the WTA of that good; if not, sellers would not sell. When WTA exceeds WTP, buyers would not buy. Some people – most notably indigenous people or those with a deep sense of connection to the place in which they live or the customs by which they live – may feel as though no amount of money can compensate for the loss of something, such as the bulk removal of *in-situ* water.

A parallel example, for oil, is that of the U'wa people of Colombia. They have stated that "there is no possible compensation" for extraction of oil from their lands. Indeed, they feel so strongly on this issue that they have vowed to commit mass suicide if oil is removed. So far, the Colombian Supreme Court has upheld their property rights in this regard, although Occidental Petroleum has been permitted to construct oil wells up to the perimeter of their tribal lands. Hanemann (1991) has demonstrated that this type of situation is more likely to occur when the "something" in question is perceived by its owner to be unique. Water and its traditional local uses (e.g., aesthetic, religious, or cultural) may be perceived as irreplaceable, and locals may therefore be unwilling to trade water for money.

4 HOW CAN WATER BE MANAGED AS A SOCIAL AND ECONOMIC GOOD?

As noted above, we currently face the challenge of finding ways to simultaneously manage water as a social and economic good. Two principles for achieving this goal are presented in this section. At the end of this paper, we offer some principles for balancing the risks and benefits of privatization of freshwater supplies. But the two principles presented in this section are not about balancing risks and benefits. They are methodological, and are therefore presented prior to our discussion of globalization and privatization.

4.1 Create valuation processes; not value estimates

Following the Dublin meeting, the UN Conference on Environment and Development, held in Rio in 1992, clearly recognized that economics must play a part in efficient water management:

> "Integrated water resources management is based on the perception of water as an integral part of the ecosystem, a natural resource, and a social and economic good…" (UN Agenda 21, Chapter 18.8).

The theory of allocating water across its competing uses, however, is difficult to implement in practice because there may be few or no practical ways to measure all of the costs and benefits associated with alternative uses of water. Estimates of the dollar value of ecosystem or cultural values, for example, nearly always involve large uncertainties, and may be misused in public policy processes. Those who approach water management from a narrow economic perspective argue that these practical difficulties can be overcome by more complete cost-benefit analyses and other quantitative efforts. Eventually, so they argue, our ability to quantify environmental, cultural, and distributional impacts will be strong enough to allow decision-makers to weigh the monetary gains from managing water as an economic good against any adverse social impacts.

Although this may be true in some instances – when volumes of data and large data-analysis budgets exist – it is also untrue in most instances. Efforts to simultaneously manage water as an economic and social good must avoid intellectual arrogance. In most policy settings, we will never have reasonable estimates of the monetized (dollar-denominated) values of ecological and cultural dimensions of water management alternatives. Fortunately, economic theory and practice do not require ex ante knowledge of these values before water can be managed simultaneously as an economic and social good.

A broad economic approach recognizes that some significant benefits and costs – especially cultural, ecological, and distributional benefits and costs – will not be quantifiable in practice. Consequently, cost-benefit analysis is treated as "interesting" and "useful" but not decisive in policy making processes. The broad economic approach involves quantifying costs and benefits when doing so is feasible and affordable, but more importantly seeks to put into place stakeholder participation processes. People know what they value and how much they value it, even when they or experts cannot quantify those values. Open and full participation is the first change in institutions (e.g., formal or informal property rights and rules) needed to eventually allow water to be allocated across competing uses in ways that are economically efficient and that are accepted as fair and equitable by stakeholders. The broad economic approach recognizes that one need not quantify the relative value of alternative uses for water in advance (*ex ante*), in order to reach desirable policy outcomes. One can establish democratic and economic processes that eventually identify and lead society to desirable outcomes; outcomes that may in fact be unforeseeable in advance.

4.2 Get the facts

Efforts to simultaneously manage water as an economic and a social good are vulnerable to misinformation and disinformation (deliberate misinformation). There are many powerful interest groups that want water managed as only a social good or as only an economic good.

Box 3. The perception of affordability problems for the poor.

An incident that occurred in the Curu Valley of Ceara, Brazil, in 1995, is an excellent example of disinformation about the affordability impacts on the poor of managing water as an economic good. In February of that year, the new president of the water utility announced to the press that the utility was planning to impose a water tariff of $25/1,000 m^3. Until that time, water had been regarded as a good that should be provided for free by the government. Public reaction to the tariff was stormy. A cartoon published in a local newspaper, reproduced below, shows the water utility president stopping a poor farmer already burdened by the water he is carrying, and maliciously asking: "Wait a minute! Have you already paid the tariff?"

Ironically, the water tariffs were proposed only for large water withdrawals (2,000 l/h or more) and the proposed tariff would have exempted the small farmer in the picture. Although it is true that small farmers and others may have been unaware of the exemption, there is evidence that the elite in this region of Brazil have in the past claimed that the region needs subsidies for the sake of the poor and the famished, while in fact subsidies have been appropriated by large landowners and politicians. Successful management of water as a social and economic good requires that the needs of the poor be addressed, and that misperceptions and disinformation be overcome (Kemper, 1996).

Source: Diário do Nordeste. February 15, 1995.

For example, rapid implementation of private-public partnerships has, in too many cases, blatantly disregarded the needs of the poor. But those who claim to defend the poor are not always doing so. In Mexico, the main opposition to raising water fees comes from powerful irrigation interests who argue that poorer farmers will suffer most (Muñoz, 2001). While this may be true in some parts of Mexico, it is certainly not always the case. Similar claims in Brazil were documented to be inaccurate, as described in Box 3.

5 GLOBALIZATION: INTERNATIONAL TRADE IN WATER

The world's water is unevenly distributed, with great natural variations in abundance. Indeed, the complex and expensive water systems that have been built over the past few centuries have been designed to capture water in wet periods for use in droughts and to move water from water-rich regions to water-poor regions. As domestic, industrial, and agricultural demands for fresh water have grown, entrepreneurs have created a wide range of markets for water, leading to various forms of international water trading and exchanges. Water has long been transferred among regions and uses via canals or pipelines, but growing demand and uncertainty of supply in the face of population growth, climate change, and other factors is motivating many states, provinces, and even individuals and corporations to examine new ways to transfer or trade water from areas with water "surplus" to areas with unmet needs. These include longer and longer pipelines, the sale of various forms of bottled water, the physical transport of liquid water in tankers or large bags towed through the ocean, and even the capture and use of icebergs.

In the past, most large-scale transfers of water occurred within national and political borders. Agreements were also common among nations that share a watershed, such as the US and Mexico over the Colorado, the Sudan and Egypt over the Nile, and many others. Now, however, proposals for bulk water transfers are being made at international, and even global, levels between parties that do not share a watershed. In recent years Alaskan, Canadian, Icelandic, Malaysian, Turkish, and other waters have been proposed as sources for international trade in bulk water. Besides the historically important environmental and socioeconomic implications of water transfers, the possibility of large-scale bulk trading of fresh water has now become an issue in international trade negotiations and disputes.

Treating water as a good to be traded has a wide range of implications. The possibility of bulk water transfers has caused concern in water-abundant regions that a global water-trading regime might lead to the requirement that their resources be tapped to provide fresh water for the rest of the world, at the expense of their own environment and people.

5.1 Current trade in water

Water can be traded as either a raw (bulk) or value-added product. Indeed, a large and rapidly growing international market already exists for various forms of processed, value-added water – particularly bottled waters. Bottled water sales worldwide in the mid-90s exceeded 50 billion liters and such sales have been increasing by nearly 10% a year since the 70s (Table 1). In 1999, the bottled water industry in the US alone generated nearly $5 billion from the sale of more than 17 billion liters – up from less than 2 billion liters annually in the mid-70s (Figure 1). Most of this is domestically produced: about 8% was imported in 1999.[5] Figure 2 shows the sources of bottled water imported into the US. Canada, France, and Italy accounted for more than 90% of the US imports in 1999.

Table 1. Global bottled water sales.[6]

Country/Region	1996 Sales (Million liters)	Projected 2006 sales (Million liters)	Annual % growth
Australasia	500	1,000	11
Africa	500	800	4
CIS	600	1,500	13
Asia	1,000	5,000	12
East Europe	1,200	8,500	14
Middle East	1,500	3,000	3
South America	1,700	4,000	7
Pacific Rim	4,000	37,000	18
Central America	6,000	25,000	11
North America	13,000	25,000	4.5
Western Europe	27,000	33,000	2.5
Total	57,000	143,800	

[6] Modified from www.soc.duke.edu/~s142tm16/World%20Markets.htm

[5] www.soc.duke.edu/~s142tm16/world.htm

Bottled water sales are also increasingly prevalent and important in poorer countries (Table 2). We believe that bottled water sales must not be considered acceptable substitutes for adequate municipal water supply. Bottled water rarely provides adequate volumes of water for domestic use, and the costs of such water are typically exorbitant. There may be circumstances when readily available (but non-potable) water for domestic uses, plus high quality and affordable bottled water for drinking, are adequate, but we could not find any examples.

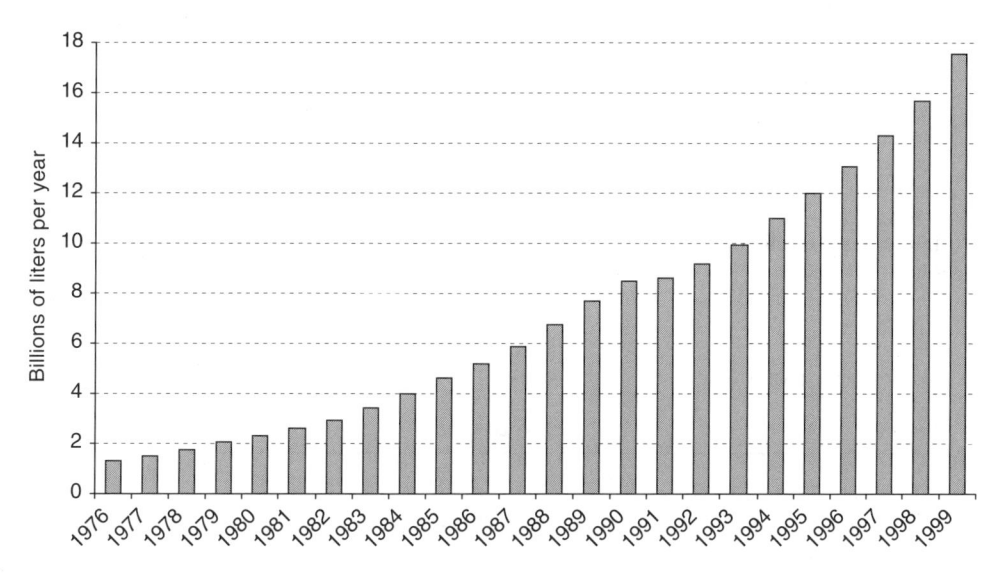

Figure 1. US Bottled water sales: increasing by 10% annually.

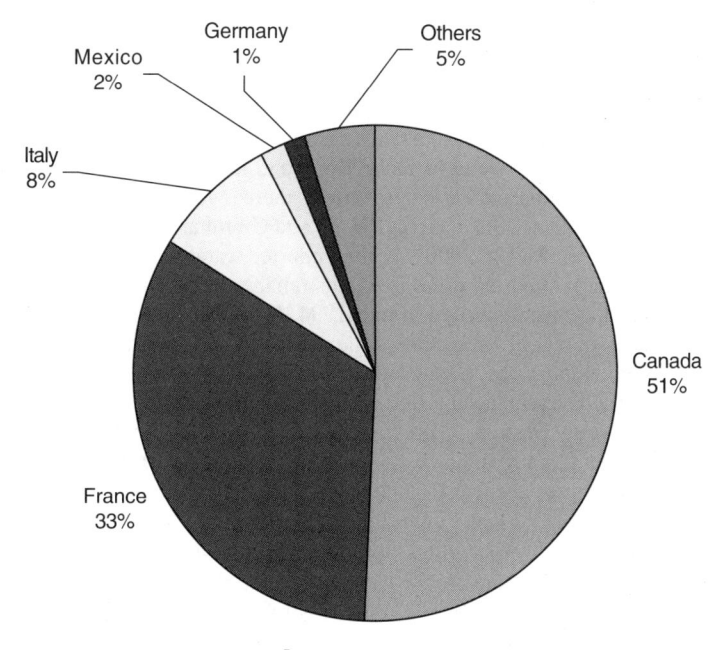

Figure 2. US Bottled water imports in 1999[7].

---

[7] www.soc.duke.edu/~s142tm16/imports.htm

Table 2. Bottled and vended water: urban and rural use (WHO, 2000).

Country	Year	Source of water	% of Population that consumes bottled or vended water Urban areas	Rural areas
Angola	1996	Tanker Truck	25.2	0.8
Cambodia	1998	Vendor	16.0	3.5
Chad	1997	Vendor	31.5	0.5
Dominican Republic	1996	Bottled Water	37.0	6.3
Ecuador	1990	Tanker Truck	16.0	7.0
Eritrea	1995	Tanker Truck	30.5	1.4
Guatemala	1999	Bottled Water	25.5	7.1
Haiti	1994	Bottled Water	26.0	0.3
Jordan	1997	Tanker Truck	1.0	10.6
Libyan Arab Jamahiriya	1995	Tanker Truck	6.8	13.9
Mauritania	1996	Vendor	53.0	0.9
Mongolia	1996	Vendor	16.0	1.0
Níger	1998	Vendor	26.4	1.9
Oman	1993	Bottled Water	39.5	42.0
Syrian Arab Republic	1997	Tanker Truck	4.1	11.3
Turkey	1998	Bottled Water	14.9	1.0
Yemen	1997	Bottled Water	14.6	0.1

Water traded as bottled or value-added water is covered by international trade rules like any other economic good. Much of the debate and concern at present is focused on proposals to trade bulk, unprocessed water across international borders, either for later processing or for use for municipal or industrial purposes. We exclude from this discussion in-basin trades or transfers of water by countries that share a watershed – such international trades occur all the time, although typically through political agreements rather than market deals.

Interestingly, there have been relatively few long-term, international, out-of-basin water trades to date, although various proposals have been put forward. On occasions, bulk water has been brought by tankers to Pacific or Caribbean islands during drought to supplement limited local supplies. In the 60s, tankers brought water to Hong Kong and loaded freshwater from the Houston River as backhaul cargo to Curaçao for use in refineries there (Meyer, 2000). In the 80s, there was a shuttle trade between the River Tees in Great Britain and Gibraltar, when Spain shut off its water supply for political purposes (Meyer, 2000). Aruba imported water by tanker from Dominica. The small island nation of Nauru imported water from Australia, New Zealand, and Fiji (Box 4). In rare circumstances, such as the severe drought in mid-1994, Japan has imported limited quantities of water by tanker to maintain refinery and automobile production. This water has come from Alaska, Vietnam, South Korea, Hong Kong, and China (Sugimoto, 1994; Jameson, 1994; AFX News, 1994; Brown et al., 1995). The provincial government of Mallorca contracted with a tanker company for shipments of water from Western Spain until desalination plants could be built (Huttemeier, 2000). In general, however, the very high cost of tankered water is a barrier to such transfers, especially if long-term supplies are needed. In such situations, the maximum amount a buyer will be willing to pay does not exceed the cost of alternatives, such as desalinated water. As a result, most of these transfers are phased out when other cheaper and more reliable solutions are found.

At present, there are few major proposals pending for large-scale, long-term transfers of water across international borders (Box 5). Turkey has offered water from the Manavgat River and has been negotiating with Israel (Ekstract, 2000; Turner, 2001). Tankers or giant bags would transfer the water to an Israeli coastal port, where it would be treated and used. The alternative for Israel would be reallocation of existing water resources within Israel (from, for example, agriculture to cities), improvements in water-use efficiency, or new supplies from elsewhere, such as desalination.

Box 4. Transfers of water out of a water shed.[8]

International, Out-of-Basin Transfers

Aruba (Netherlands Antilles) has received water from Dominica by tanker.

In the 60s, Hong Kong received some water via tanker. Prior to the return of Hong Kong to China, the city received 75% of its potable water from China. 50% was piped in from the mainland. The remaining 25% was piped from Lantau Island.

Nauru, an island nation located in the Central Pacific, has received as much as a third of its water as return cargo from ships exporting phosphate. The water is from Australia, New Zealand, and Fiji.

In the 90s Tonga regularly received water by tanker and the Canary Islands imported practically all of its potable water as bottled water, before beginning to build desalination plants for local domestic and industrial use.

Unusual Domestic, Out-of-Basin Transfers

Hong Kong's islands, Lamma and Ma Wan, receive water by submarine pipeline.

Malaysia's Penang receives some of its water from the Malaysian peninsula via submarine pipes.

China's Xiamen Island receives 50% of its supply from the mainland.

St. Thomas and St. John (US Virgin Islands) have received water from Puerto Rico transported by sea intermittently since 1955.

In the Bahamas, New Providence received an average of 21% of its total water supply from Andros Island from 1978–1987. The water was transported by barge. In 1987, 31% of the total supply was transported. The Bahamas now have about 54,000 m^3/day of desalination capacity.

Mallorca receives water by tanker from the Spanish mainland, but has drawn up plans to lay underwater pipes or build a desalination plant to avoid having to bring in drinking water by tanker each summer. Turkey sends water to Turkish Cyprus by tanker.

The smaller Fijian Islands commonly received water from the larger islands starting in the early 70s, especially during drought periods.

[8] Sources: Coffin & Richardson (1981), Water Supplies Department (1987), Zhang & Liang (1988), Fiji Country Paper (1984), Gattas (1998), Gleick (2000), Lee Yow Ching (1989), Swann & Peach (1989), Huttemeier (2000), Brewster & Buros (1985), Lerner (1986), Jacobson & Hill (1988), UNESCO (1992), Meyer (2000).

Box 5. International, out-of-basin transfers under consideration.[9]

Italy is considering importing water to its dry southern regions by building a pipeline under the Adriatic Sea to pump in supplies from Albania.

In October 2000, Austria claimed that it could supply all 370 million people in the EU with well or surface water needing treatment, just ahead of EU plans to liberalize its water industry. Austrian ministers, including the agricultural and environmental minister, argue the economic advantages outweigh the political liabilities.

Spain is considering importing water from the Rhone River with a pipeline extending from Montpellier, France to Barcelona.

Israel is negotiating to buy water from Turkey's Manavgat River. Negotiations will include an Israeli request for an annual amount of 15–25 million cubic meters of water with the possibility of doubling the amount for a period of 5–10 years. Israel sees Turkish water imports as a quick alternative to its plans to build a desalination plant. Water would be shipped from the Manavgat, in Southwestern Turkey, to the Israeli port of Ashkelon, where it could be further distributed.

[9] Sources: Boulton & Sullivan (2000), Rudge (2001), Demir (2001), Financial Times (2000).

Whether the proposed bulk transfer will be able to surmount the economic and political hurdles facing it remains to be seen, but as of this writing, solicitations are being considered and negotiations are continuing on price. Similarly, Spain is considering reviving an old proposal to import water from the Rhone River with a 320-kilometer aqueduct extending from Montpellier, France to Barcelona (Financial Times, 2000). If implemented, this would be the first trans-basin water deal in the European Union (EU), but it is unlikely, even if approved, to be completed within a decade.

5.2 The rules: international trading regimes

Rules governing international trade are complex and often contradictory. In recent years, efforts to implement standard rules have been developed in several international fora, and these rules have become increasingly sophisticated and important to the global economy. At the same time, they have become increasingly controversial, as their implications for the environment, civil society, and local economies become clearer. In this section we discuss two important agreements as they relate to the globalization of water resources: the General Agreement on Tariffs and Trade (GATT) and the 1994 North American Free Trade Agreement (NAFTA). GATT is the overriding international trade agreement, and NAFTA is an excellent example of how the international water trading debate has been influenced by a regional trade agreement. Other regional trade agreements have been signed or are being negotiated, and future assessments might consider these in more detail. For example, the General Agreement on Trade in Services (GATS), which came into force in January 1995, is the only set of multilateral rules covering international trade in services, which could be particularly relevant for efforts to privatize water, discussed later in this report (Box 6).

GATT provides the basic legal architecture that governs international trade for more than 140 member countries of the WTO. NAFTA governs trade in goods between the US, Canada, and Mexico. While the two trade regimes have many similarities, there are also several provisions, described below, that distinguish the two. For the discussion about water, NAFTA is particularly important among the regional trade regimes because it includes the US, the largest national economy in the world, and Canada, the nation that has expressed the greatest concern and taken the strongest actions with regard to international bulk water trading.

It is worth noting at the outset that there is little legal precedent pertaining directly to international trade in water, making it difficult to predict the outcomes of current and future trade disputes in this area with certainty. However, commercial pressures to export water are increasing, making resolution of these ambiguities an important goal. In addition, adverse, even virulent public sentiment over several proposed exports highlights the need to resolve and clarify issues (Barlow, 1999; FTGWR, 2001).

Box 6.   The General Agreement on Trade in Services (GATS).[10]

The GATS covers all internationally-traded services with two exceptions: services provided to the public in the exercise of governmental authority, and, in the air transport sector, traffic rights and all services directly related to the exercise of traffic rights. The GATS also defines four ways in which a service can be traded:
  – Services supplied from one country to another (e.g., international telephone calls), officially known as "cross-border supply".
  – Consumers from one country making use of a service in another country (e.g., tourism), officially known as "consumption abroad".
  – A company from one country setting up subsidiaries or branches to provide services in another country (e.g., a bank from one country setting up operations in another country), officially known as "commercial presence".
  – Individuals travelling from their own country to supply services in another (e.g., an actress or construction worker), officially known as "movement of natural persons".

---

[10] www.wto.org/english/tratop_e/serv_e/gats_factfiction1_e.htm

The degree to which countries will be able to impose controls on the exportation of water will hinge upon the determination of whether bulk water in its natural state is considered a "product", and if so, whether exemptions in the trade agreements are applicable. Treating water as a product (a good) on the global market typically implies that transfers will be governed by international trading obligations. This has raised concerns among local communities and environmental groups that water resources may suffer the fate of other global common resources: overexploitation with significant restrictions on national environmental control and restraints. For example, some trade analysts interpret WTO and NAFTA rules as requiring that water must continue to be traded in bulk once it has begun to be traded. These issues are discussed below.

5.2.1 *The General Agreement on Tariffs and Trade (GATT)*

GATT is a comprehensive international trade agreement and provides the basic legal structure that governs international trade for members of the WTO. All potential commodities are defined and described in the "Harmonized Tariff Schedule" (HTS). This schedule, used by the US and all WTO countries, includes water of all kinds (other than seawater, which is described in a separate heading) under Section 2201 (Appleton, 1994).

> "2201.90.0000: Other waters, including natural or artificial mineral waters and aerated waters, not containing added sugar or other sweetening matter nor flavored; ice and snow".

The existence of an HTS number means that there is a mechanism under which shipments of fresh water can be processed by US Customs and comparable customs organizations of other nations.

Some communities and environmental groups are concerned that if bulk water is traded as a product anywhere in the international community, GATT Article XI, "General Elimination of Quantitative Restrictions" would be interpreted to prohibit export bans by any country choosing to sell bulk water. Article XI, Section 1 states:

> "No prohibitions or restrictions other than duties, taxes or other charges, whether made effective through quotas, import or export licences or other measures, shall be instituted or maintained by any contracting party on the importation of any product of the territory of any other contracting party or on the exportation or sale for export of any product destined for the territory of any other contracting party".

Once bulk water transfers are initiated by domestic industry, Article XI plays a significant role in constraining WTO member governments' ability to establish policies, programs, or legislation that regulate, curtail, or eliminate such transfers. However, the language of the agreement is unclear as to whether these restrictions apply only to specific, actual bulk water trades once they have begun or apply to all potential bulk water trading arrangements once trade in bulk water has begun between any WTO signatories.

Under either interpretation, however, other portions of GATT appear to be relevant to the question of whether some trade in bulk water may be limited or constrained by national laws, and hence exempt from the Article XI provisions. For bulk water, the two most relevant clauses upon which a government could base an exemption and adopt measures that restrict trade are found in Article XX:

> "Article XX (b) necessary to protect human, animal or plant life or health".

> "Article XX (g) relating to the conservation of exhaustible natural resources if such measures are made effective in conjunction with restrictions on domestic production or consumption".

There is considerable debate among legal experts as to whether WTO member governments can control bulk water exports based on these resource conservation principles, and there are few legal precedents. The WTO has struck down two previous cases where national governments attempted to challenge unlimited resource trades using these exemptions (WCEL, 1999). In the 1991 Tuna-Dolphin case and the 1998 Shrimp-Turtle case, the WTO expert panels ("arbitration committees" appointed individually for each trade dispute) did not accept domestic environmental protection legislation as a valid basis for imposing trade restrictions (Box 7). Some WTO member countries such as South Africa, Angola, Ecuador, Chile, and Indonesia have publicly stated during a recent

> Box 7. The turtle and dolphin WTO cases: relevant for water?
>
> In two separate cases related to protection of ocean resources, WTO expert panels rejected protection of dolphins and turtles through import restrictions against tuna and shrimp from countries that did not require their boats to use dolphin- or turtle-safe nets. The WTO argued that import restrictions on the basis of the process of production of imported goods were in violation of GATT. Applying this reasoning to possible bulk water trade restrictions might suggest that prohibiting trades that undermine ecological values or human access to water at the area of origin would also be in violation of GATT. In the 1998 Shrimp/Turtle case, however, the US changed the method of enforcement to allow shrimp from any member country as long as they could be shown to have come from boats that do use turtle-safe nets. In June of 2001, the WTO upheld this new interpretation (Lazaroff, 2001).
>
> The tuna/dolphin and shrimp/turtle cases both involved import rather than export restrictions, which are explicitly prohibited. In both cases, the importing country was trying to influence management of natural resources beyond its borders. The concerns raised in the case of bulk water trading are associated with the power of the nation of origin to protect people or ecological functions within their own borders. It is possible that the tuna/dolphin and shrimp/turtle legal precedents would not greatly influence a case involving bulk water export controls imposed by the nation of origin, but this is speculative without an actual test case under GATT. See the further discussion in the text.

Table 3. Some heavy exploited aquifers of the world (Margat, 1996).

Region	Aquifer	Avg. annual recharge (km^3/year)	Avg. annual use (km^3/year)
Algeria/Tunisia	Saharan basin	0.58	0.74
Saudi Arabia	Saq	0.30	1.43
China	Hebei Plain	35.00	19.00
Canary Islands	Tenerife	0.22	0.22
Gaza Strip	Coastal	0.31	0.50
US	Ogallala	6 to 8	22.20
US	Selected Arizona	0.37	3.78

WTO High-Level Symposium on Trade and Environment that environmental protection measures should not stand in the way of economic development (IISD, 1999). The burden of proof that an environmentally based trade restriction is necessary rests on the country imposing the restriction (French, 1999).

Because bulk water, unlike petroleum or minerals, can be a renewable resource depending upon the way it is extracted, some may argue that it falls outside the Article XX(g) exemption. The exemption appears to have been drafted with reference to "mining" of minerals and fossil fuels, but water also has non-renewable characteristics well understood by hydrologists. As noted in Gleick (1998):

> "Freshwater resources typically are considered renewable: they can be used in a manner that does not affect the long-term availability of the same resource. *However, renewable freshwater resources can be made non-renewable by mismanagement of watersheds, overpumping, land subsidence, and aquifer contamination. Water policy should explicitly protect against these irreversible activities*"

Thus, this exemption could be interpreted to apply to stocks of water that were deposited long ago and are not being replenished at a significant rate compared with the rate of use. In such circumstances, a strong argument can be made to support an Article XX(g) exemption for bulk water resources where freshwater water resources are "non-renewable" or exhaustible through overuse or abuse, assuming domestic production or consumption is also limited to prevent non-renewable uses.

There are many examples of groundwater overdraft, where human extraction exceeds natural replenishment, sometimes by a wide margin. Table 3 lists some major groundwater aquifers where annual withdrawals approach, or even exceed, annual recharge. Similar aquifers must therefore be considered "non-renewable" just as stocks of oil are considered exhaustible. In the case of exports of water from the Great Lakes of North America, some have argued that only a tiny fraction of the lakes are "renewable" and that the vast bulk of the stored water was laid down in geologic times (Barlow, 1999). In such circumstances, exports could, if large enough, lead to the irreversible decline in lake levels. In 1999, the International Joint Commission between the US and Canada issued a report concluding that the Great Lakes are non-renewable, with an eye to ensuring that they would be subject to a GATT Article XX exemption (IJC, 2000). Cases where water stocks have been contaminated by human actions also represent the conversion of renewable water resources into a non-renewable resource – appropriate for an Article XX(g) exemption.

In some circumstances, bulk water exports could also be subject to an Article XX(b) exemption if such exports threaten human or ecosystem health. Biologists and ecologists have long understood and demonstrated that some amounts of water are needed *in situ* to protect animal and plant life and health, although the precise quantities needed to maintain adequate instream flows and *in situ* resource values are subject to study and debate. Many specific instream flow requirements have been set for particular watersheds to maintain ecosystem health. Although a specific Article XX(b) exemption has not been tested in the context of bulk water, we believe that it would also support a ban on bulk exports of water when such exports threaten ecosystem or human health.

5.2.2 *The North American Free Trade Agreement (NAFTA)*

Canada, the US, and Mexico have developed the NAFTA as a regional extension of GATT in many ways. For example, the HTS of the US and the Canadian Customs Tariff, in which "ordinary natural water of all kinds" are classified under tariff heading 22.01, also implies that water can be traded internationally as a good. Article 201(1) of NAFTA defines a "good" as a "domestic product as these are understood in the [GATT] or such goods as the Parties may agree, and includes originating goods of that Party" (Yaron, 1996). Hence the categorization of water provided above (HTS Number 2201.90.0000) is applicable under NAFTA as well as GATT.

Several other factors, however, complicate the interpretation of NAFTA rules for water. In 1993, the three NAFTA parties signed a joint declaration to provide explicit protection for *in-situ* water resources and the rights of the country of origin under NAFTA and GATT:

> "Unless water, in any form, has entered into commerce and becomes a good or product, it is not covered by the provisions of any trade agreement, including the NAFTA. And nothing in the NAFTA would obligate any NAFTA Party to either exploit its water for commercial use, or to begin exporting water in any form. Water in its natural state, in lakes, rivers, reservoirs, aquifers, water basins and the like is not a good or product, is not traded, and therefore is not and never has been subject to the terms of any trade agreement".

This joint declaration is the clearest exposition of the intent of the parties to NAFTA to protect natural waters from uncontrolled bulk withdrawals for international trade. But there is disagreement about the extent to which such declarations are binding forms of international law. Some analysts argue that the protections offered by joint declarations are not legally binding and establish no legal obligations:

> "It has long been recognized in international practice that governments may agree on joint statements of policy or intention that do not establish legal obligations ... These documents are sometimes referred to as non-binding agreements, gentlemen's agreements, joint statements or declarations". (Shrybman, 1999a).

At the same time, such declarations can fall under customary law, which can be binding. The 1993 joint declaration has received little subsequent formal support from the three governments (Shrybman, 1999a), and the sentiment expressed could be considered inconsistent with GATT and NAFTA tariff headings, "US Law", and "International Law", both of which define water as a good (Shrybman, 1999a). Official US policy in this area was further confused when then-US Trade Representative, Mickey Kantor wrote in 1993:

> "... when water is traded as a good, all provisions of the agreements governing trade in goods apply ..."

On the one hand, Kantor's statement appears to be a simple reiteration that bulk water, once it has entered trade, must be subject to the existing trade agreements. But some are concerned that Kantor's statement could be interpreted as saying that the joint declaration would no longer apply once *any* NAFTA signatory government permits sale in bulk of any water for commercial purposes (Barlow, 1999). This interpretation would put bulk water back into the realm of "goods to be traded", something environmentalists were hoping the joint declaration specifically and permanently excluded. A less extreme interpretation would be that the signers intended to exempt water in its natural state from trade agreement provisions, except for specific quantities of water that have been put into commerce with the approval of the country of origin. Under this interpretation not *all* waters become open to such trade provisions once *some* water has been traded. Under either interpretation, however, trade analysts agree that once international trade in bulk water has begun, at least the amount of water that has been authorized for trade cannot be withdrawn from trade by an action of the country of origin unless specific exemptions within GATT or NAFTA are satisfied.

The Canadian government took action to prevent bulk water trading in 1994 because it was concerned that its NAFTA obligations could impinge upon its ability to develop national water policy. On January 1 of that year, the North American Free Trade Implementation Act was proclaimed into force by the Canadian Parliament, including a rider that specified that (1) nothing in either NAFTA or its implementing legislation, except the provision on tariff elimination, applied to water, and (2) "water" in this context meant natural surface and ground water in liquid, gaseous or solid state, not including water packaged as a beverage or in tanks. As appealing as this definition is for environmental and public interest groups looking to forestall bulk exports, it is a matter of Canadian domestic legislation and may not be binding on NAFTA dispute resolution panels (Appleton, 1994). In February 1999, the government of Canada requested that each province implement a voluntary moratorium on bulk exports, but the federal government has yet to enact a national ban. Indeed, in spring 2001, the premier of Newfoundland called for lifting the ban on bulk water exports, reopening the contentious debate there (MacDonald, 2001).

As noted earlier, whether fresh water *in situ* falls under NAFTA's (or the GATT's) definition of a good has not been legally settled. If, however, bulk fresh water is considered a good under the NAFTA definition, there are three conditions of NAFTA that affect international trade. First, similar to GATT Article III, "National Treatment", each signatory country must accord businesses and investors from the other signatory countries the same preferential treatment that it accords its own businesses and investors for both goods and services. Article 1102, "National Treatment" states:

"Each Party shall accord to investors of another Party treatment no less favorable than that it accords, in like circumstances, to its own investors with respect to the establishment, acquisition, expansion, management, conduct, operation and sale or other disposition of investments"

This means that any NAFTA country cannot treat other NAFTA bulk water exporters or importers any differently than it treats its own bulk water exporters or importers.

Second, NAFTA Chapter 11 also allows investors in any signatory country to sue the government of either of the other two signatories if that government takes some future action (usually legislation) to "expropriate" that company's profits (Barlow, 1999). According to the provisions of Article 11.10, "Expropriation and Compensation":

"No Party shall directly or indirectly nationalize or expropriate an investment of an investor of another Party in its territory or take a measure tantamount to nationalization or expropriation of such an investment ("expropriation"), except:

(a) for a public purpose;
(b) on a non-discriminatory basis;
(c) in accordance with due process of law and the general principles of treatment provided in Article 1105;
(d) upon payment of compensation in accordance with paragraphs 2 to 6".

Chapter 11 issues have already been raised in the context of bulk water exports. In the fall of 1998, a Santa Barbara, California company called Sun Belt Water, Inc. sued the government of Canada under NAFTA Chapter 11. Sun Belt lost a contract to export water to California when the British

Columbia (BC) provincial government banned bulk water exports in 1995 (Shrybman, 1999b). While the Chapter 11 suit cannot overturn the BC law, it can make the government of Canada liable for the profits that Sun Belt would have made on this contract had BC not passed its export ban. This makes federal, state, and provincial governments reluctant to implement legislation regulating commerce in natural resources.

We note that while the government of Canada may be liable under these provisions of NAFTA, the profits Sun Belt might have actually received are highly uncertain – indeed, an argument can be made that any profits were unlikely. Water is very expensive to move from one place to another, and commands a high price only in the luxury form of bottled water – a form of water all participants in this debate agree is already covered by trading rules. Moreover, the amount that potential importers are willing to pay is capped at the cost of alternative sources, including desalination, making the size of possible profits highly speculative. Even assuming the sellers could command a price of $1.50\$/m^3$ – more than double what most municipalities and industries currently pay for reliable urban supplies – optimistic tankering costs for water are between 2 and $4\$/m^3$ or even higher (Bardelmeier, 1995; Huttemeier, 2000), depending on distance. This is why, historically, contracts to tanker water from one place to another have consistently given way to more local solutions, such as reallocation among end users, or desalination. Hence the size of the actual liability to Canada may be small or zero. This issue remains to be resolved.

Third, NAFTA Article 309 states that constraints on exports of any good must be shared proportionally across the signatory countries (Barlow, 1999). This means that if Canada were to start exporting water in bulk and subsequently faced a drought or other shortage, it could not reduce the amount of water exported to the US and Mexico in order to *maintain* unreduced deliveries to domestic customers. All customers must take proportional reductions; this is a de facto extension of the "national treatment" clause, wherein all customers are treated equally. These provisions reinforce the constraint on national sovereignty that arises under GATT: once bulk water is traded as a good under legally valid contracts, it must continue to be traded. This alarms many environmentalists who feel that the best method of protection for natural resources on a watershed scale is domestic legislation.

In sum, our analysis suggests that large-scale, long-term bulk exports of water across international borders are unlikely for many reasons, especially the high economic cost of moving water. Nevertheless, great uncertainty continues to revolve around the legal interpretation of international trade agreements in the context of globalizing water resources. Because of the risk of ecological damages and non-sustainable withdrawals of water for export, we urge clarification of rules governing bulk exports of water. *In particular, we recommend national water policies that explicitly protect water necessary to support human and ecosystem health and prohibit the mining and export of non-renewable water resources.*

6 PRIVATIZATION: PUBLIC-PRIVATE "PARTNERSHIPS"

"Food and water are basic rights. But we pay for food. Why should we not pay for water?" *Ismail Serageldin at the 2nd World Water Forum (2nd WWF), The Hague.*

"Water should not be privatized, commodified, traded or exported in bulk for commercial purposes". *Maude Barlow, International Forum on Globalization.*

"... the purpose of government is to make sure services are provided, not necessarily to provide services". *Mario Cuomo.*

One of the most important – and controversial – trends in the global water arena is the accelerating transfer of the production, distribution, or management of water or water services from public entities into private hands – a process loosely called "privatization". Treating water as an economic good, and privatizing water systems, are not new ideas. Private entrepreneurs, investor-owned utilities, or other market tools have long provided water or water services in different parts of the world. What is new is the extent of privatization efforts underway today, and the growing public awareness of, and attention, to problems associated with these efforts.

issue has resurfaced for several reasons: first, public water agencies have been unable to y the most basic needs for water for all humans; second, major multinational corporations have greatly expanded their efforts to take over responsibility for a larger portion of the water service market than ever before; and third, several recent highly publicized privatization efforts have failed or generated great controversy.

The privatization of water encompasses an enormous variety of possible water-management arrangements. Privatization can be partial, leading to so-called public/private partnerships, or complete, leading to the total elimination of government responsibility for water systems. At the largest scale, private water companies build, own, and operate water systems around the world with annual revenues of approximately $300 billion, excluding revenues for sales of bottled water (Gopinath, 2000). At the smallest scale, private water vendors and sales of water at small kiosks and shops provide many more individuals and families with basic water supplies than they did 30 years ago. Taken all together, the growing roles and responsibilities of the private sector have important and poorly understood implications for water and human well-being.

As a measure of the new importance of privatization, the 2nd WWF in the Hague in March 2000 gave special emphasis to the need to mobilize new financial resources to solve water problems and called for greater involvement by the private sector. Indeed, the "Framework for Action" released at that meeting called for $105 billion per year in new investment – over and above the estimated $75 billion per year now spent – to meet drinking water, sanitation, waste treatment, and agricultural water needs between now and 2025. The Framework called for 95% of this new investment to come from private sources (GWP, 2000). There was enormous controversy at this meeting about the appropriate role of governments and non-governmental organizations, and a planned public workshop and discussion on privatization and globalization of water was cancelled. In addition, the World Bank, other international aid agencies, and some water organizations like the World Water Council are increasingly pushing privatization in their efforts, but they lack a common set of guidelines and principles.

Along with the growing efforts at water privatization, there is rapidly growing opposition among local community groups, unions, human rights organizations, and even public water providers. Protests – sometimes violent – have occurred in many places, including Bolivia, Paraguay, South Africa, the Philippines, and various globalization conferences around the world. Opposition arises from concerns over the economic implications of privatizing water resources, the risks to ecosystems, the power of corporate players, foreign control over a fundamental natural resource, inequities of access to water, and the exclusion of communities from decisions about their own resources. This report reviews why efforts are growing to turn over responsibility for public provision of water and water services to the private sector, the history of privatization efforts, and the risks of these efforts. We also offer some fundamental principles that we believe are necessary to prevent inequitable, uneconomic, and environmentally damaging privatization agreements.

7 DRIVERS OF WATER PRIVATIZATION

In 1992, the summary report from the water conference in Dublin set forth four "principles", including the concept that water should be treated "as an economic good" (ICWE, 1992). This principle is, without doubt, the most important and controversial of the four. It was sufficiently vague to be accepted by the participants, and yet sufficiently radical to cause serious rethinking of water management, planning, and policy. In the years following Dublin, the concept of water as an economic good has been used to challenge traditional approaches to government provision of basic water services. Economists seized upon the idea to argue that water should be treated as a private good, subject to corporate control, financial rules, markets forces, and competitive pricing.

Various pressures are driving governments to consider and adopt water privatization. These pressures fall loosely into five categories (Neal et al., 1996; Savas, 1987):

- *Societal* (the belief that privatization can help satisfy unmet basic water needs).
- *Commercial* (the belief that more business is better).

- *Financial* (the belief that the private sector can mobilize capital faster and cheaper than the public sector).
- *Ideological* (the belief that smaller government is better).
- *Pragmatic* (the belief that competent, efficient water-system operations require private participation).

Privatization efforts in the United Kingdom (UK) and Europe were ideologically driven at first, but are increasingly characterized as commercial and pragmatic (Beecher, 1997). Privatization efforts in the US were initially pragmatic, but are now strongly ideological, as can be seen by the public policy push being given to water privatization by libertarian and free-market policy institutes. Privatization efforts in the developing world can primarily be described as financial and pragmatic, though some argue that the social benefits are significant (GWP, 2000). Interestingly, several countries with strong ideological foundations have also chosen to explore water privatization for pragmatic reasons. China and Cuba, for example, have both recently awarded contracts to private companies to develop and operate municipal water-supply systems and build wastewater treatment plants.

The provision of water to individuals, families, and communities has long been considered an essential public good, and hence a core governmental responsibility. In many countries, including the US, people expect safe drinking water to be distributed to everyone at low or subsidized prices. Yet despite intensive efforts in the 80s and early 90s, more than 1.2 billion of the six billion people on the Earth still lack access to clean drinking water. Nearly 2.5 billion do not have adequate sanitation services (WHO, 2000). This failure is one factor leading governments, companies, NGOs, and individuals to rethink their attitudes and approaches to water management worldwide. With world population eventually expected to increase to nine billion, or ten billion, or even more, water managers have realized that adequate water services cannot be provided without enormous increases in investment, improvements in the efficiency of water capture, storage, distribution, and use, and greater wastewater reclamation and reuse.

Water is an essential resource for economic development. Up to a point, greater water availability increases labor supply and reliability by reducing water-related diseases, lowers constraints on agricultural and industrial development, and provides direct employment opportunities as well. A recent analysis ranked water treatment as the second most critical infrastructure investment for emerging economies (Tan, 2000). Historically, governments have viewed increased water supply – over and above, or even instead of, meeting basic water needs – as a way to strengthen the overall attractiveness of an economy for both private investors and official development assistance. At the same time, private corporations are moving to ensure adequate water supplies for industrial development when governments cannot, or do not, address corporate concerns. Recent research indicates that increased economic growth need not always require increased water development, but the trend toward privatization still reflects this traditional belief (Gleick, 2001).

Water-supply projects can be extremely capital intensive, though estimates of future needs vary widely. The World Bank estimates that new investment required for water infrastructure over the next decade will exceed $60 billion per year. In mid-2001, the American Water Works Association released a study suggesting that $250 billion may be needed over the next 30 years just to upgrade and maintain the existing drinking water system in the US (AWWA, 2001). The Framework for Action that emerged from the 2nd WWF in The Hague in March 2000 called for an additional $100 billion annually from private sources for the next 25 years to meet basic water needs (GWP, 2000).

Whatever the actual investment required, emerging economies face significant hurdles finding the capital to expand coverage in rapidly growing urban areas, maintain existing infrastructure, and treat wastewater to even minimal quality standards. One option is for governments to turn to the private sector, with its greater access to private capital (Faulkner, 1997). Because of this, private participation in the water sector is growing especially quickly in developing countries.

Governments must, of course, also spend limited public and international financial capital to meet other social needs (Yergin & Stanislaw, 1999). By creating water systems that are self-supported

> Box 8. Privatization in Buenos Aires, Argentina (Trémolet, 2001; Tully, 2000; Muller, 1999).
>
> The 1993 privatization of water supply in Buenos Aires, Argentina led to some rapid improvements in water availability. The percentage of the population served has increased from 70% to 85%, an addition of 1.6 million customers, many of who are poor. The privatization reduced water company staff by 50%, reduced non-payment of water bills from 20% to 2%, and resulted in more modern and efficient billing and water-delivery operations. The contract governing the privatization action requires a 27% decrease in water prices over time. It is too soon to know whether privatization will achieve this goal, or eventually result in higher water prices or distributional inequities.

through private investment and by implementing water pricing that pays back the investments, developing country governments can significantly reduce their fiscal and balance of payment problems (Shambaugh, 1999). It is also difficult for government officials subject to political processes to raise water prices; privatization permits governments to give that problem to private entities.

The perception that companies are more competent and efficient than government also contributes to pressures to privatize water systems. The complexity of large water systems and their poor historical performance have encouraged the belief that the technical and managerial skills needed to improve water supply and management systems are only available, or at least more efficiently applied, through partial or complete privatization of water supply. Many developing country politicians also view introducing competition as desirable (Shambaugh, 1999). Initially favorable results from a few privatization actions have supported these beliefs (Box 8), although experience with government management of water systems in the industrialized countries demonstrates that governments are not necessarily less efficient or competent than business.

8 HISTORY OF PRIVATIZATION

Private involvement in water supply has a long history. Indeed, in some places, private ownership and provision of water was the norm, until governments began to assume these responsibilities. In the US, municipal services were often provided by private organizations in the early 1800s. Toward the latter half of that century, municipalities started to confront problems with access and service and began a transition toward public control and management. In particular, private companies were failing to provide access to all citizens in an equitable manner. Private water companies provided 94% of the US market in the 19th Century, dropping to only 15% by 2000 (Beecher et al., 1995). As Blake (1991) points out:

> "Private companies supplied water to Boston from 1796 to 1848, and to Baltimore from 1807 to 1854. As late as 1860, 79 out of 156 water works in the US were privately owned. But eventually most cities turned to municipal ownership. The profit motive was ill suited to the business of supplying water to city dwellers. Private companies were reluctant to invest enough capital; they preferred to lay their distributing pipes through the wealthier sections of the city and to hold back from carrying water into the poorer districts".

Anderson (1991) notes that the experience in Chicago and other cities in the US was similar:

> "Private companies were notorious for choosing a water source that would minimize the initial investment outlay, and for ignoring the concomitant shortcomings in water quantity and quality. Only municipal governments, so the argument goes, had the foresight and the latitude to invest large sums now in order to gain a future payoff in the form of years of excellent water".

In 19th Century France, the trend moved in the opposite direction: municipalities that previously had responsibility for providing water services began to contract services to private operators. Over the years, these operators expanded beyond the borders of France and as a result, they now have a dominant position in much of the world in providing private water services.

Major international efforts to privatize water systems and markets are still a relatively recent phenomenon, with major transfers taking place only over the past ten to fifteen years. By the end of 2000, at least 93 countries had partially privatized water or wastewater services (Brubaker, 2001), including Argentina, Chile, China, Colombia, the Philippines, South Africa, Australia, the UK, and Central Europe, but less than 10% of all water is currently managed by the private sector (LeClerc & Raes, 2001).

In South America, public monopolies were the norm until the mid-90s. As in many other regions, public water systems consistently failed to provide universal coverage, to treat most wastewater, and to find and reduce water losses that can be as high as 50%. Because of these failures, governments in South America increasingly seek private sector involvement. In some cases, such as Buenos Aires, governments have sold or leased water facilities, allowing private operators to sell services directly to the public, with government regulation. Mexico City took another approach to privatization, contracting the rights to operate parts of the city water system to multiple operators, with the goal of stimulating competition among them (Waddell, nd). By 2000, almost all countries in the region have begun to commit themselves to long-term private concessions. Chile has gone farther than most by combining the granting of concessions with private ownership of water resources.

Major cities in Asia also suffer from inadequate infrastructure, huge water losses, inadequate sewage treatment, and lack of service to large numbers of peri-urban residents. Australia, New Zealand, Malaysia, and the Philippines are all exploring various forms of privatization, and the water and wastewater utilities in almost every major city in Oceania have been taken over by private entities, or have contracted some important services. Recent efforts in Manila and Malaysia have run into political or economic controversy, causing private companies and governments to rethink the design of contracts and the conditions for concessions, but privatization efforts seem to be accelerating.

Nations in Europe have also explored a variety of different models recently, and the UK, Germany, France, and Italy all now encourage water privatization. In France and in the UK, the process is far advanced. The British, under Prime Minister Margaret Thatcher, for example, sold state-owned water operations to private investors more than a decade ago. Some of those newly privatized companies have become multinational players in privatization markets. Service providers in all four countries initially tried to keep prices low, but have recently imposed large price increases in order to upgrade their plants and distribution systems to accepted European standards. These price increases have led to growing consumer distrust, though some argue that governments would have similarly had to raise taxes or increase borrowing to make comparable improvements, or worse, would have failed to make them.

The US and Canada have moved more slowly toward privatization. The US has long had a mix of privately owned and publicly regulated water and wastewater utilities, though an estimated 85% of residences still receive water from public agencies (Box 9 for the recent experience of Atlanta, Georgia).

9 THE PLAYERS

There are a handful of major international private water companies, but two French multinational corporations dominate the sector: Vivendi SA and Ondeo (a subsidiary of Suez Lyonnaise des Eaux). These two companies own or have interests in water projects in more than 120 countries and each claims to provide water to around 100 million people (Barlow, 1999, FTGWR, 2000) (Tables 4 and 5). Vivendi's water activities are, themselves, a small part of the larger company Vivendi Universal, which was created in December 2000 when it merged with the Seagram Company to form a global media and telecommunications company. As an example of the diversity of Vivendi's activities, in spring 2001, Vivendi purchased MP3.com. The total annual revenue from the interlocking subsidiaries of Vivendi in 2000 exceeded $37 billion, of which more than 25% came from the water business (Market Guide, 2001).

Box 9. Water utility privatization in Atlanta, Georgia, USA.

Throughout much of the 80s and 90s, Atlanta's wastewater system faced growing problems, including aging infrastructure and inadequate wastewater treatment. Federal, state, and private complaints resulted in millions of dollars in fines and a consent decree specifying expensive corrective action (Brubaker, 2001). Faced with the need for almost $1 billion in capital for urgent improvements, the city government began to explore the possibility of privatization of some aspects of the local water system. The hope in Atlanta was that privatization would dramatically reduce annual operating costs, reduce the likelihood of rate increases, and free up money for new capital improvements.

In late 1998, the city signed a 20-year agreement to contract water services to United Water Services Atlanta (UWSA), a subsidiary of Suez Lyonnaise des Eaux. While the company made a number of innovative concessions, they also benefited from some significant tax breaks offered by the city. UWSA agreed to locate its regional headquarters in Atlanta, committed to hiring 20% of its workforce from the area, and offered to provide $1 million in annual funding for water research at Clark Atlanta University. The firm is benefiting from tax incentives of as much as $8,000 per employee.[11]

The 20-year agreement between the city and UWSA went into effect on January 1, 1999, covering the operations and maintenance of two water-treatment plants serving 1.5 million people, 12 storage tanks, 7 pumping stations, fire hydrants, water mains, billing, collections, and customer service. The contract set UWSA's annual operations and maintenance fee at $21.4 million – thus UWSA can count on nearly half a billion dollars in service fees to be paid by the City of Atlanta over the term of the contract. This is substantially less than the city was expected to spend running the system itself over the same period. The city will continue to spend approximately $6 million annually on power, insurance, and monitoring the contract agreements. Atlanta retained responsibility for most capital investments. The agreement also stipulated that there would be no layoffs during the life of the contract, but staff reductions due to retirements and voluntary departures substantially reduced employment costs. At the time of turnover, many of the municipal employees objected to the privatization agreement. While it is too soon to know how well the goals of the effort to privatize the city's water system will be, other US municipalities are watching closely.

[11] "Atlanta Project in Capsule" – waterindustry.org/frame-8.htm

Table 4. Population served by Vivendi water and wastewater concessions (FTGWR, 2000).

Country/Area	Population supplied in 2000 (millions)
France	25.0
Western Europe	18.5
Central and Eastern Europe	6.3
Middle East and Africa	8.5
North America	16.8
Latin America	7.8
Asia	14.6
Total	97.5

Suez is active in more than 100 countries and claims to provide 110 million people with water and wastewater services. Of the 30 biggest cities to award contracts between 1995 and 2000, 20 chose Suez, including Manila, Jakarta, Casablanca, Santiago de Chile, and Atlanta. Suez also purchases stakes or full interests in other water companies: with its $1 billion purchase of United Water Resources, it became the second largest manager of municipal systems in the US, just behind American Water Works. Suez also purchased Nalco and Calgon in the US for $4.5 billion, making it the biggest provider of water treatment chemicals for both industry and cities. In 2000,

Table 5. Population served by Suez Lyonnaise des Eaux water and wastewater concessions.[12]

Country/Area	Population supplied in 2000 (millions)
Europe and Mediterranean	43
North America	14
South America	25
Asia Pacific	23
Africa	5
Total	110

[12] www.suez.fr/metiers/english/index.htm

Suez reported profits of 1.9 billion euros on sales of 35 billion euros: of this, 9.1 billion euros (or 44%) of revenues came from their water businesses.

Other companies also have major water interests, including Thames Water (now part of the German conglomerate RWE) and United Utilities in Great Britain, Bechtel and Enron in the US, Aguas de Barcelona in Spain, and NUON in the Netherlands. To add to the complexity, however, many of these companies have interlocking directorates or partial interests in each other. For example, in spring 1999, Vivendi purchased US Filter Corporation. United Utilities of the UK has joint ventures with Bechtel. United Water Resources in the US is partly owned by Suez Lyonnaise des Eaux.

10 FORMS OF PRIVATIZATION

Despite the growing debate about privatization, there is considerable misunderstanding and misinformation circulating about what the term itself means. Privatization can take many forms. Only the most absolute form transfers full ownership and operation of water systems to the private sector. Much more common are forms that leave public ownership of water resources unaffected and include transferring some operational responsibilities for water supply or wastewater management from the public to the private sector. Privatization also does not, or should not, absolve public agencies of their responsibility for environmental protection, public health and safety, or monopoly oversight.

There are many different forms of privatization arrangements, agreements, and models. There is also a fundamental difference between public and private ownership of water assets. Private ownership involves transferring assets to a private utility. Public ownership involves keeping the assets in the public domain, but integrating the private sector in various utility operations and activities through contract (Beecher, 1997). Public or private-sector employees can perform various functions. As an illustration, Box 10 lists several functions that could be assigned to private or public employees in thousands of combinations ranging from completely public to completely private operations. These different forms have very different implications worthy of careful analysis.

The different functions in Box 10 can be combined, or broken into even more sub-functions (e.g., design of reservoirs versus design of neighborhood-scale distribution piping). The functions can also be performed privately in one geographic area and publicly in another (e.g., northern and southern halves of a metropolitan area). In the remainder of this chapter we describe fully public water systems and compare them to four variations of models of privatization. Figure 3 "locates" these models in comparison with each other and the many forms of privatization not discussed here, using ownership and management of assets as organizing principles.

10.1 *Fully public water systems*

The various private models described below must be contrasted with fully public water systems. Fully public management of water often takes place through national or municipal government

Box 10. Water system functions that can be privatized.

1. Capital improvement planning and budgeting (including water conservation and wastewater reclamation issues).
2. Finance of capital improvements.
3. Design of capital improvements.
4. Construction of capital improvements.
5. Operation of facilities.
6. Maintenance of facilities.
7. Pricing decisions.
8. Management of billing and revenue collection.
9. Management of payments to employees or contractors.
10. Financial and risk management.
11. Establishment, monitoring, and enforcement of water quality and other service standards.

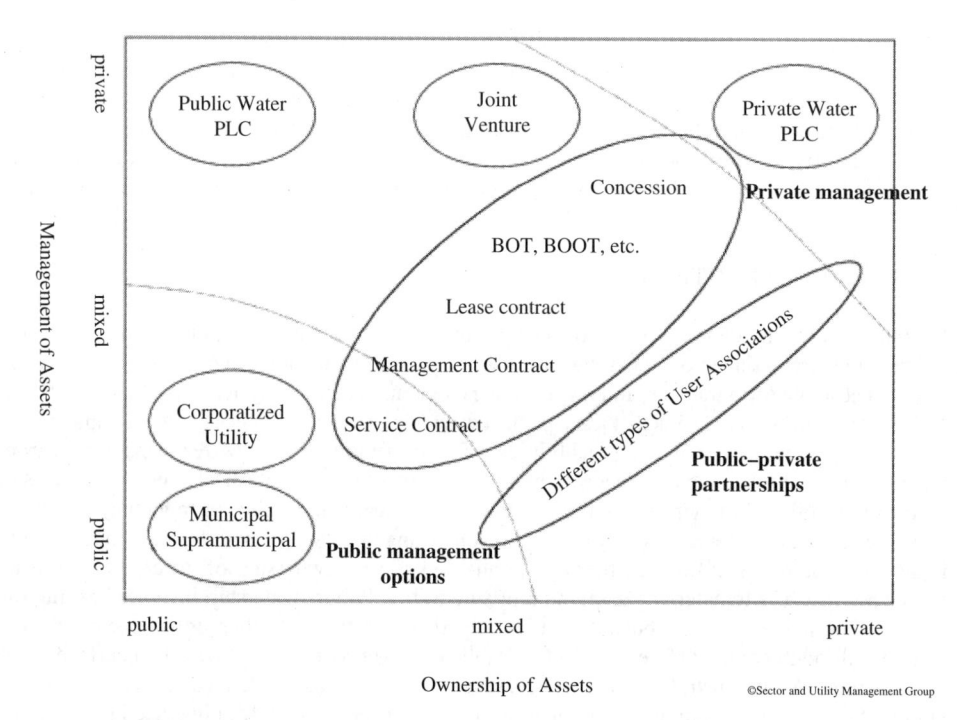

Figure 3. Types of public and private water providers, characterized by the public and private nature of the assets and management (Blokland et al., 1999).

agencies, districts, or departments dedicated to providing water services for a designated service area (in some cases, an entire country). Public managers make decisions, and public funds are used to finance construction, operation, and maintenance of facilities. Funds may be provided from general government revenues, in competition with other government investments or a water agency may be self-supporting via water charges. Governments are responsible for oversight, setting standards, and facilitating public communication and participation. Independent, special-purpose water agencies or districts can have technical and financial capacity equal to private corporations. The poor record of government in many less developed countries has caused this fact to be obscured, and is driving the privatization process more rapidly in developing countries than in, for example, the US or many EU countries.

Another form of public management involves cooperatives and user associations in local water-system governance. Typically, local users join together to provide public management or oversight. An example is the public water cooperative in Santa Cruz, Bolivia, which serves nearly one hundred thousand customers. It arose out of a history of central government neglect and, consequently, a strong belief in decentralization. Customers are split into water districts, each covering approximately ten thousand people. All customers have decision-making powers through elections for different water authorities. Elections to six-year terms are staggered, and different authorities are designed to supervise each other. The system is also externally audited each year. Despite the common belief that cooperatives create greater participation by customers, voter turnout has been low. For example, turnout was only 2.5% in the June 1998 elections (Nickson, 1998).

In 1997, the cooperative compared well to other Bolivian utilities in terms of efficiency, equity, and effectiveness. It received its share of a World Bank loan for all Bolivian water utilities and was quicker to reach the goals funded with the loan. The group incorporates the ideas of a basic water requirement and affordability for the poor through varying rate structures. It also incorporates conservation through increasing block rates, which are not applied to the very poor.

10.2 Public water corporations and corporate utilities

Private-sector participation in public water companies has a long history. In this model, ownership of water systems can be split among private and public shareholders in a corporate utility. Majority ownership, however, is usually maintained within the public sector, while private ownership is often legally restricted, for example, to 20% or less of total shares outstanding. Such organizations typically have a corporate structure, a managing director to guide operations, and a Board of Directors with overall responsibility. This model is found in the Netherlands, Poland, Chile, and the Philippines (Blokland et al., 1999).

A main benefit of the system is that it combines two potentially conflicting goals of water supply. Private owners seek to recover costs and maximize profits. Public owners may also seek to recover costs, but they are more likely to embrace concerns about affordability, water quality, equity of access, and expansion of service.

Empirical evidence suggests such models can attain a high level of operational efficiency and quality of service. Another strength is the stronger potential for public participation and protection of consumer rights. In the Philippines, the Board includes consumer rights associations and in the Netherlands consumer associations publish comparisons among similar companies. This type of customer representation encourages efficiency and discourages political exploitation of the water utility.

Public corporations may have less access to capital, though many have received multilateral development bank funds and the Netherlands system has a strong enough performance record to receive commercial bank loans. The Dutch model also holds the Managing Director personally liable for losses due to mismanagement – increasing the incentive for efficiency, quality, and protection against political intervention.

10.3 Service and leasing contracts: mixed management

In some cases, public water utilities may give private entities responsibility for operation and maintenance activities, general services contracts, or control over management of leased facilities. Ownership continues to reside in public hands. Such models do not usually address financing issues associated with new facilities, or create better access to private capital markets. They do, however, bring in managerial and operational expertise that may not available locally.

Leasing contracts may include tariff (revenue) collection responsibilities as well as operation and maintenance (Rivera, 1996). Such contracts may last for 10 to 15 years or more and arrangements are sometimes made for the private company to share some in the increases in revenues generated from better management and bill collection (Panos, 1998). Service contracts range from smaller, one-time arrangements such as meter installation or pipeline construction to longer-term comprehensive arrangements. Areas in which service contracts have proven effective include: maintenance and repair of equipment, water and sewerage networks, and pumping stations; meter installation and maintenance; collection of service payments; and data processing (Yepes, 1992).

10.4 Concession models

Much of the debate in recent years over privatization has revolved around more comprehensive concessions to the private sector. This is especially true in Latin America and Asia. The full-concession model transfers operation and management responsibility for the entire water-supply system along with most of the risk and financing responsibility to the private sector. Specifications for risk allocation and investment requirements are set by contract. To recoup heavy initial investments, concessions are usually long-term, as long as 25 to 50 years. Technical and managerial expertise may be transferred to the local municipality and community over time, as local employees gain experience.

Variations on full concessions include Build-Operate-Transfer (BOT), Build-Operate-Train-Transfer (BOTT), Build-Own-Operate-Transfer (BOOT), Rehabilitate-Operate-Transfer (ROT), and Build-Operate-Own (BOO). These arrangements can be thought of as "partial concessions" that give responsibilities to private companies, but only for a portion of the water-supply system. Ownership of capital facilities may be transferred to the government at the end of the contract. Training of local workers and managers over years prior to the transfer, with their jobs retained from some period after transfer, is a way of transferring skills along with the capital asset.

For both full and partial concessions, governments and companies are finding that responsibilities and risks must be defined in great detail in the concession contract since such contracts are for a lengthy period, and ultimately govern how the concession will perform (Komives, 2001). Case-by-case concession contract writing has led to vastly different outcomes for similar physical and cultural settings. The benefits to the public appear to be maximized when the government serves as a skillful contract negotiator.

10.5 Fully private businesses and small-scale entrepreneurs

The opposite extreme from government agency provision and management of water is supply and management by fully private actors, whether large corporations or small-scale entrepreneurs. In this model, water-quality regulation and other means of protecting "public goods", such as basic water rights or protection of environmental resources, may be non-existent. Fully private businesses and entrepreneurs are already often found where the existing water utility has low coverage or poor service. They may obtain water directly from a water utility, indirectly from the utility through customers who have utility service, or from private water sources. In some cases, early settlers of an area have privately developed piped water systems, with later settlers becoming customers of, rather than partners in, the piped system. Private providers operate most often in poor urban and peri-urban areas, but they also serve higher income groups or businesses when water is scarce or inconvenient to obtain.

Private suppliers of water also co-exist with public systems when the public system is unreliable, inconvenient, or rationed (e.g., the utility pressurizes pipes only a few hours each day). In Kathmandu, Nepal, water from privately controlled sources are sold by tanker truck to both low- and high-income areas of the city unserved by regular, reliable supply. Customers may turn to private vendors when they have more money than time for water collection. The public provision of tankered water can also involve high costs and even corruption: in parts of Mexico households dependent on public tankering services must often "tip" providers to ensure service (Muñoz, 2001).

Private businesses and small-scale entrepreneurs often operate free of regulation in less developed countries. Private water companies are usually regulated to some extent (e.g., water quality) in more developed countries. Without regulation, high prices or low water quality can cause significant social problems. Numerous studies have shown that the poor often pay much more for water from private suppliers or small-scale vendors than they would pay if a regulated community water system, piped or otherwise, were put into place. For example, in El Alto, Bolivia, where a concession was granted in the mid-90s, households with private connections spend around $2.20 per month for 10 cubic meters, while those relying on private vendors pay over $35.00 for the same account of water (Komives, 2001).

11 THE RISKS OF PRIVATIZATION: CAN AND WILL THEY BE MANAGED?

The move toward privatization of water services raises many concerns, and in some places, even violent opposition. In large part, opposition arises because of doubts about whether purely private markets can address the many different social good aspects of water, or whether some non-market mechanisms are necessary to serve social objectives. Other concerns relate to a fundamental distrust of corporate players and worries about the transfer of profits and assets outside of a community or even a country. *The greatest need for water services often exists in those countries with the weakest public sectors; yet as we shall see below, the greatest risks of failed privatization also exist where governments are weak.* The rapid pace of privatization in recent years and the inappropriate ways several projects have been implemented have compounded the worries of local communities, non-governmental organizations, and policymakers. As a result, private water companies are increasingly seeing serious and sustained public opposition to privatization proposals. This section describes the major concerns and risks of privatization of water systems.

11.1 *Privatization usurps a basic responsibility of governments*

Governments have a fundamental duty to see that basic services, such as water, sewerage, and energy, are provided to their people. The failure to satisfy such basic needs, or at least enable the means for them, must be viewed as irresponsible. Efforts of international lending agencies and development organizations have, in the past, focused on helping governments to provide these services. More recently, these organizations have begun to shift their efforts, pushing privatization as a new solution. We have serious concerns about this transfer of responsibility and the loss of control it implies.

11.2 *Privatization may bypass under-represented and under-served communities*

One of the basic goals of any proposal to provide water services (publicly or privately) should be to meet explicitly the needs of under-served communities through an expansion of access to water or wastewater services. Poor peri-urban populations have traditionally been under-served because they lack political power or representation, they come from unofficial "communities", or they may be unable to pay as much for water as residents in wealthier areas. Privatization can potentially worsen this neglect.

In the past, private companies have been reluctant to make large investments in the water sector in the poorest economies. In some cases, however, reaching underserved populations has been an explicit part of privatization concessions or contracts. The concession granted to serve La Paz-El Alto in Bolivia was designed with performance requirements to expand service to the poor. These "expansion mandates" set obligations to achieve certain levels of coverage and water quality (Komives, 2001). The mandates in Bolivia took three forms: connection requirements, coverage targets, and the requirement to connect households meeting specific criteria. The first mandate required Aguas del Illimani (the subsidiary of Lyonnaise des Eaux that won the concession), to install nearly 72,000 new water connections by the end of 2001. The second called for at least 90% coverage by 2011. The third required the company to extend services to areas that meet specific population density criteria. One problem with this particular concession is the definition of areas – "area no servida" – that the company has no formal obligation to serve.

Some multinational companies balk at provisions requiring expansion of coverage to marginal communities, stating that it is unrealistic to expect universal household connections in low-income

> Box 11. Condominial sewers in Brazil.
>
> The development of "condominial" sewers in Brazil is an example of a much lower cost method (up to 70% reduction in cost) of providing sewage coverage developed when providing coverage in poorer communities is non-negotiable (Wright, 1997). Condominial sewers are a particular form of the lateral sewers used to connect homes to main, trunk sewers. Traditionally, one lateral sewer has been provided for each home. In the condominial scheme, groups of homes share a single lateral sewer that runs from the home furthest away from the trunk sewer to the trunk sewer passing under or alongside homes. The group of homes shares maintenance of the lateral through a users group. The "headaches" of sharing a sewer lateral are considerably less "costly" than the additional capital and maintenance costs incurred if traditional, individual home lateral sewers were provided. More recently, this approach has been applied in Bolivia in the La Paz – El Alto effort.

areas in the immediate future, that lack of roads hinders expansion, and that rapid, uncontrolled peri-urban growth prevents proper water planning and service provision (Shambaugh, 1999). When meeting these unmet needs is a top priority for governments, as we believe it should be, tools for inducing concessionaires to invest in coverage in low-income areas should be part of any agreement, with provisions for mandates, quantitative performance indicators, and economic incentives. One benefit of such mandates is that they provide companies with an incentive to develop innovative, lower-cost options for residents. For example, a requirement in Brazil that mandated coverage in certain areas led to the development of condominial sewer systems (Box 11). Many private companies also request exclusivity over certain service areas, which may help increase the attractiveness of a concession, but exclusivity may suppress competition and the provision of equitable service.

11.3 *Privatization can worsen economic inequities and the affordability of water*

One of the greatest concerns of communities and individuals is that privatization will lead to increases in the cost of water to consumers. Water pricing is a complicated issue – indeed, we note three major different types of pricing and affordability questions associated with privatization:

- Are price increases to consumers necessary?
- Are subsidies appropriate, and if so, what kind and how big?
- How should rates for service be designed?

11.3.1 *Are price increases necessary?*

The problem of water rates is perhaps the most controversial issue around privatization efforts. One of the leading arguments offered by proponents of privatization is that private management or ownership of water systems can reduce the water prices paid by consumers. Ironically, one of the greatest concerns of local communities is that privatization will lead to higher costs for water and water services. The actual record is mixed – both results have occurred. Significant price increases for some groups of water users may also take place even when overall prices do not rise.

Savings from privatization can result from reduction in system inefficiencies, overhead, labor costs, and management expenses. Economies of scale may also exist for developing new infrastructure. Conversely, water supply is a costly business, and substantial improvements in water systems can lead to increased need for revenue from rates. In addition, the requirement that private companies make a profit may drive up rates in systems where government subsidies were the norm. Private utilities may have better access to capital than some public systems, but they may also have to pay a higher cost for that capital, as well as pay taxes. Finally, a move toward full-cost pricing may improve overall economic efficiency, but contribute to rate increases as subsidies are removed or reduced.

There is abundant evidence that people – even those with low incomes – are willing to pay for water and sanitation when the services are reliable and the cost of delivering services is reasonably

Box 12. WTP depends on the type and reliability of water services (World Bank, 1992).

A water supply project in Northeastern Thailand was intended to provide protected water at the lowest possible cost because people in the area were poor. Because groundwater is abundant in the region, the technology chosen was hand pumps. After about five years most of the hand pumps were not working, and water use habits were largely unchanged. In a follow-up phase, motor pumps provided piped water at community standpipes. Again, the project failed. Five years after implementation 50% of the systems were not working at all and another 25% operated intermittently.

The failures were attributed to technologies that were too complex to maintain and to the inability of the villagers to pay for improved water supplies. Gradually, however, it became apparent that the main problem was not the capabilities of the villagers but the fact that the service being offered was not what they wanted. They did not want hand pumps, which were not considered an improvement over the traditional rope and bucket system. And standpipes, being no closer than their traditional sources, offered no obvious benefits. Only piped water to yard taps could meet people's aspirations.

In the next project yard taps were allowed, with the users paying the full costs of connection. Five years later the verdict was in: 90% of the systems were functioning reliably, 80% of the people were served by yard taps; meters had been installed, and locally adapted charging systems had been developed. Not only were the systems well maintained, but because the services was so popular, many systems had extended distribution lines to previously unserved areas.

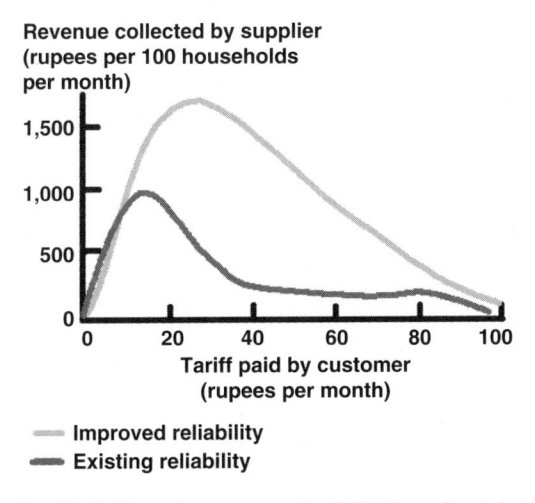

Figure 4. Data from Punjab, Pakistan demonstrates that WTP depends not just on the type of water service, but also on the reliability of that service. As reliability improves, so do acceptable tariffs and revenues to service providers.

transparent and understandable to customers (Box 12). Experience also suggests that people and businesses will pay more for water without significant resistance when they receive new or improved services that they desire. In the context of privatization, this suggests that dissemination of detailed information about the improvement in services, and the capital investments needed to create those improvements, is essential to public acceptance of increases in overall water prices. The new or improved services should be clearly described and rate changes should be phased in together with strong education and information programs describing the changes and their reason.

Problems with setting and defining rates are a leading source of controversy over privatization proposals. When the basic guidelines described here are not followed, rapid and large increases in water rates cause strong social and political reactions. Public protests and political demonstrations over price increases have taken place in such diverse settings as Cochabamba, Bolivia; Tucuman,

Box 13. Failed privatizations: Tucuman, Argentina and Cochabamba, Bolivia.[13]

Aguas del Aconquija, a subsidiary of Vivendi and local Argentinean companies, won a 30-year concession contract in July 1995 to run the water-supply system for the 1.1 million people of Tucuman, Argentina. Aguas del Aconquija doubled water tariffs within a few months time in order to meet aggressive investment requirements specified in the concession. A new governor took office around the same time and objected to the privatization. He and his supporters encouraged residents to stop paying bills. Soon afterward, delivered water turned brown, an incident Aguas del Aconquija assured was not harmful and was attributable to the naturally high sediment content of the city's water. The explanation did not convince residents: approximately 80% stopped paying their bills. In October 1998 the concession was terminated, but on the condition that the company continue to operate the water system for 18 months. Vivendi agreed, but quickly filed a $100 million suit against the government, and joined several other companies who had filed complaints against Argentina with the World Bank Arbitration Panel.

An outbreak of violence resulting from a proposal to privatize the public water system in Cochabamba, Bolivia's third largest city, exemplifies the severe problems that can result from rushing toward management of water as an economic good while disregarding its social good aspects. In 1999, the Bolivian government privatized the water system of Cochabamba, partly in response to pressures from the World Bank to make structural adjustments to its economy. The government granted a 40-year concession to run the water system to a consortium led by Italian-owned International Water Limited and US-based Bechtel Enterprise Holdings. The consortium also included minority investment from Bolivia. The newly privatized water company immediately modified the rate structure, putting in place a tiered rate and rolling in previously accumulated (but not recovered) debt. As a result, many local residents received (or anticipated receiving) increases in their water bills. Aguas de Tunari maintained that the rate hikes would have a large impact only on industrial customers; however, the poor peasants of the town claimed that some residents saw increases as high as 100%. Water collection also required the purchase of permits, which threatened access to water for the poorest citizens.

Local farmers had already expressed concerns about privatization. In October 1998 3,000 farmers organized a march protesting a draft law to charge for water that they believe they own, and for the lack of government attention to a drought that was ruining farmers and livestock owners. After the contract was signed, local groups, including rural farmers, community organizers, and even wealthier groups with an interest in maintaining the parallel private water tanker/trucking market held several protests to demand that the water system stay under local public control. During the protests, the Bolivian army killed as many as nine (reports range from one to nine killed), injured hundreds, and arrested several local leaders. The government also reportedly cut off drinking water to Villa Tunari during this time. Martial law was declared on April 8, 2000 but in late April the government gave in and canceled its contract with Aguas de Tunari. While the cancellation of the contract and the violence have helped put a spotlight on problems with privatization elsewhere, nearly 60% of the population still are not served with any water other than expensive water from private tankers. Moreover, these segments of the population are likely to remain the long-term losers from the continued failure to provide adequate clean water.

[13]Sources: Brook (1999a), Mandell-Campbell (1998), Hudson (1999), Pilling (1996), FTGWR (2000), Goldman Foundation (2001), Business Wire (1999), International Press Service (1998), Earth Island Journal (2000).

Argentina; Puerto Rico, US; and Johannesburg, South Africa. In Argentina and Bolivia, rate concerns along with other factors led to privatization efforts being cancelled (Box 13). Across Southeastern Asia, disputes over water tariffs are raging. In Malaysia, rate increases just prior to privatization led to protests. In Manila, Maynilad Water Services has been lobbying for a rate adjustment to cover losses caused by currency fluctuations, and threatening to return the concession if its petition is refused by the Philippine government (FTGWR, 2001). These experiences fuel public skepticism over arguments that water prices will decline as a result of privatization. In the La Paz/El Alto privatization agreement, the General Manager of Aguas del Illimani mollified the public by stating that he would not raise rates in the first five years of the contract, even if the company's costs rise (Komives, 2001).

> Box 14. Subsidies in water system: balancing social and market needs.[14]
>
> The traditional approach to subsidizing low-income water users has been to sell a first "block" of water to each user at a subsidized rate. This approach encourages conservation of water use in excess of the "block quantity", but ensures the provision of a minimum amount of water for all users. As a practical matter, the first block is usually made available to all water customers, not just those with a demonstrable financial need.
>
> Another way to subsidize low-income water users begins from the common practice of charging water users periodic fixed fees for repayment of fixed costs of the water supply system (e.g., a bi-monthly fee of $8.00). These costs, such as amortization of investments in reservoirs and pipelines, do not vary with the amount of water purchased. They can be waived (or even made negative) for low-income households without encouraging wasteful use of water. Wealthier households must pay a higher periodic fee to support this subsidy, but doing so does not distort the incentive for efficient water use. Since the fixed costs of water supply can be quite significant – sometimes more than half the total cost of water supply – this approach has great promise.
>
> In Santiago, Chile, officials have introduced a "water stamps" scheme that covers part of the cost of water purchases for the poorest residents. As a result, the private firm running the city's water has a direct incentive to serve the poor. This program is similar to the US food stamps program, a program that has been found to be superior to food price subsidies because it more tightly targets the people it intends to help and it gives companies the incentive to serve the poor.
>
> [14] Sources: World Commission on Water for the 21st Century (2000), The Economist (2000), Empresa Metropolitana de Obras Sanitarias S.A. (1995).

11.3.2 *Are water subsidies appropriate and desirable?*

Subsidies – especially water subsidies – have been a controversial topic for many years (Myers & Kent, 2001). On the one hand, economic theory acknowledges that they can be socially desirable and economically efficient in some circumstances. On the other hand, they are often applied as policy favors or social gifts far more widely than necessary to meet critical social goals. Many groups claim they deserve subsidies. Businesses threaten bankruptcy or job cuts if water prices increase. Other users argue that their products or water uses are socially critical or particularly beneficial.

Water-pricing systems often already include some subsidies. Government policies often keep water tariffs low to benefit public welfare. "Lifeline" rates for basic water needs are sometimes available for the lowest income groups in a community. Most governments offer substantial subsidies for agricultural water use by farmers and the poorest urban users. A 1997 rate study from the World Bank notes that agricultural water users may pay as little as 20% of the total costs of providing irrigation water and may never fully repay capital costs of projects that benefit them (Dinar & Subramanian, 1997). When properly designed, subsidies can satisfy social goals without causing serious problems for the overall market (Box 14).

The results of subsidies, however, have often been unsatisfactory. The quality of water services and coverage are inadequate in many countries. Subsidies directed at the poor often end up benefiting the wealthier populations, while many poor remain unconnected to the system.

One of the potential benefits of privatization is elimination of inappropriate subsidies, a point not lost on those who argue for increased private control. The public sector is often sensitive – some might say too sensitive – to calls for subsidies from various interest groups. Shifting responsibility to the private sector can lead to prices that better reflect costs and allow governments to discontinue subsidies, while letting private providers take most of the heat for price increases. This can be a clear advantage to privatization.

We also note, however, that lack of water subsidies in some cases can have disastrous results, especially when combined with pressures to recover costs. In South Africa in 2000, a massive outbreak of cholera occurred in the KwaZulu-Natal region when the local water agencies began requiring repayment of fees for water services. This led some of the poorest communities to abandon clean utility services and switch to free, but contaminated water, from other sources.

We believe that there has been inadequate attention given in privatization negotiations and debates to identifying the difference between appropriate and inappropriate subsidies. When water systems or operations are privatized, it may be desirable to protect some groups of citizens or businesses from paying the full cost of service, perhaps permanently. For example, an affordable supply of water sufficient to meet basic needs may be a fundamental human right or may be socially desirable for other reasons, and hence worthy of a subsidy (Gleick, 1996 and 1999). In such cases, it might be appropriate to promote the use of tradeable water stamps, as has been done in Chile, or to provide direct subsidies to reduce water prices for some economic classes because of the positive externalities of higher water use. Water-dependent industries that are critical to local employment patterns or long-term economic growth may also be worth subsidizing, either with revenue from other water users or with general tax revenues.

There are more complicated examples of subsidies, such as when companies bid low on contracts or overpay for water concessions in order to get a foothold in a region. In this case, later and higher bids are used to subsidize the development of a regional presence. For example, the French giants Suez Lyonnaise and Vivendi have been accused of overpaying for initial water concessions. After winning the bid for Germany's first privatization, one Vivendi executive stated, "Berlin was a flagship contract – a symbol. We will use that to make progress municipality by municipality" (Gopinath, 2000).

Obtaining contracts at loss, with expectation of making a profit later, is dangerous for both the private company and the host community. An example of the danger to the private company is provided in Box 15. The dangers to the host community are numerous, all related to pressure by the company to increase prices and make a formerly unprofitable contract profitable. Even when prices are controlled by the local community or specified in the privatization contract, companies have claimed that changed conditions render agreed-upon prices inadequate, leading to litigation, reduction in services, or bankruptcy.

The competitive efforts of international companies trying to get a foot in the door or establish a dominant international position have driven immediate profit margins down in many developing country privatization proposals. Some recent requests for proposals explicitly announce that the

---

Box 15. The risk of accruing present losses in the hope of futures profits.

Positioning a company for future markets while keeping it afloat in the present is a difficult balance. While some analyses suggest that privatization can lead to significant savings, other suspect that low bids for projects represent an effort to get control of a market in the hope of larger returns later. A representative of one large water company was quoted as saying that most of the dramatic savings claimed by private water companies represent losses for the bidders. In Atlanta, for example, he commented, "It all boils down to who wanted to lose the most money for the longest time" (Brubaker, 2001).

To the extent that firms so highly value an opportunity to establish themselves that they are willing to underbid their competitors at a loss to themselves, savings will be more modest in the future. Enron Corporation's water utility spin-off, Azurix, planned to become the next Suez or Vivendi in international markets. Azurix seemed ready to expand rapidly. International markets opened more slowly than expected, they failed to obtain several important contracts, and several that they did obtain proved to be uneconomic. In June 1999, Azurix paid $440 million for 90% of the concession company serving two of the three regions of Buenos Aires, Argentina. It subsequently invested an additional $94 million.

Some analysts have argued that the price paid was excessive (FTGWR, 2001, 114, p.8), and in January 2001 Azurix took a one-time charge of $470 million in an acknowledgement that the contract had serious problems. In particular, Azurix concluded that the terms of the contract prevented them from adequately raising capital or receiving an appropriate return on investment. In the quarter ending September 30, 1999, Azurix brought in a net income of $18.8 million on revenues of $170.5 million. One year later, they were carrying a $3.6-million loss on revenues of $183.7 million. The share price of Azurix – one measure of confidence that revenues and return on investment would be attractive – fell from $19 per share at the time of the company went public to under $4 per share. In 2001, Enron took Azurix private again (Rossa, 2001).

contract will be awarded to the bid offering the lowest water rate. Although this type of competition, and legitimately improved management in some cases, have led to lower water prices following some privatization actions, such as in Buenos Aires and Manila, it is reasonable to wonder whether privatization will provide water services that are affordable and fairly priced in the long run. Indeed, in Manila, serious problems are now surfacing for both of the water concessionaires and residents due to unexpected currency fluctuations, problems with non-revenue customers, and the size of the debt burden (FTGWR, 2001). As a result, one of the concessionaires, Maynilad, currently loses two pesos for every peso billed – an unsustainable economic situation.

11.3.3 *How should rates be designed?*

Public acceptance of efforts to privatize water services often hinges on decisions about the design and size of rates. One of the problems with privatization is that the incentives for companies to put in place innovative rate structures often conflict with the incentives for companies to generate revenues. When income is a function of how much water a company sells, rate structures that encourage efficient use and conservation may simply reduce overall income. Similarly, inequitable rate structures that favor one class of user over another may be economically beneficial to a company, but socially undesirable.

Rates depend on a wide range of factors, including the balance between one-time system connection fees, fixed fees, and volumetric water prices. Water can be expensive on a volumetric basis, but affordable for low-income families if connection charges and periodic fixed fees are set at zero for these families. Conservation-oriented rate structures, however, are still not well understood or consistently applied even in more developed parts of the world. For example, in a recent survey of California water utilities, more than half of all rate structures used either flat or declining block rates, which are usually less effective at encouraging efficient use of water than increasing block rates (Figure 5).

In the La Paz/El Alto concession, a progressive rate structure was developed that subsidized low-volume residential users and imposed an increasing four-block rate – the more water used, the higher the tariff. Industrial customers pay a single rate, equal to the long-run marginal cost. Two tiers were set for commercial users (Table 6).

Significant customer concerns about changes in water prices are to be expected. As a result, transparent and reasonable explanations for proposed changes are essential for public acceptance. When subsidies from general tax revenues are eliminated, causing rates to increase in general and

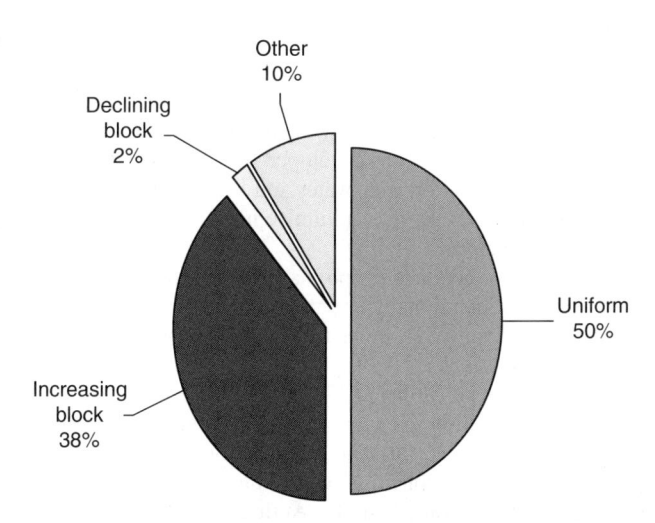

Figure 5. Only 38% of water utilities in California reported increasing block rates in 2000 (Black & Veatch, 2001).

Table 6. Tariff structure for Aguas del Illimani, Bolivia (Komives, 2001).

Tariff ($/m^3)	Residential (m^3)	Commercial (m^3)	Industrial
0.2214	1 to 30		
0.4428	31 to 150		
0.6642	151 to 300	1 to 20	
1.1862	Above 300	Above 20	All water

Notes:
1. 99% of all residential customers use less than 150 m^3/month.
2. The long-run marginal cost is estimated at 1.18$/m^3.

rate structures to be changed, detailed information about the alternative use of general tax revenues may also be essential to public acceptance.

Privatization efforts have also been opposed when rate changes have occurred rapidly and without public education. Rapid rate increases, for any reason, tend to engender opposition and protests, even in publicly operated water systems. Phasing in such increases allows people and businesses to adjust to price changes if the schedule of change is communicated in advance and people believe that it will actually be implemented. In many instances, measures to reduce water use can be adopted before price changes take place, which reduces the financial burden of the changes to consumers. This adjustment effect will yield less revenue than would occur with more rapid rate changes, but revenue "losses" due to phasing should not be seen as losses – they would occur anyway as people and businesses seek to avoid higher water prices. In fact, phased changes in water prices and rates are not only less burdensome for customers, they create greater revenue stability for the water supplier and make financial projections less difficult and burdensome.

11.4 *Privatization agreements may fail to protect public ownership of water and water rights*

Privatization of water management can, under some circumstances, lead to the loss of local ownership of water systems, which in turn can lead to neglect of the public interest. Many of the concerns expressed about privatization relate to the control of water rights and changes in water allocations, rather than explicit financial or economic problems. In part, this is the result of the deep feelings people have for water. It is also the result, however, of serious neglect of these issues by some who promote privatization.

11.5 *Water rights and control*

Control of water has enormous implications for any society or culture. Many different forms of water "rights" exist – this is not the place to review them (Bruns & Meinzen-Dick, 2000). But each of these cultural, social, or legal controls has developed over time to address some aspect of public ownership, control, and participation over water and water policy. Among other things, they may ensure equitable access to water service, minimize impacts on downstream water users, protect water quality, or resolve disputes.

While some privatization contracts and proposals do not lead to any formal change in water rights, a growing number either intentionally or unintentionally change the status quo. Some even explicitly transfer ownership of water resources from public to private entities. For example, the Edwards Aquifer Authority in the central US has considered selling water rights for either a limited period of time (e.g., one year) or in perpetuity (EAA, 2001) (Box 16). Granting perpetual withdrawal rights would reduce the public's ability to ensure that the aquifer is managed as a social good.

Despite numerous legal challenges, this and other actions to establish public ownership of underground water in Texas have been upheld. Most strikingly, the Supreme Court of Texas rejected a claim that action creating the Edwards Aquifer Authority deprived landowners of a property right vested to them by the Texas Constitution. Establishment of the Edwards Aquifer Authority is an excellent example of the type of changes in property rights and rules that are necessary if water

> Box 16. Establishing public property rights for *in-situ* water.
>
> The Edwards Aquifer of South Central Texas is the sole source of drinking water for 1.5 million people in parts of eight counties, including all of San Antonio, the ninth largest city in the nation (according to the 2000 US Census). The aquifer provides 300 million cubic meters of irrigation water annually for about 34,000 hectares of agricultural land. It also supports an extremely diverse wildlife population in surface springs and underground. At least nine endangered species rely on springflows for their survival, baseflow in the Guadalupe and San Antonio rivers depends in part on the aquifer, and its subterranean aquatic ecosystem is believed to be the most diverse in the world.
>
> Historically, Texas law granted complete ownership of groundwater to the landowner above it.[15] This common law rule was replaced long ago in most other US states. Several serious droughts (1984 and 1996), legal decisions to enforce the Endangered Species Act (between 1990 and 1996), and citizen action that raised public understanding of the importance of the aquifer, led the Texas legislature to gradually impose public control over (and hence partial public ownership of) water in this and other aquifers in Texas. In 1993 the Texas Legislature created an Edwards Aquifer Authority to limit water pumping, penalize violators, issue permits, control the transfer of water rights, and institute water quality programs.
>
> [15] Unless the groundwater is flowing in an underground stream or river, in which case the laws governing surface water apply.

is to be managed effectively as both a social and an economic good (EAA, 2001). However, the existence of such public bodies does not ensure sound water management. The Edwards Aquifer Authority itself has allowed some water rights holders to sell those rights in perpetuity, thereby reducing the public's ability to ensure that future water from the aquifer is managed as a social good. Full implementation of public ownership of water at the source requires that ownership cannot be permanently transferred to private hands.

Changes in access and water rights may also occur without explicit agreement. One of the causes of tensions in Bolivia over the proposal to privatize the water systems in Cochabamba were efforts to restrict unmonitored groundwater pumping by rural water users and to bring them into the private system. While this may make sense from a purely economic and efficiency perspective, it imposed a fundamental change in the historical use rights in the region.

Another challenge associated with privatization is the degree to which the process of privatization leads to the transfer of government or public assets into the hands of those who are friends of government, or already wealthy. When privatization results in a redistribution of wealth in an inequitable way, there will be strong pressure to oppose or cancel reforms. Confidence in the fairness of the process, in turn, depends on both the design and the transparency of the rules and legal system (Yergin & Stanislaw, 1999).

11.6 *Privatization agreements often fail to include public participation and contract monitoring*

Oversight and monitoring of public-private agreements are key public responsibilities. Far more effort has been spent trying to ease financial constraints and government oversight, and to promote private-sector involvement, than to define broad guidelines for public access and oversight, monitor the public interest, and ensure public participation and transparency.

Weaknesses in monitoring progress can lead to ineffective service provision, discriminatory behavior, or violations of water-quality protections. In the late 80s, Guinea had one of the least developed urban water-supply systems in West Africa. Fewer than 40% of urban residents had access to piped water, services were irregular, and water quality was unreliable. In 1989, the government of Guinea entered into a lease arrangement for the capital and sixteen other cities and towns. Considerable improvements have resulted (Brook, 1999b), but problems with weak monitoring and enforcement have led to fewer gains to consumers than expected.

Scorecard for Water and Sewerage Companies in England and Wales
1999–2000 Score

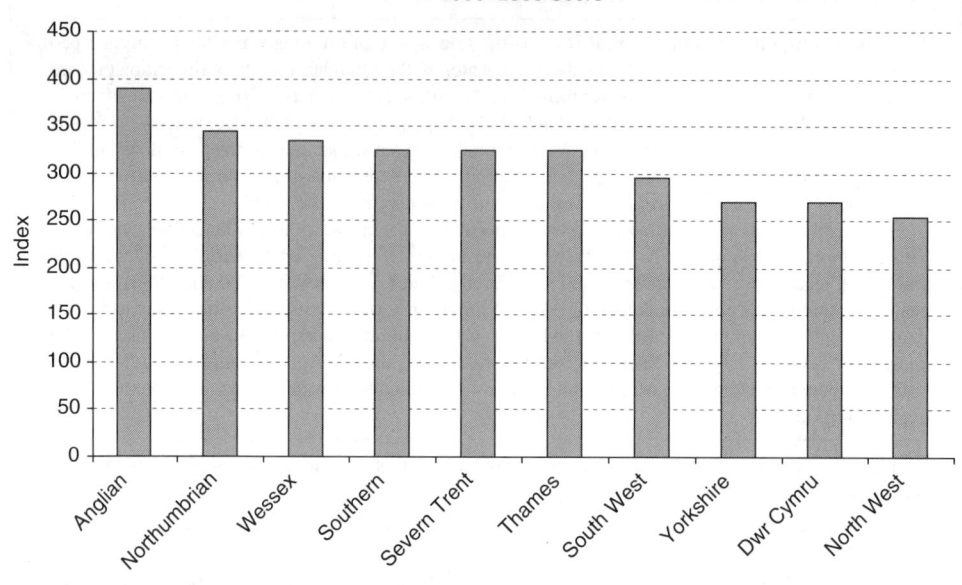

Figure 6. OFWAT is responsible for oversight and monitoring of private water agencies in the UK. They produce an annual report addressing the performance of water companies. The higher the score, the better the performance.

One option is to have regulators set and monitor explicit indicators of service performance – "benchmarking". Benchmarking can focus attention on service quality and provide incentives for long-term performance. Performance benchmarking has become standard practice in the water-sector reforms in England and Wales. OFWAT, the public regulator, collects and publishes sets of indicators on an annual basis from water and sewerage companies. These scorecards help pressure the worst providers to improve service and boost the reputations of the best providers (Kingdom & Jagannathan, 2001). Figure 6 shows the "scorecard" OFWAT produced in 2000, based on variety of performance criteria, including customer service, water pressure, billing factors, public complaints, supply interruptions, water quality, and more.

In Sao Paulo, Brazil, the introduction of pollution tests and public reporting has led 95% of polluting industries to install waste-treatment units to avoid paying fines and seeing their names published. Procuraduria Federal de Proteccion al Ambiente, Mexico's environmental enforcement agency, will shortly publish information on the environmental performance of industries in an effort to encourage improvements in environmental quality (Kingdom & Jagannathan, 2001).

There are many barriers to public reporting, including inadequate data, vested interests that block exposure of poor practices, conflicts of interest among agencies that both provide and regulate services, and costs. Nevertheless, the clear advantage of performance monitoring is a strong argument for more universal programs to collect and disseminate benchmarks as a basic part of privatization efforts. The World Bank recently launched an international water benchmarking network to help provide information and cross-country comparisons, but input from consumer groups, local communities, and others must be sought more actively.

11.7 *Inappropriate privatization efforts ignore impacts on ecosystems or downstream water users*

Many privatization contracts include provisions to encourage the development of new water supplies, often over a long time period. If privatization contracts do not also guarantee ecosystem

Box 17. The cost of failure to define minimum in-stream flows.

Purchase or lease of existing water rights has been used to increase in-situ water resources (e.g., water in a river), and is likely to be used much more extensively in the future. For example, the California-Federal analysis of water-management options (CALFED, 1999) that would, in part, increase "environmental flows" through the San Francisco Bay-Delta ecosystem considers payments to farmers to fallow their farmland during dry years. They suggest that $142-284 million per year would be sufficient to fallow enough farmland to achieve a legally mandated target of an additional 710,000 acre-feet of instream flow. This is a modest amount given the ecological benefits expected. The market tool of paying farmers to fallow land during a drought can be an effective way to improve management of water as an economic good.

The direction of payment, however, suggests that the public in general must dig into their pockets to pay owners of water rights that their government representatives gave away for free in the first place. The historic neglect of instream flows and other environmentally valuable uses of water have created a situation in which the public may have to pay again to restore a public good already lost to them.

water requirements, development of new supply options will undermine ecosystem health and well-being (for both public and private developments). Famous examples of this problem include the Aral Sea in central Asia and the San Francisco Bay-Delta ecosystem in central California. Similarly, the largest lake in Mexico, Lake Chapala, is shrinking due to overextraction of groundwater, strong expansion of irrigated areas, reduced flows to the Lerma River, and unchecked urban water demand in the watershed. Once *in-situ* flows fall below minimum levels, significant and costly ecosystem damage occurs or society in general is required to purchase water rights from those who have obtained them for free. Authorities in Mexico are now trying to buy back water from agricultural producers (Muñoz, 2001).

Decisions about water supply and system operations affect natural flows of water and ecosystem health. Water withdrawals and use come at the expense of riparian and riverine ecosystems. Timing and magnitude of flows may change. Private operators have little incentive to operate reservoirs to maintain minimum downstream flows required for ecosystem health, fishing or recreational interests, and so forth. Balancing ecological needs with water supply, hydroelectric power, and downstream uses of water is a complex task involving many stakeholders. In addition to our growing understanding of the ecological impacts of water development, there has been new attention given in recent years to the economic impacts of these environmental changes as well. We will not review here the growing literature on quantifying the ecological benefits of water systems in economic terms (for example, Postel & Carpenter, 1997; Daily, 1997), but we point out the growing economic costs – typically billions of dollars – being spent to restore previously degraded systems such as the Everglades in Florida and California's Sacramento-San Joaquin Delta (Box 17).

11.8 Privatization efforts may neglect the potential for water-use efficiency and conservation improvements

Selling water itself is much easier than selling water conservation and efficiency improvements. One of the greatest concerns of privatization watchdogs is that efficiency programs are typically ignored or even cancelled after authority for managing public systems is turned over to private entities. Improvements in efficiency reduce water sales, and hence may lower revenues. As a result, utilities or companies that provide utility services may have little or no financial incentive to encourage conservation. In addition, conservation is often less capital intensive and therefore creates fewer opportunities for investors. Consequently, it may be neglected in comparison with traditional, centralized water-supply projects such as new reservoirs.

Where water scarcity is an important issue, or when new sources of supply are expensive, water-use efficiency improvements may be particularly cost effective. Many of the benefits of such improvements, however, may not be easily or directly measured, including improvements to ecosystem health, energy savings, and reduction in wastewater treatment costs. Capturing those improvements may also be a challenge, requiring policies ranging from proper rate design and pricing to rebates to education and information transfers. Water prices are important tools to encourage improvements in water conservation and use efficiency.

11.9 *Privatization agreements may lessen protection of water quality*

Private suppliers of water have few economic incentives to address long-term (chronic) health problems associated with low levels of some pollutants. In addition, private water suppliers have an incentive to understate or misrepresent to customers the size and potential impacts of problems that do occur. As a result, there is widespread agreement that maintaining strong regulatory oversight is a necessary component of protecting water quality. Concerns about the ability of private water providers to protect water quality led the National Council of Women of Canada, a non-partisan federation of organizations, to adopt a policy in 1997 of opposition to the privatization of water purification and distribution systems (NCWC, 1997). The Water Environment Federation (WEF) in the US supports "national policy to encourage public/private partnerships (privatization)" but with appropriate public oversight (WEF, 2000).

When strong regulatory oversight exists, privatization can lead to improvements in water quality. For example, Standard and Poor's notes that water and wastewater quality have improved in the UK after water privatization (S&P, 2000). Indeed, prior to privatization, there was a distinct reluctance of government agencies to monitor and fine other government water providers who were violating water quality standards – a classic conflict of interest. Governments that own, operate, and finance water and wastewater utilities have shown that they cannot always properly regulate them, too. Privatization has the potential to reduce those conflicts and permit governments to regulate. In the UK, government regulators have greatly increased their successful prosecutions for violations (Orwin, 1999).

In Buenos Aires, privatization in the early 90s led to rapid improvements in wastewater treatment. Aguas Argentinas increased the capacity of water-treatment plants and brought on-line wastewater plants that were previously inoperable (Idelovich & Ringskog, 1995). In Chile, municipal water companies have been run by concessions for many years, with different regional companies granting concessions for water supply and treatment, water distribution, operation of sewers, and sewage treatment. Starting in 1997, the Chilean government began to privatize wastewater treatment as well. All operators are kept under close scrutiny by the Superintendencia de Servicios Sanitarios, an autonomous government agency (Orwin, 1999).

11.10 *Privatization agreements often lack dispute-resolution procedures*

Public water companies are usually subject to political dispute-resolution processes involving local stakeholders. Privatized water systems are subject to legal processes that involve non-local stakeholders and perhaps non-local levels of the legal system. This change in who resolves disputes, and the rules for dispute resolution, is accompanied by increased potential for political conflicts over privatization agreements. Public–private partnerships have not often developed clear mechanisms for open participation in dispute resolution, and contracts are often ambiguous in this area. Carefully worded contracts can avoid some such problems. But the water market is relatively new, and some problems are likely to occur even with carefully developed contract language.

It is becoming clear that governments in developing countries are not experienced in negotiating often very complex contracts that specify level and quality of service, monitoring and success indicators, water quality protection, and so forth, in the midst of difficult-to-estimate growth in demand for water. Contracts also have cultural contexts that differ widely and should be accounted for in specific contract language, such as that related to dispute resolution (Calaguas, 1999).

Some have called for voluntary "codes of conduct" by which companies would acknowledge their social responsibilities in providing access to water services. Most recently, the ISO has been presented with a proposal for a set of standards that would apply to all privatization agreements. While we strongly support the concept of standards, benchmarks, and clear contract agreements, such standards must be negotiated in an open, transparent process, with input from all parties, not just water companies.

11.11 *Privatization of water systems may be irreversible*

When governments transfer control over their water system to private companies, the loss of internal skills and expertise may be irreversible, or nearly so. Many contracts are long term – for as much as 10 to 20 years. Management expertise, engineering knowledge, and other assets in the public domain may be lost for good. Indeed, while there is growing experience with the transfer of such assets to private hands, there is little or no recent experience with the public sector re-acquiring such assets from the private sector.

11.12 *Privatization may lead to the transfer of assets out of local communities*

In the past, revenues generated from local sales of water and services went to local agencies for reinvestment in the community. Because of the multinational character of most water privatization companies, some opponents of privatization fear the loss of a wide range of assets that could be transferred out of local communities. These assets include jobs that may go to outside parties and the profits from operations that go to corporate entities in other countries.

12 PRINCIPLES AND STANDARDS FOR PRIVATIZATION

Despite the vociferous, and often justified, opposition to water privatization, proposals for public–private partnerships in water supply and management are likely to become more numerous in the future. There are many forms of water privatization, or public–private partnerships, making unilateral support for, or opposition to, privatization illogical. We do not argue here that privatization efforts must stop. We do, however, argue that all privatization agreements should meet certain standards and incorporate specific principles. Consequently, we conclude this section with suggested Principles and Standards for privatization of water-supply systems and infrastructure that are now primarily public in character.

We believe that the responsibility for providing water and water services should still rest with local communities and governments, and that efforts should be made to strengthen the ability of governments to meet water needs. As described above, the potential advantages of privatization are often greatest where governments have been weakest and failed to meet basic water needs. Where strong governments are able to provide water services effectively and equitably, the attractions of privatization decrease substantially. Unfortunately, the greatest risks of privatization are also where governments are weakest, where they are unable to provide the oversight and management functions necessary to protect public interests. This contradiction poses the greatest challenge for those who hope to make privatization work successfully.

12.1 *Continue to manage water as a social good*

12.1.1 *Meet basic human needs for water. All residents in a service area should be guaranteed a basic water quantity under any privatization agreement*

Contract agreements to provide water services in any region must ensure that unmet basic human water needs are met first, before more water is provided to existing customers. Basic water requirements should be clearly defined (Gleick, 1996 and 1999).

12.1.2 *Meet basic ecosystem needs for water. Natural ecosystems should be guaranteed a basic water requirement under any privatization agreement*

Basic water-supply protections for natural ecosystems must be put in place in every region of the world. Such protections should be written into every privatization agreement, enforced by government oversight.

12.1.3 *The basic water requirement for users should be provided at subsidized rates when necessary for reasons of poverty*

Subsidies should not be encouraged blindly, but some subsidies for specific groups of people or industries are occasionally justified. One example is subsidies for meeting basic water requirements when that minimum amount of water cannot be paid for due to poverty.

12.2 Use sound economics in water management

12.2.1 *Water and water services should be provided at fair and reasonable rates*

Provision of water and water services should not be free. Appropriate subsidies should be evaluated and discussed in public. Rates should be designed to encourage efficient and effective use of water.

12.2.2 *Whenever possible, link proposed rate increases with agreed-upon improvements in service*

Experience has shown that water users are often willing to pay for improvements in service when such improvements are designed with their participation and when improvements are actually delivered. Even when rate increases are primarily motivated by cost increases, linking the rate increase to improvements in service creates a performance incentive for the water supplier and increases the value of water and water services to users.

12.2.3 *Subsidies, if necessary, should be economically and socially sound*

Subsidies are not all equal from an economic point of view. For example, subsidies to low-income users that do not reduce the price of water are more appropriate than those that do because lower water prices encourage inefficient water use. Similarly, mechanisms should be instituted to regularly review and eliminate subsidies that no longer serve an appropriate social purpose.

12.2.4 *Private companies should be required to demonstrate that new water-supply projects are less expensive than projects to improve water conservation and water-use efficiency before they are permitted to invest and raise water rates to repay the investment*

Privatization agreements should not permit new supply projects unless such projects can be proven to be less costly than improving the efficiency of existing water distribution and use. When considered seriously, water-efficiency investments can earn an equal or higher rate of return to that earned by new water-supply investments. Rate structures should permit companies to earn a return on efficiency and conservation investments.

12.3 Maintain strong government regulation and oversight

12.3.1 *Governments should retain or establish public ownership or control of water sources*

The "social good" dimensions of water cannot be fully protected if ownership of water sources is entirely private. Permanent and unequivocal public ownership of water sources gives the public the strongest single point of leverage in ensuring that an acceptable balance between social and economic concerns is achieved.

12.3.2 *Public agencies and water-service providers should monitor water quality. Governments should define and enforce water-quality laws*

Water suppliers cannot effectively regulate water quality. Although this point has been recognized in many privatization decisions, government water-quality regulators are often under-informed and under-funded, leaving public decisions about water quality in private hands. Governments should define and enforce laws and regulations. Government agencies or independent watchdogs should monitor, and publish information on, water quality. Where governments are weak, formal and explicit mechanisms to protect water quality must be even stronger.

12.3.3 *Contracts that lay out the responsibilities of each partner are a prerequisite for the success of any privatization*

Contracts must protect the public interest; this requires provisions ensuring the quality of service and a regulatory regime that is transparent, accessible, and accountable to the public. Good contracts will include explicit performance criteria and standards, with oversight by government regulatory agencies and non-governmental organizations.

12.3.4 *Clear dispute-resolution procedures should be developed prior to privatization*

Dispute resolution procedures should be specified clearly in contracts. It is necessary to develop practical procedures that build upon local institutions and practices, are free of corruption, and difficult to circumvent.

12.3.5 *Independent technical assistance and contract review should be standard.*

Weaker governments are most vulnerable to the risk of being forced into accepting weak contracts. Many of the problems associated with privatization have resulted from inadequate contract review or ambiguous contract language. In principle, many of these problems can be avoided by requiring advance independent technical and contract review.

12.3.6 *Negotiations over privatization contracts should be open, transparent, and include all affected stakeholders*

Numerous political and financial problems for water customers and private companies have resulted from arrangements that were perceived as corrupt or not in the best interests of the public. Stakeholder participation is widely recognized as the best way of avoiding these problems. Broad participation by affected parties ensures that diverse values and varying viewpoints are articulated and incorporated into the process. It also provides a sense of ownership and stewardship over the process and resulting decisions. We recommend the creation of public advisory committees with broad community representation to advise governments proposing privatization; formal public review of contracts in advance of signing agreements; and public education efforts in advance of any transfer of public responsibilities to private companies. International agency or charitable foundation funding of technical support to these committees should be provided.

13 CONCLUSION

As the 21st Century unfolds, complex and new ideas will be tested, modified, and put in place to oversee the world's growing economic, cultural, and political connections. One of the most powerful and controversial will be new ways of managing the global economy. Even in the first years of the new century, political conflict over the new economy has been front and center in the world's attention.

This controversy includes how fresh water is to be obtained, managed, and provided to the world's people. In the water community, the differing notions of water as an economic good and as a social good have become the focal point of contention. In the last decade, the idea that fresh water should be increasingly subject to the rules and power of markets, prices, and international trading regimes has been put into practice in dozens of ways, in hundreds of places, affecting millions of people. Prices have been set for water previously provided for free. Private corporations are taking control of the management, operation, and sometimes the ownership of previously public water systems. Sales of bottled water are booming. Proposals have been floated to transfer large quantities of fresh water across international borders, and even across oceans.

These ideas and trends have generated enormous controversy. In some places and in some circumstances, treating water as an economic good can offer major advantages in the battle to provide every human with their basic water requirements, while protecting natural ecosystems. Letting private companies take responsibility for managing some aspects of water services has the potential to help millions of poor receive access to basic water services. But in the past decade, the trend toward privatization of water has greatly accelerated, with both successes and spectacular failures. Insufficient effort has been made to understand the risks and limitations of water globalization and privatization, and to put in place guiding principles and standards to govern privatization efforts.

There is little doubt that the headlong rush toward private markets has failed to address some of the most important issues and concerns about water. In particular, water has vital social, cultural, and ecological roles to play that cannot be protected by purely market forces. In addition, certain management goals and social values require direct and strong government support and protection, yet privatization efforts are increasing rapidly in regions where strong governments do not exist. We strongly recommend that any efforts to privatize or commodify water be accompanied by formal guarantees to respect certain principles and support specific social objectives. Among these are the need to provide for basic human and ecosystem water requirements as a top priority, independent monitoring and enforcement of water quality standards, equitable access to water for poor populations, inclusion of all affected parties in decision making, and increased reliance on water-use efficiency and productivity improvements. Openness, transparency, and strong public regulatory oversight are fundamental requirements in any efforts to share the public responsibility for providing clean water to private entities.

Water is both an economic and social good. As a result, unregulated private market forces can never completely and equitably satisfy social objectives. Given the legitimate concerns about the risks of this "new economy of water", efforts to capture the positive characteristics of the private sector must be balanced with efforts to address its flaws, gaps, and omissions. Water management is far too important for human and ecological well-being to be placed entirely in the private sector. The proper balance requires that new water management policies and mechanisms be developed that make it possible to manage water as both a social and an economic good. Whether that balance will be achieved remains to be seen.

REFERENCES

AFX News. 1994. "Mitsubishi Oil to import water from Alaska". August 24, 1994.
AWWA. 2001. Reinvesting in Drinking Water Infrastructure. American Water Works Association. Denver, Colorado (May). Available at http://www.awwa.org/govtaff/infrastructure.pdf.
Anderson, L. 1991. "Water and the Canadian City". Water and the City. Public Works Historical Society, Chicago.
Appleton, B. 1994. Navigating NAFTA: A Concise User's Guide to the North American Free Trade Agreement. Carswell Publishing, Scarborough, Ontario, Canada. 214 pp.
Bardelmeier, W. 1995. "Water is too heavy a burden to carry". Lloyd's List, September 22, 1995.
Barlow, M. 1999. Blue Gold: The Global Water Crisis and the Commodification of the World's Water Supply. International Forum on Globalization. Sausalito, California.
Beecher, J.A., Dreese, G.R. & Stanford, J.D. 1995. *Regulatory Implications of Water and Wastewater Utility Privatization*. The National Regulatory Research Institute, Columbus, Ohio (July).
Beecher, J.A. 1997. Water utility privatization and regulation: Lessons from the global experiment. Vol. 22, No.1, pp. 54–63.
Black & Veatch. 2001. *California Water Charge Survey 2001*. Black and Veatch Corporation, Irvine, California.
Blake, N.P. 1991. Water and the City: Lessons from History. *Water and the City*. Public Works Historical Society, Chicago.
Blokland, M., Braadbaart, O. & Schwartz, K. (eds). 1999. *Private Business, Public Owners: Government Shareholdings in Water Enterprises*. Ministry of Housing, Spatial Planning and the Environment, The Hague, the Netherlands.
Boulton, L. & Sullivan, R. 2000. Thirsty markets turn water into valuable export: Drinking water is ever more scarce in the parched lands around the Mediterranean. *Financial Times (London)*. November 7, 2000.
Briscoe, J. 1996. Water as an economic good: The idea and what it means in practice. *World Congress of the International Commission on Irrigation and Drainage*. Cairo, Egypt (September 1996). Available at http://www-esd.worldbank.org/rdv/training/icid16.htm.)
Briscoe, J. 1997. Managing Water as an Economic Good: Rules for Reformers. Keynote Paper. *International Committee on Irrigation and Drainage Conference on Water as an Economic Good*. Oxford, United Kingdom. September 1997.

Brewster, L. & Buros, N. 1985. Non-conventional water resources: economics and experiences in developing countries. (I). *Natural Resources Forum*. Vol. 9, No. 2, pp. 133–142.

Brook, P.J. 1999a. Bail out: The global privatization of water supply. *The Urban Age Magazine*. Winter. Urban Development. The World Bank, Washington, D.C.

Brook, P.J. 1999b. Lessons from the Guinea Water Lease. *Public Policy for the Private Sector, Note*. No. 78, The World Bank Group, Washington, D.C. April 1999.

Brown, L.R. & associates. 1995. *State of the World 1995*. Norton Press, New York. via http://www.worldwatch.org/pubs/sow/sow95/ch01.html.

Brubaker, E. 2001. *The Promise of Privatization*. Energy Probe Research Foundation, Toronto, Ontario (April). Available in full at http://www.environmentprobe.org/enviroprobe/pubs/Ev548.htm.

Bruns, B.R. & Meinzen-Dick, R.S. (eds). 2000. *Negotiating Water Rights*. International Food Policy Research Institute. Vistaar Publications, New Delhi, India.

CALFED. 1999. *Economic evaluation of water management alternatives California-Federal Bay Delta Program*. Sacramento, California.

Calaguas, B. 1999. Private Sector Participation. *Vision 21 process, Export Group Meeting*. Wageningen, The Netherlands, April.

Coffin & Richardson, Inc. 1981. *Water Conservation under Conditions of Extreme Scarcity: the U.S. Virgin Islands*. Office of Water Research and Technology. National Technology Information Service. Washington, D.C.

Daily, G.C. (ed). *Nature's Services: Societal Dependence on Natural Ecosystems*. Island Press, Washington, D.C.

Demir, M. 2001. Turkey, Israel sign water deal. *The Jerusalem Post*. January 29, 2001.

Dinar, A. & Subramanian, A. 1997. *Water Pricing Experiences: An International Perspective. World Bank Technical Paper No. 386*. The World Bank, Washington, D.C.

EAA. 2001. *Website of the Edwards Aquifer Authority*. www.e-aquifer.com

Ekstract, J. 2000. Turkey and Israel set to complete water import deal. *Jerusalem Post*. July 5. Available at http://www.jpost.com/Editions/2000/07/05/News/News.9164.html.

Empresa Metropolitana de Obras Sanitarias S.A. 1995. *Memoria 95*. Santiago, Chile.

Faulkner, J. 1997. Engaging the private sector through public–private partnerships. In bridges to sustainability. *Yale Bulletin Series*. No. 101. Yale University Press. New Haven, Connecticut.

Fiji Country Paper. 1984. Technical Proceedings (Part 3). *Regional Workshop on Water Resources of Small Islands*. Suva, Fiji, 1984. Commonwealth Science Council Technical Publication Series No 182, pp. 1–10.

Financial Times. 2000. *Thirsty markets turn water into valuable export*. Financial Times (London).

Financial Times. 2000. Available at http://www.internetional.se/toft/toft20011.htm#water. November 7.

FTGWR. 2000. *Bolivia: The Cochabamba crisis*. Financial Times Global Water Report. 14 April 2000. Issue 93. pp. 1–3.

FTGWR. 2000. *International Water's rebuttal: Bolivia*. Financial Times Global Water Report. 28 April 2000. Issue 94. pp. 4–6.

FTGWR. 2000. *French giants slug it out*. Financial Times Global Water Report. 28 April 2000. Issue 94. pp. 10–12.

FTGWR. 2000. *Azurix writes down its activities in Argentina*. Financial Times Global Water Report. 26 January 2001. Issue 114. pp. 8.

FTGWR. 2000. *Maynilad pull-out threat in Manila*. Financial Times Global Water Report. 26 January 2001. Issue 114. pp. 1–3.

FTGWR. 2001. *Maynilad fights on in face of government opposition*. Financial Times Global Water Report. 12 March 2001. Issue 117. pp. 6–7.

FTGWR. 2001. *US eyes Canadian water supplies*. Financial Times Global Water Report. 6 August 2001. Vol. 127, pp. 3.

French, H. 1999. Challenging the WTO. *Worldwatch*. November/December 1999, pp. 22–27.

Garn, M. 1998. Managing water as an economic good. *Conference on Community Water Supply and Sanitation*. May 5–8 1998, Washington, D.C. The World Bank, via http://www.wsp.org/english/focus/conference/managing.html.

Gattas, N. 1998. Environment-Cyprus: Tourists get first sip of water shortages. *Inter Press Service*. November 3.

Gleick, P.H. 1996. Basic water requirements for human activities: Meeting basic needs. *Water International*. Vol. 21, pp. 83–92.

Gleick, P.H. 1998. Water in crisis: Paths to sustainable water use. *Ecological Applications*. Vol. 8, No. 3, pp. 571–579.

Gleick, P.H. 1999. The human right to water. *Water Policy*. Vol. 1, pp. 487–503.

Gleick, P.H. 2000. *The World's Water 2000–2001: The Biennial Report on Freshwater Resources*. Island Press, Washington, D.C.

Gleick, P.H. 2001. Global water: Threats and challenges facing the United States. Issues for the new U.S. administration. *Environment.* Vol. 43, No. 2, pp. 18–26.

Gleick, P.H., Wolff, G., Chalecki, E.L.& Reyes, R. 2002. *The New Economy of Water. The Risks and Benefits of Globalization and Privatization of Fresh Water.* Pacific Institute. February, 2002.

Global Water Partnership (GWP). 2000. Toward Water Security: A Framework for Action to Achieve the Vision for Water in the 21st Century. Global Water Partnership, Stockholm, Sweden.

Goldman Foundation. 2001. *Goldman Prize winners 2001.* Oscar Olivera. http://www.goldmanprize.org.

Gopinath, D. 2000. *Blue Gold.* Institutional Investor International Edition. February.

Graz, L. 1998. Water source of life. *FORUM: War and Water.* International Committee of the Red Cross, Geneva, Switzerland, pp. 6–9.

Hanemann, M. 1991. *Willingness to pay and willingness to accept: how much can they differ?* The American Economic Review, Vol. 81, No. 3, pp. 635–47.

Hudson, P. 1999. Muddy waters; Argentina, the Latin American leader in private water works, struggles to create a transparent concession system. *Latin Trade.* March. 1999.

Huttemeier, J. 2000. Personal communication via email. Director Maersk Tankers. Copenhagen, Denmark. December 7, 2000.

ICWE. 1992. The Dublin Principles. International Conference on Water and the Environment. Available in full at http://www.wmo.ch/web/homs/icwedece.html.

IISD. 1999. Report on *WTO's High-Level Symposium on Trade and Environment.* 15–16 March 1999. International Institute for Sustainable Development. As found at http://www.wto.org/wto/hlms/sumhlenv.htm.

IJC. 2000. *Final Report to the Governments of Canada and the United States, Protection of the Waters of the Great Lakes.* International Joint Commission. Available at http://www.ijc.org.

IUCN. 2000. Vision for water and nature: A world strategy for conservation and sustainable management of water resources in the 21st century. *The International Union for the Conservation of Nature, World Water Forum*, The Hague, Netherlands.

Idelovitch, E. & Ringskog, K. 1995. *Private Sector Participation in Water Supply and Sanitation in Latin America.* The International Bank for Reconstruction and Development/The World Bank (May). Washington, D.C. http://www.worldbank.org/html/lat/english/papers/ewsu/ps_water.txt.

Jacobson & Hill. 1988. Hydrogeology and Groundwater Resources of Nauru Island, Central Pacific Ocean. *Bureau of Mineral Resources.* Record No 1988/12, Australian Government.

Jameson, S. 1994. With Japan's relentless heat wave, it's boom or bust. *Los Angeles Times.* August 1994.

Kemper, K.E. 1996. *The cost of free water: water resources allocation and use in the Curu Valley, Ceara, Northeast Brazil.* Linkoping University. Linkiping, Sweden.

Kingdom, B. & Jagannathan, V. 2001. *Utility Benchmarking: Public Reporting of Service Performance.* The World Bank Group, Note Number 229. World Bank, Washington, D.C. March 2001.

Komives, K. 2001. Designing pro-poor water and sewer concessions: Early lessons from Bolivia. *Water Policy.* Vol. 3, No. 1, pp. 61–80.

Lazaroff, C. 2001. WTO Upholds U.S. Right to Protect Sea Turtles. *Environment News Service.* 19 June 2001. As found at www.ens-news.com/ens/jun2001/2001|-06-19-07.html.

LeClerc, G. & Raes, T. 2001. *Water: A World Financial Issue.* PriceWaterhouseCoopers, Sustainable Development Series, Paris, France.

Lee Yow Ching. 1989. Development of Water Supply in Penang Island, Malaysia. Proc. Seminar on Water Management in Small Island States. *Cyprus Joint Tech. Council and Commonwealth Engineer's Council.*

Lerner, 1986. Leaking pipes recharge ground water. *Groundwater.* Vol. 25, No. 5, pp. 654–662.

Linton, J. 1993. *NAFTA and water exports.* Submission to the Cabinet Committee on NAFTA of the Government of the Province of Ontario. April 1993, 15 pp.

MacDonald, M. 2001. *Brian Tobin has no sway in debate over water exports.* http://ca.news.yahoo.com/010329/6/3u8a.html.

Mandell-Campbell, A. 1998. Argentina sell-off sparks litigation. *The National Law Journal.* July 20, 1998.

Market Guide. 2001. Vivendi Universal. http://yahoo.marketguide.com/mgi/MG.asp?nss=yahoo&rt=ageosegm&rn=A26C7

Margat, J. 1996. Comprehensive assessment of the freshwater resources of the world: Groundwater component. *Comprehensive Global Freshwater Assessment.* Contribution to Chapter 2. United Nations, New York.

McNeill, D. 1998. Water as an economic good. *Natural Resources Forum.* Vol, 22, No. 4, November 1998.

Meyer, T.A. 2000. Personal communication to Peter Gleick, December 9, 2000 (via email).

Muller, M. 1999. Efficiency yes, but jury is still out on equity. *Business Day.* October 6, 1999. South Africa.

Muñoz, C. 2001. Personal communication.

Myers, N. & Kent, J. 2001. *Perverse Subsidies: How Tax Dollars Can Undercut the Environment and the Economy*. Island Press, Covelo, California.

NCWC. 1997. Privatization Of Water Purification And Distribution Systems. Policy. 97.14EM. (available at http://www.ncwc.ca/policies/water_waterquality.html). National Council of Women of Canada

Neal, K., Maloney, P.J., Marson, J.A. & Francis, T.E. 1996. *Restructuring America's Water Industry: Comparing Investor-owned and Government Water Systems.* Reason Public Policy Institute, Policy Study No. 200, January.

Nickson, A. 1998. *Organizational Structure and Performance in Urban Water Supply: the case of the SAGUAPAC co-operative in Santa Cruz, Bolivia.* International Development Department. The University of Birmingham, United Kingdom.

Orwin, A. 1999. *The Privatization of Water and Wastewater Utilities: An International Survey.* Environment Probe, Canada. http://www.environmentprobe.org/enviroprobe/pubs/ev542.html.

Panos. 1998. *Liquid Assets: Is Water Privatization the Answer to Access?* Panos Briefing, London, United Kingdom.

Perry, C.J., Rock, M. & Seckler,V. 1997. *Water as an Economic Good: A Solution, or a Problem?* International Irrigation Management Institute, Research Report 14.

Pilling, D. 1996. Generale des Eaux in row over Argentine water. *Financial Times (London).* February 13, 1996.

Postel, S. & Carpenter. S. 1997. Freshwater ecosystem services. In G.C. Daily (ed.) *Nature's Services: Societal Dependence on Natural Ecosystems.* Island Press, Washington, D.C., pp. 195–214.

Rivera, D. 1996. *Private Sector Participation in the Water Supply and Wastewater Sector: Lessons from Six Developing Countries.* The World Bank, Washington, D.C.

Rogers, P., Bhatia, R. & Huber, A. 1998. *Water as a social and economic good: how to put the principle into practice.* Global Water Partnership/ Swedish International Development Cooperation Agency, Stockholm, Sweden.

Rossa, J. 2001. Azurix holders OK merger with indirect Enron unit. *Dow Jones Newswires.* March 16, 2001.

Rudge, D. 2001. Delegation going to Turkey to discuss importing water. *The Jerusalem Post.* January 17, 2001.

Savas, E.F. 1987. *Privatization: The Key to Better Government.* Chatham House, Chatham, New Jersey.

Shambaugh, J. 1999. *Role of the Private Sector in Providing Urban Water Supply Services in Developing Countries: Areas of Controversy.* Research Paper, United Nations Development Program/Yale University Research Clinic, New Haven, Connecticut.

Shrybman, S. 1999a. *Water export controls and Canadian international trade obligations.* Legal opinion commissioned by the Council of Canadians. Available at http://www.canadians.org.

Shrybman, S. 1999b. *Opening the floodgates.* Western Environmental Law Center, available at http://www.wcel.org/wcelpub/1999/12756.html

S&P. 2000. European water: Slow progress to increase private sector involvement. Standard & Poor's. http://www.sandp.com/Forum/RatingsCommentaries/CorporateFinance/Articles/eurowater.html. September 12, 2000.

Sugimoto, M. 1994. Drought forces Japan to import Alaskan water. *South Bend Tribune.* August 28, 1994.

Swann and Peach. 1989. Status of Groundwater Resources Development for New Providence [Bahamas]. *Interregional Seminar on Water Resources. Management Techniques for Small Island Countries.* UNDTCD, Suva, Fiji. ISWSI/SEM/4.

Tan, N. 2000. *Rebuilding Malaysia Inc.* In-depth Report. October 2. Merrill Lynch, New York.

The Economist. 2000. *Nor any drop to drink.* March 25, 2000, p. 69.

Trémolet, S. 2001. *Putting the poor at the heart of privatization.* Financial Times Global Water Report (FTGWR), Issue 114.

Tully, S. 2000. Water, water everywhere. *Fortune.* 15 May 2000.

Turner, W. 2001. *Water export from Manavgat, Turkey.* Available at http://www.waterbank.com/Newsletters/nws23.html.

UN. 1997. *Comprehensive Assessment of the Freshwater Resources of the World.* United Nations, New York.

UNESCO. 1992. *Small Tropical Islands-Water Resources of Paradises Lost.* IHP Humid Tropics Programme Series No.2, UNESCO, Paris, France.

Waddell, J.A. no date (nd). *Public Water Suppliers Look to Privatization.* PricewaterhouseCoopers. http://Waddell.UtilitiesProject.com.

WEF. 2000. *Financing water quality improvements.* Water Environment Federation. http://www.wef.org/govtaffairs/policy/finance.jhtml.

Water Supplies Department. 1987. *Hong Kong's Water.* Land and Work's Branch. Hong Kong Government: 32 pp.

WCEL. 1999. West Coast Environmental Law. http://www.wcel.org/wcelpub/1999/12926.html.

World Bank. 1992. *The World Development Report, 1992.* Oxford University Press, New York.

World Commission on Water for the 21st Century. 2000. *A Water Secure World: Vision for Water, Life, and the Environment*. World Water Vision Commission Report. The Hague, The Netherlands.

WHO. 2000. *Global Water Supply and Sanitation Assessment: 2000 Report*. World Health Organization and United Nations Children's Fund. Report. Available in full at http://www.who.int/water_sanitation_health/Globassessment/GlobalTOC.htm.

Wright, A.M. 1997. *Toward a Strategic Sanitation Approach*. UNDP-World Bank Water and Sanitation Program. Washington, D.C.

Yaron, G. 1996. *Protecting British Columbian waters: The threat of bulk water exports under NAFTA*. University of British Columbia, Canada. April 1996, 61 pp.

Yepes, G. 1992. *Alternativas en el manejo de empresas del sector agua potable y saneamiento*. Departamento Nacional de Planeación. Seminario Internacional Eficiencia en la Prestación de los Servicios Públicos de Agua Potable y Saneamiento Básico. Bogotá, Colombia. pp. 111–116.

Yergin, D. & Stanislaw, J. 1999. *The Commanding Heights: The Battle Between Government and the Marketplace That Is Remaking the Modern World*. Simon and Schuster, New York.

Zhang, Z. & Liang, Z. 1988. Study of Water Supply to Xiamen Island. *South East Asian and the Pacific Regional Workshop on Hydrology and Water Balance of Small Islands*. UNESCO/ROSTSEA, Nanjin, China. pp. 134–140.

A structure for water administration in the 21st Century. The case of Australia

B. Martin
Office of Water Regulation, Australia

ABSTRACT: The provision of water services in Australia is a State responsibility. Each of the eight separate States and Territories is responsible for its own water industry, and each State runs a regulatory system. Water resources are shared between the States of Queensland, New South Wales, Victoria and South Australia because each has access to the major Australian river system, the Murray Darling System. The Commonwealth Government does not have a direct role in the water industry. However, by virtue of its taxation and funding powers, the Commonwealth has been able to impose reform conditions on the State water industries. As a result the water industry, and across Australia, has undergone significant change over the past 10 years. Major reorganisations have occurred in most States, which have separated the three functions of service provision, resource management, and regulation of the industry. Mechanisms for the regulation of the water industry vary considerably from state to State. In fact, Australia has been described as a laboratory of experimentation for different regulatory arrangements. This paper describes the industry models applying in the States of Australia and assesses these against Best Practice Benchmarks.

1 OVERVIEW OF THE AUSTRALIAN WATER INDUSTRY

The provision of water services in Australia is a State responsibility. Each of the 6 separate States and 2 Territories (Figure 1) is responsible for its own water industry, and each jurisdiction is responsible for its own regulatory system.

Australia is an extremely dry continent with the availability of water defining opportunities for agricultural and most other activity, other than mining. There are many river basins and underground water resource systems. The largest of these are the Murray Darling River and the Great Artesian Basin, both of which cross several States (Figure 2). Because of its economic importance and the sensitive level of the exploitation of the Murray Darling Basin, a special Commission of Management – the Murray Darling Basin Commission – has been established to manage the resource.

There are a wide range of climatic conditions and water resource availability, which is a feature of the vastness of the Australian continent.

Some idea of the importance of water to the development and population of Australia can be seen from the distribution of its population. The people of Australia are very much located around the major city centres, and the regional areas that are well watered. The majority of the population is also located in coastal areas. The centre of the continent is largely desert, and has a low population density.

Table 1 shows the dominant position of agriculture as a user of the water resources of Australia. Its share is 70%. This compares with about 8% for household consumption which is a little more than the usage in the treatment of water and wastewater. Manufacturing uses only 3%, and mining even less, of the total water used.

Australia has experience relatively rapid growth of its population and its industry since the 50s. There was a heavy emphasis on migration in the decades following the Second World War, resulting

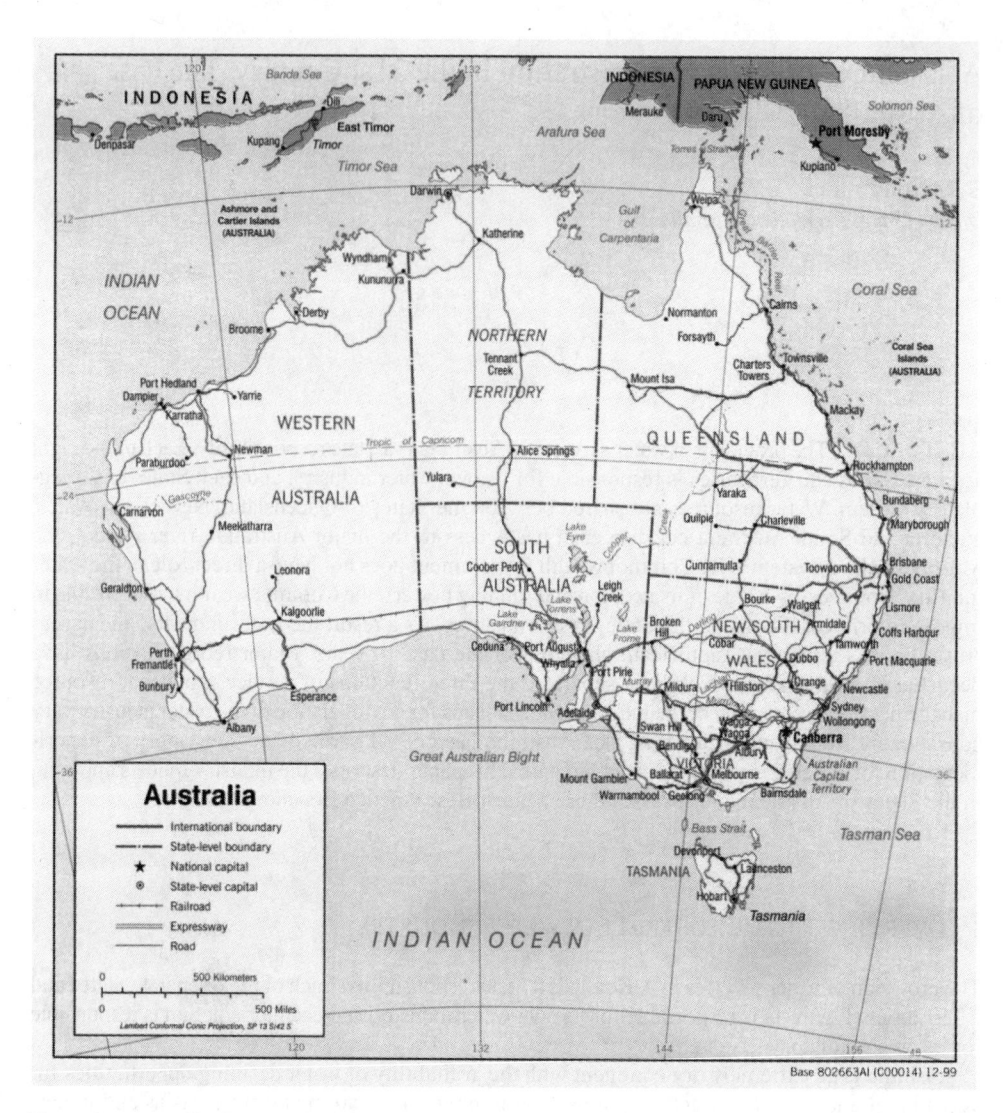

Figure 1. Australia: States and Territories.

in sustained growth in population of the order of 3% to 4%. This, together with the development of new industry and the boom in mining and minerals processing has led to large increases in demand for water services.

The water industry has a history of working to high engineering standards, and there has been a relatively high level of usage per capita. More recently, these demands on the water supply industry have been added to with increasing attention to environmental and health standards. There has been an increasing community awareness of the value of the water resource and the need to manage water resources with care. A lot of attention has been given to assessing sustainable yields from surface and underground resources, and protecting both the resource and the environment that it supports. In a number of areas across Australia, water resources have come to be under stress. In some cases this has been a result of over-allocation of the available resource. In others, climate change has led to a downward revision of the sustainable yield available from resources.

Water industry providers have traditionally been public works departments and there is still a high level of government ownership of infrastructure assets. Water utility operators are now a mixture of

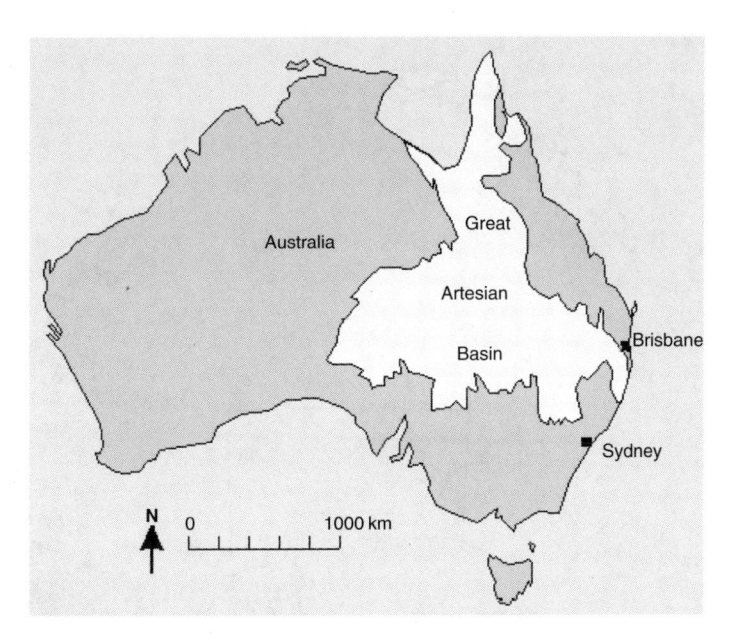

Figure 2. Australia: the Great Artesian Basin.

Table 1. Water usage by industry for Australia, 1997–98 (ABS, 2000 and 2002).

Industry	Self extraction (gigalitres)	Recirculated use (gigalitres)	Total use (gigalitres)
Agriculture, forestry and fishery	7,162	8,360	15,522
Mining	545	25	570
Manufacturing	217	511	728
Electricity and gas	1,262	46	1,308
Water/wastewater treatment	1,707	0	1,707
Residential	33	1,796	1,829
Other	104	419	523
Total	11,030	11,157	22,187

corporations, statutory authorities and local government municipalities. There is very little competition between service providers.

Pricing of water has traditionally been set by government policy with water prices often being used to further other policy objectives of governments such as regional development, establishment of new agricultural industries, and assistance to relatively disadvantaged groups in the community. As a result water pricing has become a politically sensitive issue, although there does seem to be a growing community awareness that water has been under-priced.

During the 90s there has been a period of significant change for the industry. Many water utilities have been commercialised or corporatised, and this has assisted with the clarification of roles for all parties involved, and improved the focus on outcomes to be achieved. Roles of the regulators have developed and become more specialised, and more effective. Standard setting has become more explicit. Licensing of utilities has been introduced, and these have established formal standards of performance and standards for the delivery of services to customers. Economic regulation has been established in some States.

The structure and organisation of the water industry varies considerably from State to State. In fact, there is a range of ownership and management arrangements. While some irrigation schemes have

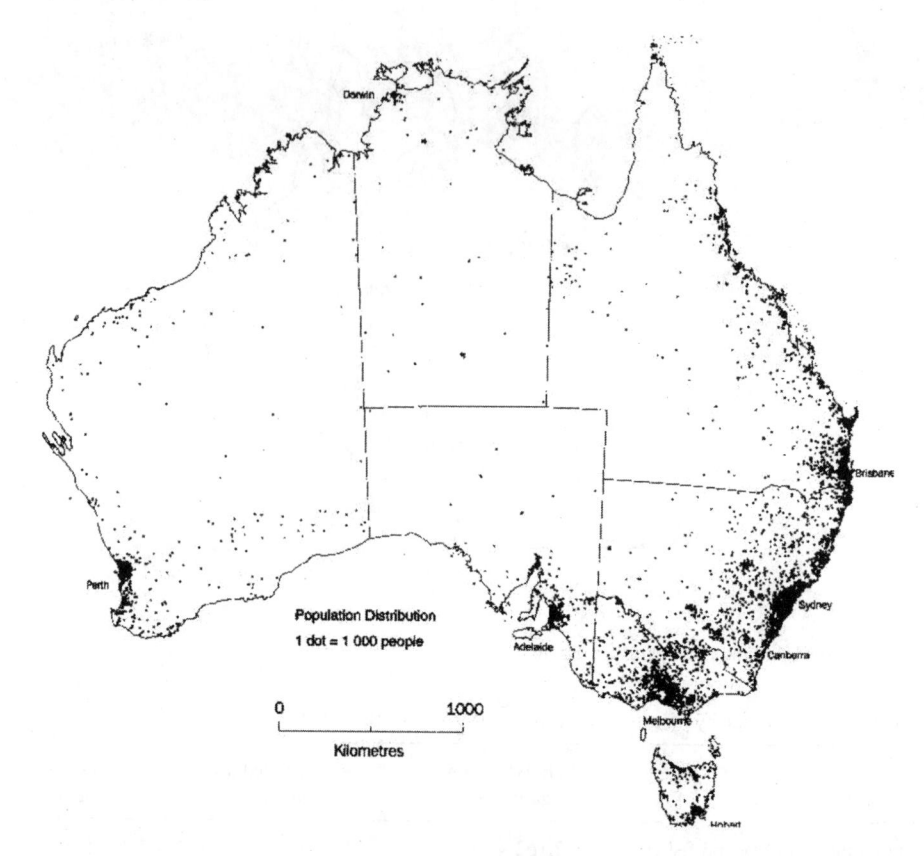

Figure 3. Australia: population distribution and density (ABS, nd). Data are based on the population of Statistical Local Areas (SLA, generally equivalent to Local Government Areas). One dot is allocated for every 1,000 people within the SLA. The dot density presentation of the population distribution for very large area SLA such as Kalgoorlie-Boulder (WA) and Mt Isa (Qld) can give the impression that the population is spread over a vast area when in fact it is mostly concentrated in the major regional centre within the SLA.

been privatised (in the sense of asset sale), Australia has stopped short of fully privatising any of its urban water systems, which are predominantly in public ownership. However, a wide range of private sector participation arrangements apply. Regulatory arrangements, also, vary considerably from State to State.

The provision of water services in Australia has traditionally been a responsibility of Government agencies. Public Works Departments and urban water departments and agencies have built and managed water infrastructure in both the major cities and in regional towns. Similarly, irrigation schemes have been developed by Governments. Governments have provided the capital, built the infrastructure and then funded and operated the irrigation schemes on behalf of irrigator farmers. In the urban water sector, the private sector has become increasingly involved in delivery arrangements over the last 40 years, and particularly in the last 10 years, through a range of mechanisms.

2   REFORM IN THE AUSTRALIAN WATER INDUSTRY

The water supply industry in Australia has served population and economic growth well for over 150 years. The main emphasis has been on building infrastructure and operating it through Government owned monopolies. This was particularly the case in the period from 1950 to the 80s. The focus tended to be on supply oriented delivery rather than on demand side reform and efficiency.

In 1994 and 1995 the Commonwealth and State Governments meeting as the Council of Australian Governments initiated a major program of reform for all industries. It gave particular attention to sectors that are dominated Government owned monopolies. This had significant implications for the water industry, and was a major milestone in the history of the industry in Australia.

The general objectives of the program were for a more competitive commercial environment for all industry, and thereby the achievement of better economic, environmental, resource management and social outcomes.

The general aims of the reforms were to:

– Reduce anti-competitive conduct.
– Reform legislation which restricts competition.
– Reform State-owned monopolies to facilitate competition.
– Provide for third party access to essential infrastructure to promote competition and efficient use of spare capacity.
– Remove abuse of monopoly power in pricing.
– Foster competitive neutrality.

The policy was a major step forward for micro-economic reform in Australia and set an important reform agenda for State governments and government owned utilities.

For the water industry the Council of Australian Governments agreed to a national approach to reform, which was comprehensive.

(a) In terms of institutional arrangements, this reform called for:
 – Separation of the roles of owner, service provider, water resource manager, and services regulator.
 – A higher level of public consultation by government agencies and service deliverers where change and/or new initiatives are contemplated involving water resources.
 – A review of water legislation to ensure that any anti-competitive elements are removed.
 – Education of the community on the value of water, and demand management.
(b) In terms of the modernisation of government-owned utilities, the reform called for:
 – Service providers, particularly in metropolitan areas, to have a commercial focus. This might be achieved by contracting out, corporatisation or privatisation.
 – Service providers should no longer be involved in the setting of prices or the provision of policy advice to governments.
 – The greater use of stormwater and re-use of wastewater. There was also agreement to improve water quality monitoring and catchment management policies and community consultation and awareness.
 – More open and transparent reporting of information so that the public might be made more aware of costs and cross-subsidies where they occur.
 – Pricing reforms, including the adoption of pay-for-use pricing and two part tariffs with a fixed connection fee and volumetric charges.
 – Removal of cross-subsidies (or at least transparency of reporting so that such subsidies would be apparent).
(c) The reform paid particular attention to irrigation schemes, and also to other rural services. It called for:
 – Achievement of full cost recovery – a measure specifically designed for irrigation schemes, most of which recovered only a small share of costs.
 – Proposed new irrigation schemes to be subject to a rigorous appraisal of their economic viability and ecological sustainability.
 – Irrigators to take greater responsibility in the management of irrigation areas, for example through operational responsibility being devolved to local bodies.
 – The introduction of a comprehensive systems of water allocations or entitlements backed by separation of water property rights from land title and clear specification of entitlements in terms of ownership, volume, reliability, transferability and, if appropriate, quality.
 – The setting of allocations or entitlements, including allocations for the environment as a legitimate user of water.

- Trading arrangements in water allocations or entitlements to be instituted once the entitlement arrangements had been settled.
- Trading arrangements to be consistent and facilitate sales across State borders where this is socially, physically and ecologically sustainable.

(d) In terms of price setting, the reform also called on States and Territories to consider establishing independent sources of price oversight advice, which would:
- Be independent from the Government business enterprise whose prices are being assessed.
- Have as a prime objective the efficient allocation of resources.
- Recognise defined community service obligations imposed on a business enterprise.
- Apply to all significant Government business enterprises that are monopoly, or near monopoly, suppliers of goods or services (or both).
- Permit submissions by interested persons, through an open and consultative process.
- Publish its pricing recommendations, and the reasons for them.

These represented significant reforms for the Australian water industry. Progress against the reform agenda has been carefully monitored and substantial payments from the Commonwealth to the States are dependent on satisfactory progress. However, the pace of reform has varied considerably. Some States have instituted all of the reforms, while others are yet to fully introduce water metering – so that two-part tariffs are not yet possible. The result is that some States have much more advanced models for water services provision and regulation, than do others.

There has been considerable progress across Australia in the reform of irrigation industries. Water entitlements, separate from land tenure, have been established. Water entitlements have been made tradeable. Most irrigation schemes have been devolved from Government and are now owned and managed by cooperatives of irrigator farmers.

3 NEED FOR REGULATION OF THE WATER INDUSTRY

Large parts of the water industry are natural monopolies, because the high capital cost of much of the infrastructure is not economic to duplicate by a second provider. This is the case for many source developments, and for transport systems for water and for sewerage. The sources of water are also natural monopolies because they either cannot be duplicated or can only be duplicated by going to more distant sources and hence at much greater cost.

Other components of the water industry are potentially competitive, although under current industry arrangements, opportunities are limited. Protection from competition means there is little incentive to improve performance, and means that there is a risk the service may not be provided at a price or standard acceptable to customers.

Industries such as water services are usually regulated by government to address "market failure". Where markets are seen to not deliver efficient or equitable outcomes they are said to "fail". The solution is to apply forms of regulation, which ideally should be designed to address the root cause of the market failure. There are four potential sources of "market failure":

- *Monopoly*. Monopoly can be a natural outcome of the nature of the industry or it can be created by artificial barriers to entry in the form of patents, agreements or exclusive licences. The essence of a natural monopoly is significant economies of scale over an extensive range of output. This means that regardless of whether an established provider is efficient or inefficient, it still has large cost advantages over an alternative provider.
- *Public goods*. These are goods that, once provided, are freely available to everyone. It is not possible to exclude anyone from consumption of the good or service. Therefore, without regulation, it is difficult to collect revenue to pay for their provision. Some aspects of the water industry, such as drainage and groundwater resources function as public goods.
- *Externalities*. Externalities are costs or benefits arising from an economic transaction that fall on a third party. There are many examples in the water industry. Over-extraction of water resources has

impacts on the environment and on the quality of people's lifestyle as stream flows and ground water tables are lowered. Drainage can release contaminants into streams and water bodies. Outfalls from sewerage and water treatment works can similarly have adverse impacts on the quality of water bodies and the wider environment. Poor quality services can result in public health costs that go beyond the individual consumer. These sorts of externalities can only be addressed by regulation. Regulators in Western Australia and elsewhere set standards which protect environmental values and public health values, and also represent the interests for good management of the water resources.

– *Inequalities*. Inequality is a natural outcome of the market economy. This is so because not all people have the same resources, and some have resources that are less valued than others. The results are inequalities of income, opportunity and location. When inequalities become unacceptable, governments act to reduce them by redistributing income. In an ideal world this is best done through the taxation and social security systems, rather than through subsidisation of the costs of access to, or consumption of, water services.

Regulation is seen as a form of countervailing force to the monopoly forces in an industry. It has been the normal experience in Australia that regulatory mechanisms have been introduced at the time public works style departments or urban water statutory authorities are corporatised.

In the State of Western Australia, for example, there have for some years been dedicated regulators for the protection of the environment (Department of Environmental Protection), and public health (Health Department of Western Australia). Formal regulation of water service providers was introduced in January 1996, at the same time as the Water Corporation was corporatised. This was most appropriate as the principal supplier (the Water Corporation) was obliged under the *Water Corporation Act 1995* and Corporations Law to maximise its profits and to grow its business. It now had a clear set of objectives and could no longer be guided by other considerations, such as community or social impacts, other than by the force of its regulators or by explicit Government direction.

The Second Reading Speech for the *Water Services Coordination Act 1995* (which establishes the regulatory framework for the service providers) clearly indicated the intention for the new regulatory arrangements. They were to set the groundwork for competition in the provision of water services within the State, with the aim of bringing about downward pressure on the charges consumers pay for these services, and to ensure that consumers' needs became of paramount importance to providers.

It notes that "... the Coordinator of Water Services will protect the interests of customers in respect to the levels of service provided and prices charged by water service utilities including the Water Corporation, the Water Boards and other licensed water providers".[1]

To this end, the new legislation had the objective of enabling other providers of water services to enter the water services market. The new licensing regime was to facilitate this objective. It also foreshadowed that certain of the rights and powers vested in the Water Corporation under other water legislation would be conferred by regulations on private providers.

States may want to continue to regulate their water industries for a variety of other reasons, which are very real but can make for a model of regulation which is less than ideal. For example, the water industry has strong links to regional growth and development. It plays an important role in supporting the development of primary industries from mining and downstream processing of minerals to the provision of irrigation services and drainage in some country areas.

While it is not an ideal model, a pragmatic world recognises that water services are often used as a mechanism to offset social and economic disadvantage. The role of the regulator in this context is to assist with development of policies that are well targeted and efficiently address the objectives being decided upon by governments.

In addition, water has a special significance because of its unique character and importance in setting a quality of life. Access to clean water is regarded as a basic human right, and this consideration is significant in developing policy settings for the water industry.

---

[1] Taken from the Second Reading Speech, Hansard (28/9/95)

A further reason for regulation is to protect the interests of Government, as owner of the assets there is a need to ensure long-lived industry assets are well planned for, efficiently used and properly maintained.

From a customer point of view, there is need for protection. There is, for example, the risk that customers may have inadequate information in relation to their water services, especially drinking water quality.

4 INDUSTRY STRUCTURES ACROSS AUSTRALIA

The structure of the water industry varies enormously between States.

4.1 *Western Australia*

In Western Australia the State-owned utility was corporatised in 1996. It provides services to the capital city, Perth, to most regional centres and small towns, and also to farms in the South West for domestic and livestock use. A number of relatively small utilities, which are owned and operated by either the State Government or by Local Governments, provide services in some non-metropolitan centres.

Irrigation services are provided by cooperatives of irrigation farmers, or are in the process of being transferred to irrigator-managed cooperatives.

4.2 *New South Wales (NSW)*

New South Wales (NSW) has city-based utilities. Sydney Water (a corporatised entity) provides services to the Sydney, Illawarra and Blue Mountains areas and Hunter Water Corporation provides services to the City of Newcastle and surrounding areas. Throughout regional NSW there are Local Government water schemes. In Sydney there has been a separation of bulk water provision, with this service being provided to Sydney water by the Sydney Catchment Authority. Irrigation services throughout the State are owned and managed by irrigator cooperatives.

Sydney Water Corporation is supplied with catchment and bulk water services by the Sydney Catchment Authority, a statutory authority.

4.3 *Victoria*

In Victoria there are large urban utilities, in Melbourne, and also regional water utilities. From 1995, the former vertically integrated monopoly, the Melbourne Water Corporation, owned by the State Government, was broken into a wholesale corporatised entity and three corporatised retail and distribution companies.

The wholesale business, Melbourne Water, is responsible for Melbourne's bulk water and bulk sewerage services. This includes management of the dams, reservoirs and sewage treatment plants for the Melbourne area. Although Melbourne Water is not regulated by operating licence, requirements for water quality and quantity are laid down in its contracts with each of the three metropolitan water companies.

The three retail and distribution companies, City West Water, Yarra Valley Water, and South East Water, provide water and sewerage services to commercial, industrial and residential customers in Melbourne and the Mornington Peninsula.

Outside Melbourne, water and wastewater services are the responsibility of 15 statutory urban water authorities and 4 statutory rural water authorities. In addition, there are six Alpine Resorts Management Boards for ski resorts, Parks Victoria supplies services to the National Parks, and other organisations provide water to remote towns, caravan parks and resorts.

4.4 South Australia

In South Australia the water utility was corporatised in 1994 to create the South Australian Water Corporation. In 1995 Adelaide's water supply and wastewater treatment services were outsourced through a management contract to a private company. In 1996 a contract was awarded to the private sector for water treatment in rural areas. Most rural water supplies are provided by the South Australian Water Corporation.

In Queensland and Tasmania water services are provided by Local Government Councils including water services in Brisbane which are provided by the City Council. In Tasmania there are three bulk water suppliers, who deliver water to the Local Government service providers.

In the Australian Capital Territory (ACT) there is a single water provider in the form of ACT Electricity and Water. This is a State owned entity. ACT Electricity and Water has entered a public-private partnership with AGL, after a competitive selection process. The structure of this relationship is a hybrid, with the public sector retaining the assets, but also entering a joint venture with the private sector.

In the Northern Territory, the Power and Water supplies water and sewerage services to the Territory's four major urban areas, as well as to a number of rural and remote communities.

5 BENCHMARKING AND PERFORMANCE EVALUATION

Unlike the many collaborative metric benchmarking developments being undertaken by regulatory agencies, research centres, academic institutions and peak water industry bodies throughout the world, in Australia there has been no similar coordinated approach to benchmarking or comparative performance reporting. Seemingly, it is an area in which Australian regulators could/should be doing more but this assumes that the regulatory circumstances in all states and territories have reached some minimum threshold of uniformity – which is simply not the case at the present time.

Even where the (variable) functions and specific objectives of Australia's regulators may be comparable across different jurisdictions, the choice of regulatory model and complementary organisational structures differ markedly due to other factors. Examples of these factors include; the degree of regulatory independence conferred, scope of regulation permitted, maturity of local industry reform, the participation of other entities in regulatory affairs, the number and mix of network industries subject to regulation by a single entity, and so on. Whilst recognising that these circumstances and factors are not unique to Australia, the fact remains – Australia needs to further minimise its "regulatory diversity" before any common approach to benchmarking and comparative performance reporting could reasonably be pursued.

Rather than wait until this occurs the Office of Water Regulation (OWR) some time ago began following significant benchmarking developments globally. Of particular interest has been the work of the IWA and its partners (including Instituto Tecnológico del Água (ITA) and Valencia Polytechnic University) in developing the SIGMA Performance Benchmarking Software and of the World Bank/PURC in establishing the International Benchmarking Project – that are now rapidly coming to maturity. Following publication of our first water services benchmarking report in July 2001, the OWR accepted a World Bank invitation to participate in its project and as a node on the global benchmarking network.

Our objective is to achieve a high level of parity between OWR's local benchmarking approach and the approaches being pursued by pre-eminent national and international benchmarking organisations and partnerships.

In doing so, the OWR is signalling its commitment both to contributing to international benchmarking projects and participating in benchmarking initiatives/partnerships with water industry regulators elsewhere. The potential benefits to Western Australian industry stakeholders of this commitment are manifold and include:

– Recognition for Western Australia as an enthusiastic, valued contributor to world-class thinking on water utility performance benchmarking.

- Access to external synergies, expertise, methods and information to improve the standard and credibility of local benchmarking efforts.
- Publication of reports that more reliably identify and evaluate significant aspects of WA water and wastewater service provider performance.
- Enhancement of "competition by comparison" by benchmarking local utility performance with that of best practice utilities in Australia and elsewhere in the world.

It is clear that the popularity of benchmarking has increased to the point where it is no longer seen as simply a growing trend, rather it has become an established tool for promoting and improving best practice within the water industry, evaluating industry performance and contributing to pricing and economic regulation methodologies.

The major trends, issues and questions that we see as important in Western Australia are:

- *What to benchmark* is one of the most pressing challenges for regulators. The extent to which universal indicators are applicable in a Western Australian context is being thoroughly investigated by the OWR. The picture emerging is that the IWA's approach is increasingly becoming apparent as the preferred approach and that the use of the related SIGMA software will facilitate its use. The development of locale-specific benchmarks while still enabling competitive comparison with other like utilities however must also be addressed by the new Economic Regulator, in particular financial and service related benchmarks.
- *The benefits of benchmarking* in a regulatory context include the use of benchmarking data to effect quality of service and pricing changes, with the publishing of benchmarking data encouraging utilities to increase their efficiency and target areas of performance where improvement is needed. This data is also being used in determining pricing paths by a number of regulators, including the Independent Pricing and Regulation Tribunal (IPART) in NSW.
- *Information needs* and the problem of an information asymmetry is just as relevant to benchmarking as it is to other areas of regulatory responsibility, with good quality, well constructed data the foundation to successful benchmarking.
- *Web-based benchmarking* has become a major trend in recent times, with increasing use being made of the Internet not only to share benchmarking information, but also to facilitate web-based benchmarking.

6 REGULATION OF THE WATER AUSTRALIAN INDUSTRY

The way in which the water industry is regulated, again, varies enormously between the States. Regulation of the water industry is currently subject to significant review and change in Victoria, NSW, South Australia and Western Australia.

6.1 *Health standards*

At the same time, there are significant similarities between States. In every State the quality standards for drinking water are set by the State Health Department (or equivalent) on the basis of the National Health and Medical Research Council guidelines, which are in turn based on World Health Organisation standards.

There have been occasional water quality scares in Australia; the most recent one of significance was in 1998 when Sydney consumers were instructed to boil drinking water in response to a suspected guardia and cryptosperidium outbreak.

6.2 *Environmental standards*

Another similarity across States is with respect to environmental conditions. Conditions for the management of environmental impacts in the form of water or wastewater treatment discharges are set by an Environment Protection Agency (EPA) and administered by a Department of Environmental Protection.

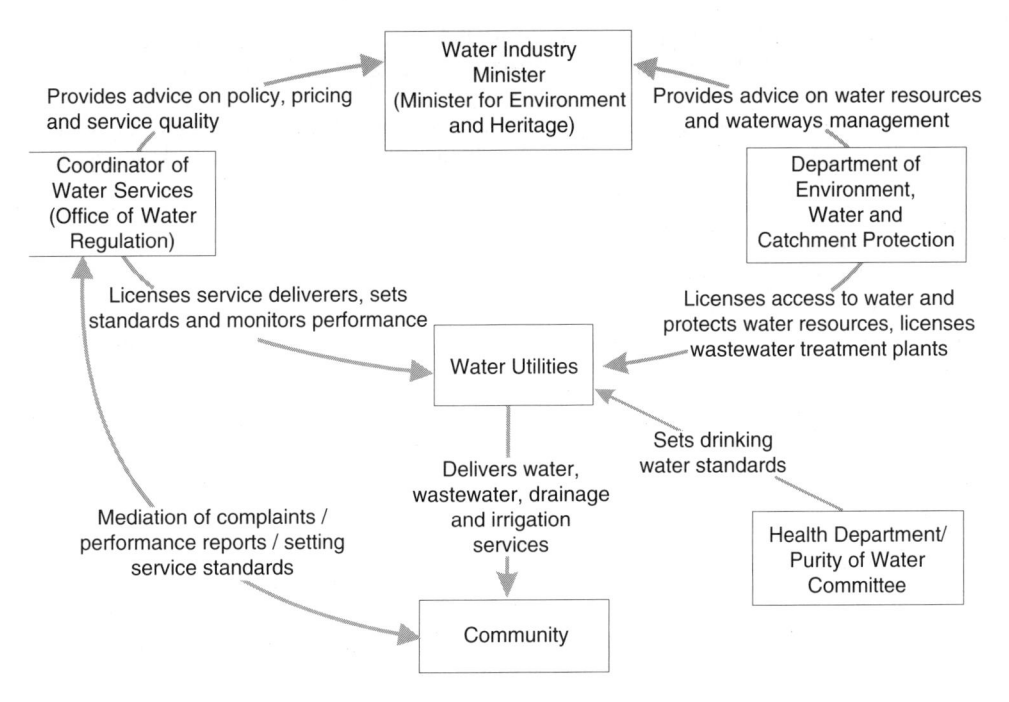

Figure 4. Relationships between regulators and their responsibilitites.

6.3 *Resource management*

Again, all States have largely the same arrangements for the protection of water resources. In each State there is a Resource Manager which is a Department of Government. This department assesses sustainable yields from all the resources, allocates access to water (entitlements) and provides for the trading of water entitlements.

Most States have adopted integrated catchment management as a standard practice for the better management of their water and land resources.

6.4 *Licensing of service providers and standard setting*

6.4.1 *Western Australia*
In Western Australia the main regulatory instrument is licensing, with licences setting standards of service to be achieved by service providers. Licences also require that utilities demonstrate they are maintaining and managing their assets and are meeting the needs of customers for quality service.

The relationships between regulators and their respective responsibilities, is shown in Figure 4.

6.4.2 *New South Wales (NSW)*
In NSW, licences, similar to those applying in Western Australia, have been administered by the Department of Land and Water Catchment, but have recently been made the responsibility of the independent pricing authority, IPART.

IPART will assess and monitor performance of the operating licences of Sydney Water, Sydney Catchment Authority and Hunter Water, against their five-year operating licences. Other water service providers have not been issued operating licences, as this is not part of the regulatory framework which applies to them.

For the three major utilities, the operating licences are very similar to those applying in Western Australia. Licences set out requirements for outcomes such as water quality, customer service standards for water and wastewater (continuity and pressure, etc.), complaint handling, and customer consultation. Some of these requirements are cross-linked to the requirements of other regulators (e.g., NSW Health and EPA). Sanctions can apply for non-achievement of the licence requirements.

The Local Government Councils in regional areas are subject to some, but not all, of the rigours of regulation of the three major utilities. They are not issued operating licences in the same way as these larger utilities. Many are supplied with bulk water by State Water, a division of the Department of Land and Water Conservation.

6.4.3 Victoria
In Victoria the new Essential Services Commission (ESC) has replaced the previous Office of the Regulator General. The ESC has responsibility for both licensing and price determination in the metropolitan areas. Non metropolitan services are licensed by the Department of Natural Resources and Environment.

6.4.4 South Australia
In South Australia, Queensland and Tasmania there is no licensing regime. Any person can provide a water service, provided they have the infrastructure, and can access water from either their own source or a bulk provider. They also must be able to meet the necessary health and environmental consideration (and provided they can obtain a water supply).

6.4.5 Northen Territory
In the Northern Territory the Utilities Commission, established in 2000, is implementing new arrangements for the regulation of the water industry. It issues licences and establishes and monitors standards of service.

6.5 Prices setting

6.5.1 Western Australia
In Western Australia prices for the government owned water utilities are set by Government. The State has a policy of almost uniform prices for water services across the whole State.

Regulatory arrangements for the water industry will change when the water industry is incorporated into the Economic Regulation Authority (ERA), which is currently proposed to be created on 1 July 2003. The ERA is to be a multi-industry economic regulator, covering the gas, rail, electricity and water industries in Western Australia.

This will have a significant impact on the way the water industry is regulated and on the Water Services Coordination Act, which is the principal legislative instrument for regulating the industry.

6.5.2 New South Wales (NSW)
In NSW prices have been determined by the IPART for the past decade. IPART has the power to regulate prices for all government monopolies, including water utilities (major urbans and municipal). This regulation is activated by the Governor declaring a utility as a monopoly. The only organisations that have been declared so far are the three major utilities plus two councils (Gosford and Wyong).

Price regulation through IPART is an established, thorough, public and open process. Submissions are received in public hearings, including submissions from Government agencies, such as Treasury, EPA, and the Department of Urban Affairs and Planning.

Earlier in this year, IPART published its determination for prices of the regulated utilities for the three years from July 2000.

The determination process also incorporated an independent review of capital and operational expenditures over the period, coupled with assessment of "reasonable" rates of return.

IPART also developed and applied a methodology for determining developer charges based upon a net present value approach, starting in 1993–94. This is aimed at sending economic signals to urban planners and the development industry by pricing developer charges to reflect the environmental and economic costs of development of urban areas.

The regulatory framework for assessment of performance against operating licence, and for price submissions, is demanding on utilities in NSW. They must provide substantial information, and are subject to detailed scrutiny and assessment. The operating licence review reports are tabled in Parliament, the price determinations are recommended to government, and are reported widely in the media.

6.5.3 *Victoria*
In Victoria the Essential Services Commission has recently been made responsible for price determination in the metropolitan areas. Pricing for non-metropolitan services are regulated indirectly by the Department of Natural Resources and Environment, through a mechanism of approvals of business plans.

The Minister for Environment and Conservation also has effective control of pricing for the non-metropolitan and rural water authorities. However, water prices for the non-metropolitan and rural sectors are set using a different mechanism. In these sectors, services are provided by statutory authorities. For these organisations, corporate planning guidelines are the main tool for regulation. Water prices must be set in accordance with the corporate plans of each authority. The Department of Natural Resources and the Environment (DNRE) is responsible for reviewing the corporate plans and the proposed water prices (tariffs). DNRE is able to advise the Minister for Environment and Conservation to veto pricing proposals made by the authorities.

6.5.4 *South Australia*
In South Australia, prices are proposed by South Australian Water and these require the approval of the Minister. The South Australia Water Corporation, with ultimate responsibility for all metropolitan and most rural services, also manages and administers the two private contractors, including the monitoring of their technical and contractual performance, and their price reviews. It operates its own water quality regulatory processes. It can impose penalties upon the private operators. These contracts also have objectives and financial incentives for the private contractors to achieve economic development for South Australia. This is unique in Australia. It also has significant difficulties in implementation and measurement.

The South Australian price setting arrangements are about to be changed with pricing responsibilities going to the South Australian Independent Industry Regulator, which until now has had responsibility for the regulation of the gas and rail sectors. South Australia, like Western Australia, has a policy of uniform water prices across the State.

6.5.5 *Queensland*
In Queensland price regulation is undertaken by the Queensland Competition Authority which sets broad principles for price setting to be administered by Government agencies.

6.5.6 *Tasmania*
In Tasmania prices are set every three years by the Government Prices Oversight Committee.

6.5.7 *Australian Capital Territory (ACT)*
In the ACT pricing is established by the Independent Competition and Regulatory Commission which conducts periodic reviews, and allows prices to increase between reviews on the basis of some predetermined function.

7 A BEST PRACTICE REGULATORY SYSTEM

In July 1999, the Australian Competition and Consumer Commission issued on behalf of the Utility Regulators Forum a discussion paper on the issue of "Best Practice Utility Regulation" which provides a guide for assessing the existing regulatory framework and proposals for change.

These principles are:

- *Communication and consultation*: Effective communication and consultation assists all stakeholders to understand regulatory initiatives and needs. Regulators should establish processes that provide relevant and comprehensive information that is also accessible, timely and inclusive of all stakeholders.
- *Consistency*: Consistency of treatment of participants across service sectors, over time and across jurisdictions provides confidence in the regulatory regime. Regulators should act in a consistent and fair way that does not adversely affect the business performance of a specific participant.
- *Predictability*: Predictability of regulation enables utilities to confidently plan for the future and be assured that their investments will not be generally threatened by unexpected changes in the regulatory environment. The principle is particularly important in the utility sector, which is characterised by major infrastructure works with long investment time horizons.
- *Flexibility*: Flexibility involves the use of a mix of regulatory tools and the ability to evolve and amend the regulatory approach over time as the external environment changes.
- *Independence*: Regulatory decisions should be free from undue influences that could compromise regulatory outcomes.
- *Effectiveness and efficiency*: Best practice regulation requires an assessment of the cost effectiveness of the proposed regulation, and an assessment of alternative options.
- *Accountability*: Accountability involves regulators taking responsibility for their regulatory actions. This requires regulators to establish clearly defined decision-making processes and provide reasons for decisions. Effective appeal mechanisms and adherence to principles of natural justice and procedural fairness are also required.
- *Transparency*: Transparency requires regulators to be open with stakeholders about their objectives, processes, data and decisions. Regulators should establish visible decision-making processes that are fair to all parties and provide rationales for decisions, including rules about treatment of information and about what information will be regarded as confidential, or to which access will be restricted.

These principles provide a guide for assessing the existing system of regulation and a basis for legislative reform.

8 ASSESSMENT AGAINST BEST PRACTICE PRINCIPLES

Australian regulators have yet to take up benchmarking in any coordinated way. The service utilities, especially the major ones, participate in their own benchmarking mechanism which is publicly reported.

8.1 New South Wales (NSW)

NSW has the most rigorous and comprehensive regulatory system in Australia for the water sector. The independence and rigour of price setting, and the price paths, provide incentives for the regulated utilities to improve efficiency. The regulatory system is robust, institutionally strong, and is evolving. There is separation of functions, coupled with transparency and public scrutiny.

With the recent changes to IPART, pricing and customer service standard setting is undertaken by a single regulator. For the major utilities, many aspects of regulation are also brought together in the operating licence. However, the community trade-offs between environmental standards and price are not efficiently handled.

The regulatory system does not yet have consistent coverage across the State, and mechanisms to introduce competition are hampered by planning processes. Different combinations of mechanisms (pricing, operating licences, customer contracts, audits, etc.) are applied to the major and minor utilities. Hence the burden of regulation sometimes falls unevenly.

The NSW system of regulation rates the most highly against the principles of best practice regulation of any Australian State.

8.2 Victoria

The Victorian regulatory system is one of the more mature regulatory systems in Australia. However, it is difficult to draw conclusions and to make comparisons with best practice principles at this time because of the number of changes underway.

Within Melbourne, there is separation between the resource manager, service provider and regulator roles. However, if the State is viewed as a whole, then the regulatory system begins to look fragmented and over-complicated. There is a need for rationalisation of roles within function, and this is now being implemented.

There appears both the scope and a strong desire for further reform within a modified version of the current regulatory model. The changes proposed appear likely to propel Victoria in the direction of better regulatory practice.

A key strength of the regulatory system in Victoria is the benchmark performance system, which operates not only amongst the metropolitan service providers, but also amongst regional providers. This allows the regulator to establish relative efficiencies and performance with relative ease.

The licensing system applies only in the capital city and regional providers are not licensed. Regulatory arrangements probably need to be made more transparent. There does not appear to be any efficient means for government to make judgements about the balance and trade-offs between service and environmental standards and prices.

8.3 Western Australia

Western Australia has a robust licensing system and has developed strong linkages between customer complaints, customer priorities and the standards specified in licences. There is a strong interest in development of benchmarking.

However, the Western Australian regulatory system is not well developed. Price setting lacks regour, transparency and accountability – shortcomings which are currently being addressed.

8.4 South Australia

South Australia has a regulatory system that is not rigorous and not well defined. Role definition and separation is not clear.

In practice, there is confusion between regulation and service provision. There is imperfect separation of roles and a lack of transparency. It is a relatively weak regulatory framework. This is clearly recognised within the State, and significant changes are now being made to the regulatory system.

8.5 Queensland

The Queensland regulatory functions have been separated from service provision. Developments of regulation are occurring in both the urban and irrigation sectors. The pricing principles have only just been established.

The model is reactive, with intervention only when there is a failure (customer complaint). The regulatory system as a whole is in a state of development.

8.6 Tasmania

The Tasmanian regulator, the Government Prices Oversight Committee sets prices every three years for all water utilities. It not regulate the quality of service provided and water service suppliers are not licensed.

8.7 Australian Capital Territory (ACT)

The ACT Territory regulatory mechanism is modelled on the NSW system, and accordingly rates well against best practice principles this is particularly so in the areas of role separation, protection of customers' interests, transparency, defined coverage, efficiency, accountability, and promotion of economic efficiency.

8.8 Northern Territory

The Northern Territory regulatory system is relatively new. Water service providers require an operational licence, and price and service quality are regulated. The new regulatory system is in its early days, and it is not possible to make a considered assessment.

9 CONCLUSION

The structure of the water industries and their regulation varies considerably between Australian States, but national policies have produced a degree of consistency. For example, separation of roles of resource manager, service provider and economic regulator has been achieved in most States.

While the models employed in NSW and Victoria would seem to come closest to satisfying best practice principles, there is no clear evidence yet that anything like an ideal model has been identified, much less applied.

There has been some maturation of licensing and pricing mechanisms, and a start toward benchmarking and performance comparisons. The need for quality information from the industry has been a priority need. There is also general agreement on the need for open and transparent processes of regulatory decision making. The need for incentives for improved performance by utilities has probably been given less than adequate attention.

One lesson that seems to have been learned in most States is that good regulation is not just a matter of good pricing mechanism, or a good licensing mechanism, or good benchmarking and customer consultation. All of these functions are so closely inter-related that to address one without simultaneously attending to the others is to fall well short of an ideal regulatory environment.

There is a real need to balance service standards and costs, in the pricing function.

Another ongoing source of contention is the level of information required by the regulator to conduct its work effectively. States differ markedly in the capacities of their regulators to obtain information. Some regulators are unable to obtain adequate information because of the inadequacy of their legislation. Others have strong powers to obtain information and are accused by the industry of imposing excessive compliance costs on industry members.

For most States, consistency of application of regulatory mechanisms within the State boundary is a real source of concern. Again, this seems to be a problem that has been recognised and is being addressed in some of the more recent proposals for development of regulatory systems.

Another problem which is being addressed is the problem of regulatory overlap. The industry has complained that regulators are duplicating areas of regulation and are requiring similar information to be reported, but often in different forms. Regulators appear to be responding to these concerns of industry.

Australia has only one example of a multi-utility in the water industry. This is the ACT Energy and Water Corporation. Regulation of multi-utilities poses an emerging problem for the regulation industry.

REFERENCES

ABS (nd). *Population by Age and Sex*. Australian Bureau of Statistics Cat. nos 3235.1–.8. Canberra. Australia.
ABS. 2000. *Water Account for Australia – May 2000*. Australian Bureau of Statistics no. 4610.0. Canberra. Australia.
ABS. 2002. *Water Account for Australia 1993–94 to 1996–97*. Australian Bureau of Statistics 2002. Cat. no. 4610.0. Canberra. Australia.
CAG. 1994. *Report on the Working Group on Water Resource Policy*. Council of Australia Governments. Feb. 1994. Hobart. Australia.
CAG. 1995. *Communique*. Council of Australian Governments. April 1995 Canberra. Australia.
URF. 1999. Best Practice Utility Regulation. *Utility Regulators Forum 1999*. Office of Water Regulation, Perth. Australia.

*Practical aspects and
sustainable water management*

Effects of climate change on hydraulic resources

L.W. Mays
Department of Civil and Environmental Engineering, Arizona State University, United States

ABSTRACT: The Pacific Institute has compiled a comprehensive bibliography of peer-reviewed literature dealing with climate change and its effects on water resources and water systems. As of December 2001 the bibliography included over 920 citations and can be found at the Pacific Institute website. The International Panel on Climate Change (IPCC) has been active in compiling information concerning future climate change (IPCC, 1990, 1996a, b, c, 1998, 2001a, b, c). Other reports have been published that contain our state-of-the-art knowledge on the topic of global climate change and water resources. Gleick (1993) edited the book *Water Crisis: A Guide to the World's Fresh Water Resources* in which he and several authors presented summaries on several fresh water resources topics related to our resources on Earth. The Conference on Climate and Water held in Helsinki (Finland) in 1989 and the Second International Conference on Climate and Water held in Espoo (Finland) in 1998 provides another source of information on this topic, particularly for the European Region. This paper is a basic review of the literature on the topic of climate change, hydrology and water resources.

1 INTRODUCTION

1.1 *The climate system*

The *climate system* (Figure 1) as defined by the IPCC (2001a) is "an interactive system consisting of five major components: the atmosphere, the hydrosphere, the cryosphere, the land surface, and the biosphere, forced or influenced by various external forcing mechanisms, the most important of which is the Sun." The effect of human activities on the climate system is considered as external forcing. The Sun provides the ultimate source of energy that drives the climate system.

The *atmosphere*, which is the most unstable and rapidly changing part of the system, is composed (Earth's dry atmosphere) mainly of nitrogen (78.1%), oxygen (20.9%), and argon (0.93%). The atmospheric distribution of ozone in the lower part of the atmosphere, the troposphere, and lower stratosphere acts as greenhouse gas. In the higher part of the stratosphere there is a natural layer of high ozone concentration. This ozone absorbs ultra-violet radiation, playing an essential role in the stratosphere's radiation balance and at the same time filters out this potentially damaging form of radiation.

The *hydrosphere* comprises all liquid surface and subterranean water, both freshwater, including rivers, lakes and aquifers, and saline water of the oceans and seas. The *cryosphere*, which includes the ice sheets of Greenland and Antarctica, continental glaciers and snow fields, sea ice and permafrost, derives its importance to the climate system from its reflectivity (albedo) for solar radiation, its low thermal conductivity, its large thermal inertia. The cryosphere has a critical role in driving the deep-water circulation. Ice sheets store a large volume of water so that variations in their volume are a potential source of sea level variations. *Marine* and *terrestrial biospheres* have major impacts on the atmosphere's composition through biota influencing the uptake and release of greenhouse gases and through the photosynthetic process in which both marine and terrestrial plants (especially

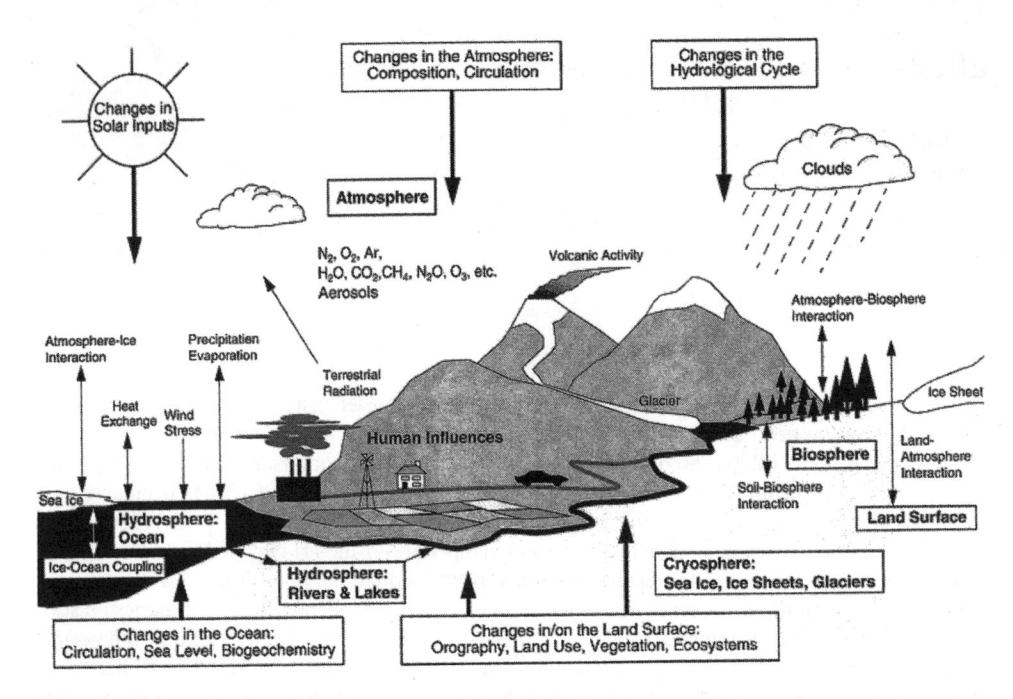

Figure 1. Schematic view of the components of the global climate system, their processes and interactions and some aspects that may change (IPCC, 2001a).

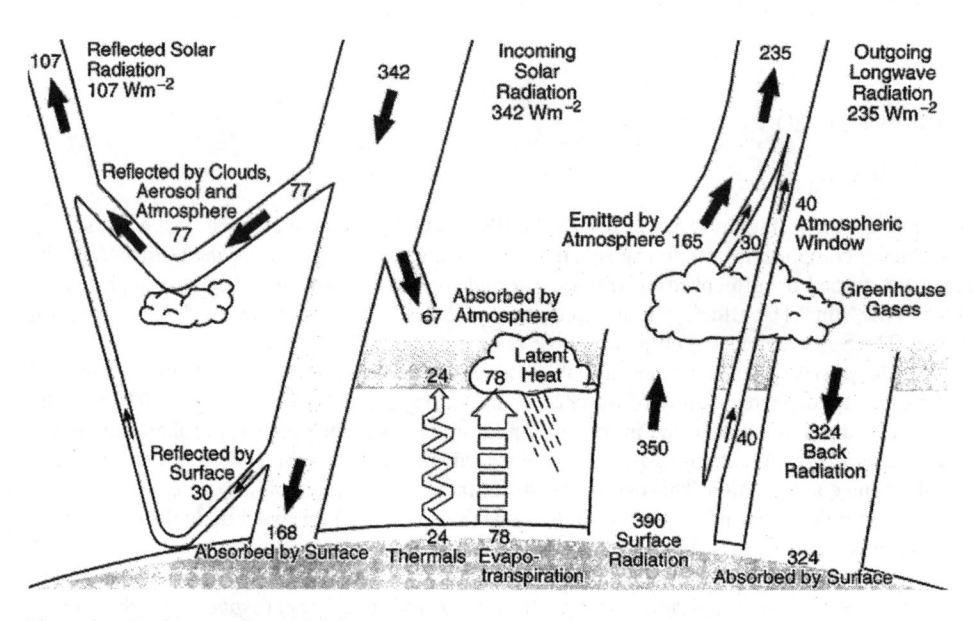

Figure 2. Earth's annual and global mean energy balance (Kiehl & Trenberth, 1997).

forests) storing significant amounts of carbon from carbon dioxide. Interaction process (physical, chemical, and biological) occur among the various components of the climate system on a wide range of space and time scales. This makes the climate system, as shown Figure 1, extremely complex.

The mean energy balance for the Earth is illustrated in Figure 2: of the incoming solar radiation, 49% (168 W/m^2) is absorbed by the surface. That heat is returned to the atmosphere as sensible heat,

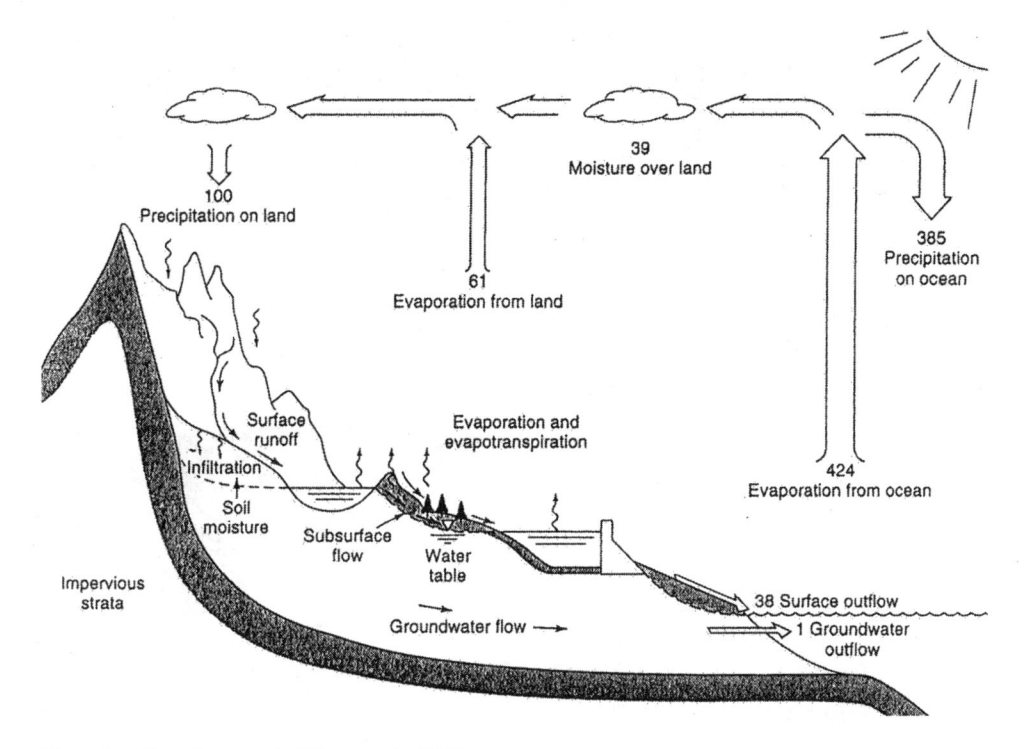

Figure 3. Hydrologic cycle (Chow et al., 1988).

as evapotranspiration (latent heat) and as thermal infrared radiation. Most of this radiation is absorbed by the atmosphere, which in turn emits radiation is absorbed by the atmosphere, which in turn emits radiation both up and down. The radiation lost to space comes from cloud tops and atmospheric regions much colder than the surface. This causes a greenhouse effect. Figure 2 shows on the left-hand side what happens with the incoming solar radiation and on the right hand side how the atmosphere emits the outgoing infrared radiation. A *stable climate* must have a balance between the incoming radiation and the outgoing radiation emitted by the climate system. On average the climate system must radiate 235 W/m^2 back into space. Climate variations may result from the radiative forcing, but also from internal interactions among components of the climate system; therefore a distinction is made between externally and internally induced natural climate variability and change.

1.2 *Hydrologic cycle*

To consider the effects of climate change on the hydraulic resources of the earth, we must begin with a definition of the hydrologic cycle. The central focus of hydrology is the *hydrologic cycle* consisting of the continuous processes shown in Figure 3. Water evaporates from the oceans and land surfaces to become water vapor that is carried over the earth by atmospheric circulation. The water vapor condenses and precipitates on the land and oceans. The precipitated water may be intercepted by vegetation, become overland flow over the ground surface, *infiltrate* into the ground, flow through the soil as subsurface flow, and discharge as surface runoff. Evaporation from the land surface comprises evaporation directly from soil and vegetation surfaces, and transpiration through plant leaves. Collectively these processes are called *evapotranspiration*. Infiltrated water may percolate deeper to recharge groundwater and later become springflow or *seepage* into streams to also become *streamflow*. Physical, chemical, and biological processes operate on the working media within the system; the most common working media involved in hydrologic analysis are water, air, and heat energy.

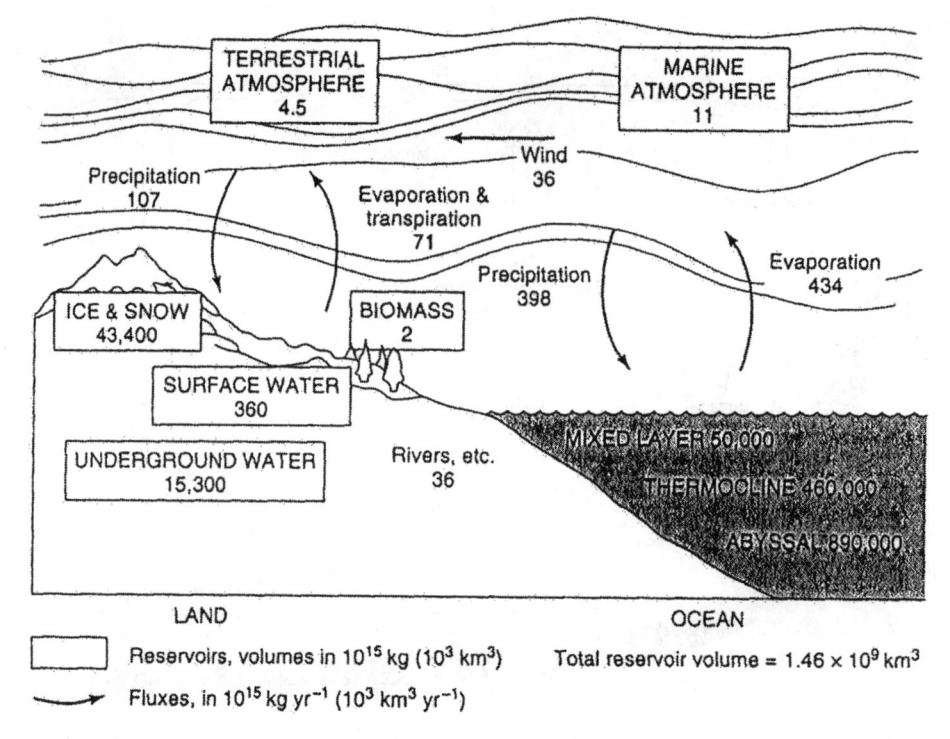

Figure 4. The hydrologic system at a global scale.

A *hydrologic system* is defined as a structure or volume in space, surrounded by a boundary, that accepts water and other inputs, operates on them internally, and produces them as outputs (Chow et al., 1988). The structure (for surface or subsurface flow) or volume in space (for atmospheric moisture flow) is the totality of the flow paths through which the water may pass as throughput from the point it enters the system to the point it leaves. The boundary is a continuous surface defined in three dimensions enclosing the volume or structure. A *working medium* enters the system as input, interacts with the structure and other media, and leaves as output. An example is the storm-rainfall-runoff process on a watershed, which can be represented as a hydrologic system. The input is rainfall distributed in time and space over the watershed and the output is streamflow at the watershed outlet. The boundary is defined by the watershed divide and extends vertically upward and downward to horizontal planes.

The *global hydrologic* cycle can be represented as a system containing three subsystems: the atmospheric water system, the surface water system, and the subsurface water system as shown in Figure 4. Eagleson (1986) discussed the emergence of global-scale hydrology.

The hydrologic cycle is intimately connected to all major processes that define the character of the various systems on the Earth. One example is the hydrologic cycle of the Artic system (Vorosmarty et al., 2002), which has key linkages among land, ocean, and atmosphere. The Artic hydrologic cycle is inextricably connected to all biological and chemical processes occurring in the biosphere, atmosphere, and cryosphere. The hydrologic interactions with terrestrial and aquatic ecosystems and their biogeochemistry control all life in the pan-Artic region. Figure 5 is a conceptual diagram of the Artic system showing the linkages among the atmosphere, land surface, and ocean components.

1.3 Hydrologic budgets

A *hydrologic budget*, *water budget*, or *water balance* is a measurement of continuity of the flow of water, which holds true for any time interval and applies to any size area ranging from local-scale

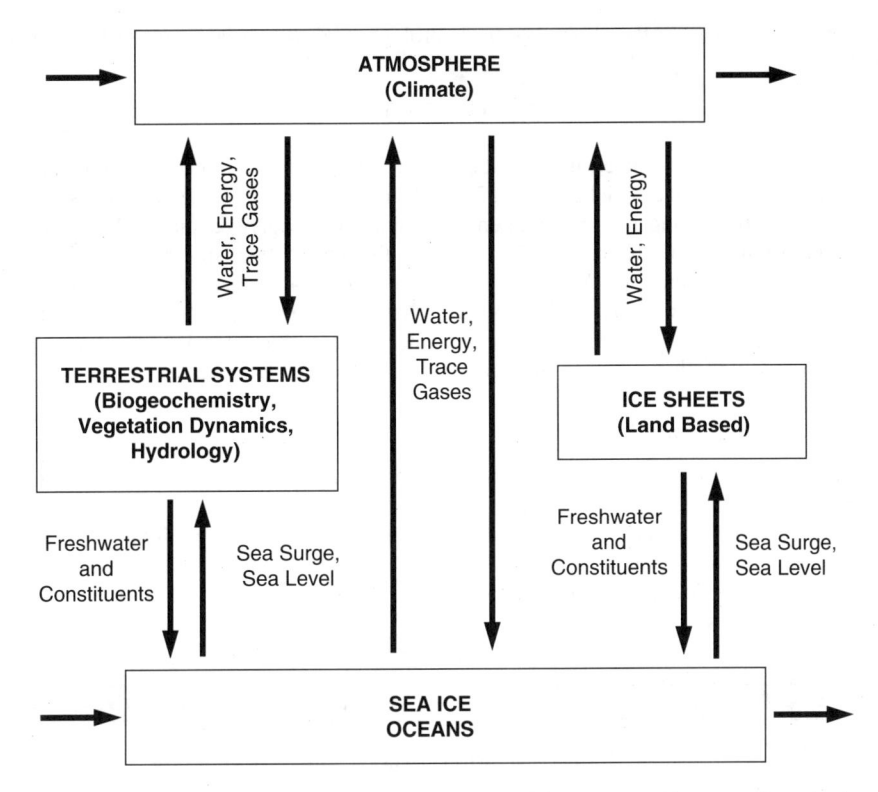

Figure 5. A conceptual diagram of the Artic system showing linkages among the atmosphere, land surface, and ocean components (Walsh et al., 2001).

areas to regional-scale areas or from any drainage area to the earth as a whole. The hydrologists usually must consider an open system, for which the quantification of the hydrologic cycle for that system becomes a mass balance equation in which the change of storage of water (dS/dt) with respect to time within that system is equal to the inputs (I) to the system minus the outputs (O) from the system. The water balance equation can be expressed in units of volume per unit time separately for the surface water system and the groundwater system. The *surface water system hydrologic budget* can be expressed as shown by Equation 1.

$$P + Q_{in} - Q_{out} + Q_g - E_s - T_s - I = DS_s \tag{1}$$

where P is the precipitation, Q_{in} is the surface water flow into the system, Q_{out} is the surface water flow out of the system, Q_g is the groundwater flow into the stream, E_s is the surface evaporation, T_s is the transpiration, I is the infiltration, and DS_g is the change in water storage of the surface water system. *The groundwater system hydrologic budget* can be expressed shown by Equation 2.

$$I + G_{in} - G_{out} - Q_g - E_g - T_g = DS_g \tag{2}$$

where G_{in} is the groundwater flow into the system, G_{out} is the groundwater flow out of the system. The *system hydrologic budget* is developed by adding the above two budgets together (Equation 3).

$$P - (Q_{out} - Q_{in}) - (E_s + E_g) - (T_s + T_g) - (G_{out} - G_{in}) = D(S_s + S_g) \tag{3}$$

Using net mass exchanges the above system hydrologic budget can be expressed as shown by Equation 4.

$$P - Q - G - E - T = DS \qquad (4)$$

Water balance models have been used in looking at the hydrologic effects of climate change. Schaake (1990) and Nemec & Schaake (1982) used simple water balance models to estimate the sensitivity of runoff to changes in precipitation and evaporation. Wigley & Jones (1985) and Karl & Riebsame (1989) used water balance models to look at the sensitivity of streamflow to the changes of temperature and precipitation. Vaccaro (1992) used a soil-water balance model to look at the sensitivity of groundwater recharge. The Soil and Water Assessment Toll (SWAT) is a hydrologic model based on a water balance equation for soil water content (Arnold et al., 1998). Arnold et al. (1999) performed a continental scale simulation of the hydrologic balance.

1.4 What is climate change?

The Earth's temperature is affected by numerous causes including: (a) the incoming solar radiation that is absorbed by the atmosphere and the Earth's surface, (b) by the characteristics (emissivity) of the matter that absorbs the radiation, and (c) by the fact that part of the long-wave radiation emitted by the surface is absorbed by the atmosphere, and then re-emitted as long-wave radiation either in the upward or downward direction. Figure 6 is a schematic of observed variations of various temperature indicators.

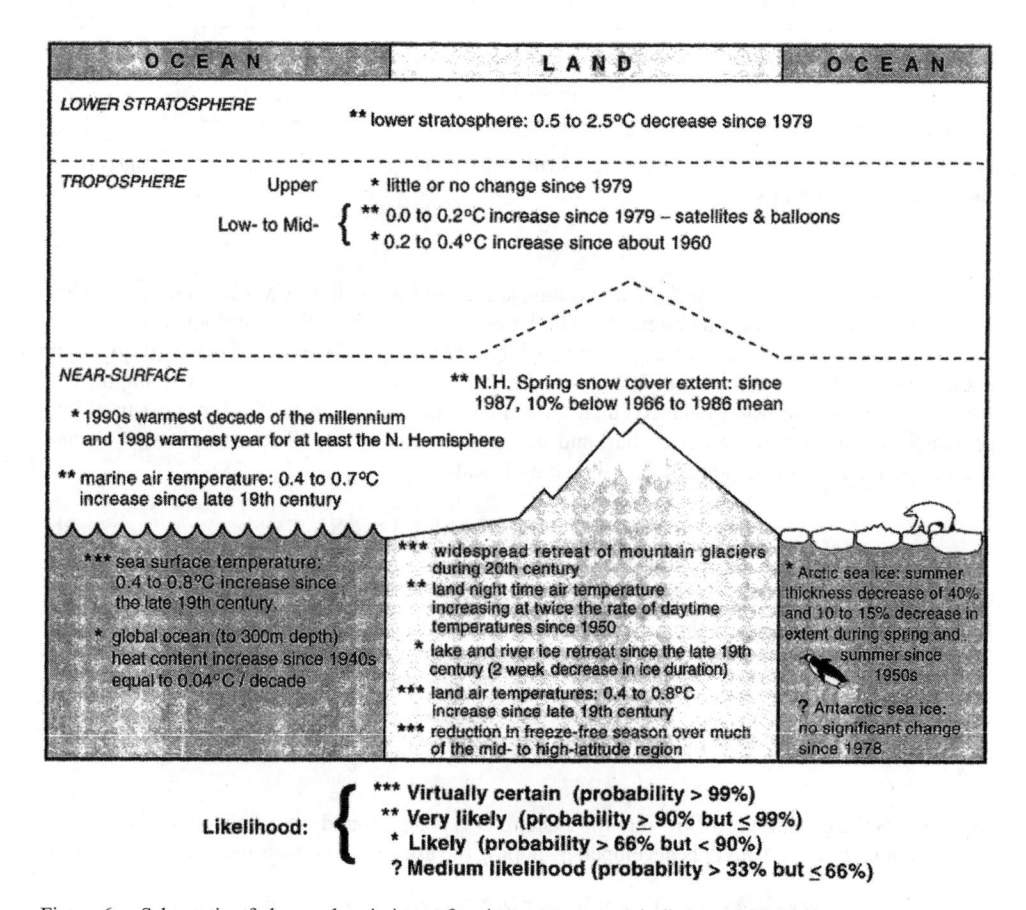

Figure 6. Schematic of observed variations of various temperature indicators (IPCC, 2001a).

The so-called *greenhouse effect* is caused by the net change of the internal radiation balance of the atmosphere due to the continued increased emission of greenhouse gases, resulting in both the atmosphere and the Earth's surface being warmer (Mitchell, 1989; Mitchell et al., 1995). Magnitude of the greenhouse effect is dependent on the composition of the atmosphere, with the most important factors being the concentrations of water vapor and carbon dioxide, and less importantly on certain trace gases such as methane. There is mounting evidence that global warming is under way (Houghton et al., 1990; Levine, 1992; Mitchell et al., 1995; Santer et al., 1996; Tett et al., 1996; Harris & Chapman, 1997; Kaufmann & Stern, 1997). Scientists agree that a global temperature rise of 1° to 2°C is likely by 2050; however, a precise knowledge of the global-average temperature change is of little predictive value for water resources. On a global scale the warming would result in increased evaporation and increased precipitation. What is more important is how the temperature will change regionally and seasonally, affecting the precipitation and runoff, which is by no means clear. The articles by Karl et al. (1995) and Karl & Quayle (1988) on climate change may be of interest to the reader.

In general the hydrologic effects are likely to influence water storage patterns through out the hydrologic cycle and influence the exchange among aquifers, streams, rivers, and lakes. Chalecki & Glieck (1999) provided a bibliography of the impacts of climate change on water resources in the United States. Climate change can have a significant effect on the organisms living in these aquatic systems (Fisher & Grimm, 1996; Firth & Fisher, 1992). Other literature concerning lakes, ecosystems, and other related topics includes Burkett & Kusler (2000), Cohon (1987), Hostetler & Small (1999).

1.5 Climate change prediction

Climate-change predictions are mostly based on computer simulations using *general circulation models* (GCMs) of the atmosphere. These models solve the conservation equations (in discrete space and time) that describe the geophysical fluid dynamics of the atmosphere. GCMs have the same general structure as *numerical weather-prediction models* (NWPMs) that are used for weather prediction; however have a much larger spatial distribution and are run for much longer time periods. Scenarios of changes in atmospheric variables (such as temperature, precipitation, and evapotranspiration) that effect the surface and subsurface hydrology are determined using methods that range from prescribing hypothetical climate conditions to the use of GCMs. Lettenmaier et al. (1996) discussed the advantages and disadvantages of using three methods: historical and spatial analogues; output from GCMs; and prescribed climate-change scenarios. They further discussed the various types of climate-change scenarios based on GCM simulations that include: scenarios based directly on GCM simulations; ratio and difference methods; scenarios based on GCM atmospheric circulation patterns; stochastic downscaling models; and nested models. Lettenmaier et al. (1996) felt that perhaps the most useful scenarios of future climatic conditions are those developed by using a combination of methods.

Figure 7 shows the results of models that were used to make projections of atmospheric concentrations of greenhouse gases and aerosols, and future climate, based upon emissions from the IPCC Special Report on Emission Scenarios (SRES) as discussed in the IPCC (2001a) report. A great deal of uncertainty lies in the estimate of climate change, which are not focused upon in this paper. Refer to Dickinson (1989) and the IPCC reports for more information on this topic.

1.6 Droughts

Droughts can be classified into different types: *meteorological droughts* refer to lack of precipitation, *agricultural droughts* refer to lack of soil moisture, and *hydrological droughts* refer to reduced streamflow or ground-water levels. *Drought analysis* is used to characterize the magnitude, duration, and severity of the respective type of drought in a region of interest. Figure 8 illustrates the progression of droughts and their impacts.

The *Palmer Drought Severity Index (PDSI)* is a measure of agricultural drought and the severity of droughts, expressed by the Equation 5.

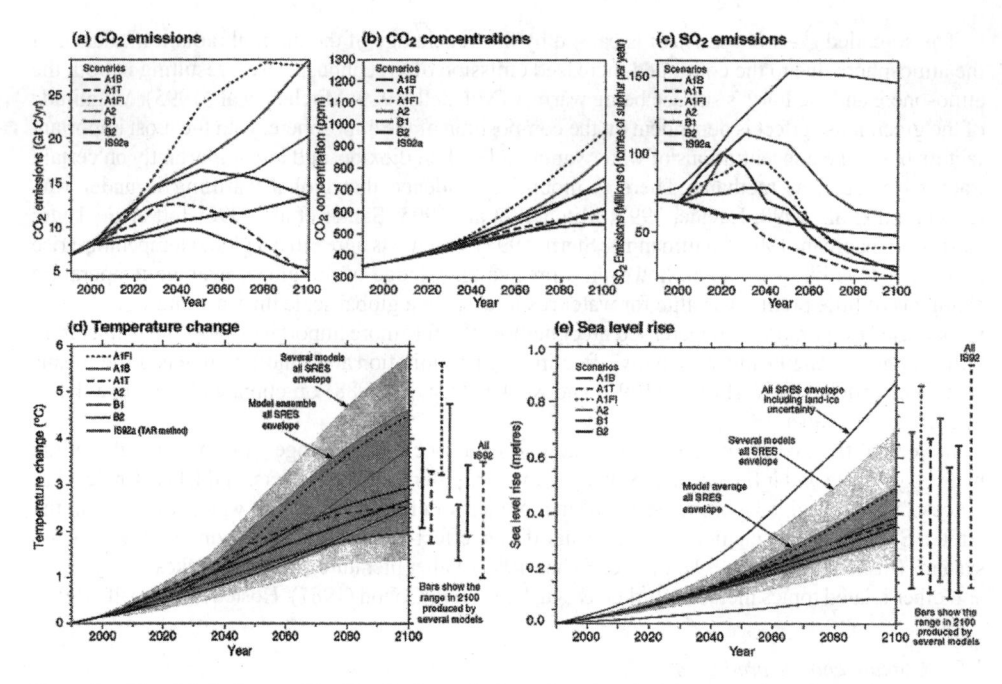

Figure 7. The global climate of the 21st Century will depend on natural changes and the response of the climate system to human activities (IPCC, 2001a).

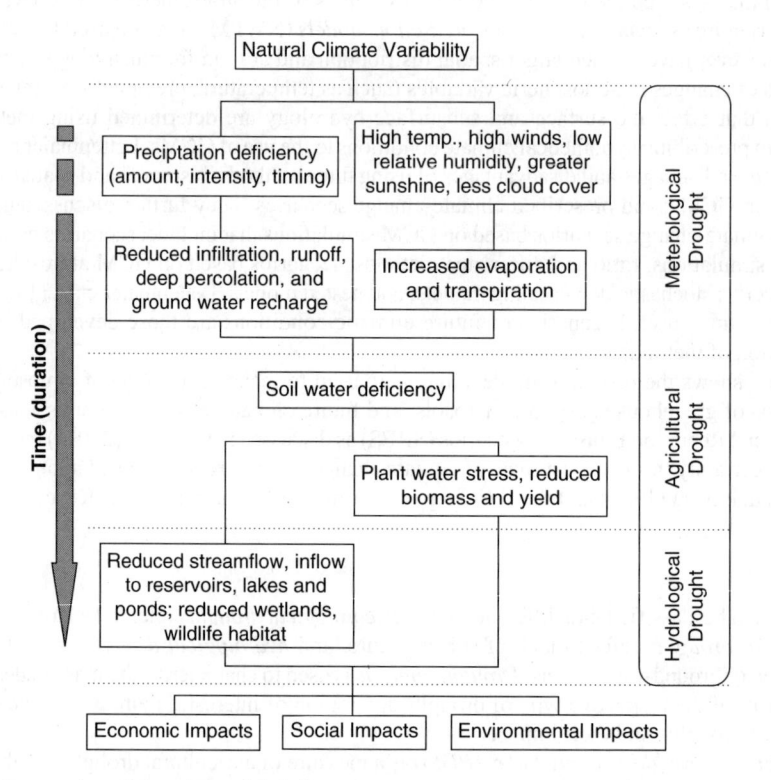

Figure 8. Progression of droughts and their impacts.

$$PDSI_i = 0.897 PDSI_{i-1} + \tfrac{1}{3} Z_i \tag{5}$$

where *PDSI* is the Palmer Drought Severity Index and *Z* is an adjustment to soil moisture for carryover from one month to the next, defined as shown by Equation 6.

$$Z_i = k_j \left[PPT_i - \left(\alpha_j PE_i + \beta_j G_j + \gamma_j R_i - \delta_j L_i \right) \right] \tag{6}$$

where the subscript *j* represents one of the calendar months and *i* is a particular month in a series of months. *PPT* is the precipitation, PE_i is the potential evapotranspiration, G_i is the soil moisture recharge, R_i is the surface runoff (excess precipitation), and L_i is the soil moisture loss for month *i*. The coefficients α_j, β_j, γ_j, and δ_j are the ratios for long-term averages of actual to potential magnitudes for *E*, *G*, *R*, and *L* based on a standard 30-year climatic period. The values are cumulated and compared with climatic normals, with numbers ranging from −6 for extreme droughts to +6 for extreme wetness.

2 CLIMATE CHANGE EFFECTS

2.1 *Global perspective on water use*

Global water use has been analyzed (Falkenmark & Lindh, 1993) using the metric of per capita water availability, A(t), for a region or country at time t,

$$A(t) = [P - ET + Q_{in}] / N(t) \tag{7}$$

where P is average precipitation, ET is average evapotransition, Q_{in} is the average rate of inflow of river and groundwater from surrounding regions, and N(t) is the population of the region or country. Falkenmark & Lindh (1993) compared water demand (use) with water availability for several countries and two regions of the United States.

Vorosmarty et al. (2000) compared the relative effects on water availability of projected climate change due to global warming and population growth in year 2025. These investigators combined a global runoff model, a global climate model, and population projections, using as a metric the dimensionless-water stress, W, defined as

$$W = U / [P - ET + Q_{in}] \tag{8}$$

where U is water use. Current (1995) world populations living under the four levels of water stress were reported, as presented in Table 1. The models forecasted an overall reduction of global runoff of 6% and a 4% increase in water stress due to climate change alone. However they forecasted a

Table 1. World population in 1995 living under various levels of water stress.

Water stress	Stress level	Population (billions)
<0.1	Low	1.72
0.1 to 0.2	Moderate	2.08
0.2 to 0.4	Medium-high	1.44
>0.4	High	0.46

50% increase in stress due to population growth and economic development. Obviously they concluded that over the next 25 years population and economic growth will produce a much greater stress on water resources than climate change.

2.2 Hydrologic effects

2.2.1 Evaporation and transpiration

The two main factors influencing evaporation from an open water surface are the supply of energy to provide the latent heat of vaporization and the ability to transport the vapor away from the evaporative source (Chow et al., 1988). Solar radiation is the main source of heat energy and the ability to transport vapor away from the evaporative surface depends on the wind velocity over the surface and the specific humidity gradient in the air above it. As temperatures rise the availability of energy for evaporation increases and the atmospheric demand for water in the hydrologic cycle increases.

Evaporation from the land surface comprises evaporation directly from the soil and vegetation surface, and transpiration through plant leaves, in which water is extracted by the plant's roots, transported upwards through its stem, and diffused into the atmosphere through tiny openings in the leaves called stomata (Chow et al., 1988). The processes of evaporization from the land surface and transpiration from vegetation are collectively termed *evapotranspiration*, which is influenced by the two factors described previously for open water evaporation, and a third factor, the supply of moisture at the evaporative surface. Potential evapotranspiration is that which occurs from a well-vegetated surface when moisture supply is not limiting. The actual *evapotranspiration* drops below its potential level as the soil dries out.

2.2.2 Soil moisture

A major part of the hydrologic cycle is subsurface water flow, which can be discussed as three major processes: infiltration of surface water into the soil to become soil moisture, subsurface (unsaturated) flow through the soil, and groundwater (saturated) flow through soil or rock strata. Climate changes that alter the precipitation patterns and the evapotranspiration regime will directly affect soil moisture storage, runoff processes, and groundwater recharge dynamics (Rind et al., 1990; Vinnikov et al., 1996). Decreases in precipitation may significantly decrease soil moisture. In regions where precipitation increases along with increases in evaporation due to higher temperatures, or the timing of the precipitation or runoff changes, the soil moisture on the average or over certain periods may decrease. In regions where precipitation increases significantly, soil moisture will probably increase and in some locations may significantly increase.

Sensitivity studies by Gregory et al. (1997) reported reduced soil-moisture conditions in mid-latitude summers in the northern hemisphere as temperature and evaporation rise and winter snow cover and spring runoff decline. Wetherald & Manabe (1999) investigated the temporal spatial variation of soil moisture in a coupled ocean-atmosphere model, with results that showed both summer soil moisture dryness and winter wetness in middle and high latitudes of North America and Southern Europe. The study results showed a large percentage reduction of soil moisture in summer and a soil moisture decrease for nearly the entire year in response to warming.

2.2.3 Snowfall and snowmelt

Temperature increases are expected to be greater in the higher latitude regions because of the dynamics of the atmosphere and feedbacks among ice, albedo, and radiation. Research has established that higher temperatures will lead to dramatic changes in snowfall and snowmelt dynamics in watersheds with substantial snow. Higher temperatures will have several major effects including (Gleick, 2000):

– Increase the ratio of rain to snow.
– Delay the onset of the snow season.
– Accelerate the rate of spring snowmelt.
– Shorten the overall snowfall season.

– More rapid and earlier seasonal runoff.
– Cause significant changes in the distribution of permafrost and mass balance of glaciers.

One of the most persistent and well-established findings on the impacts of climate change is that higher temperatures will affect the timing magnitude of runoff in watersheds with substantial snow dynamics. Both climate models and theoretical studies of snow dynamics have long projected that higher temperatures will lead to a decrease in the extent of snow cover in the northern hemisphere. More recently field studies have corroborated these findings.

2.2.4 *Storm frequency and intensity*

In general global climate change will result in a wetter world. Many model estimates indicate that for much of the northern mid-latitude continents summer rainfall will decrease slightly and winter precipitation will increase. In other studies precipitation changes in mid-latitudes are highly variable and uncertain. Unfortunately general circulation models poorly reproduce detailed precipitation patterns. As pointed out in Gleick (2000), potential changes in rainfall intensity and variability are difficult to evaluate because intense convective storms tend to occur over smaller regions that global models are able to resolve. Figure 9 presents a summary of hydrological and storm-related indicators.

2.2.5 *Runoff: floods and droughts*

A major challenge for the hydrologic community in the study of climate change is to establish the linkage between local-scale and global-scale processes. Wigley & Jones (1985) used a simple water balance models to conclude:

– Changes in runoff are more sensitivity to changes in precipitation than to changes in evaporation.
– Relative change in runoff is always greater than the relative change in precipitation.

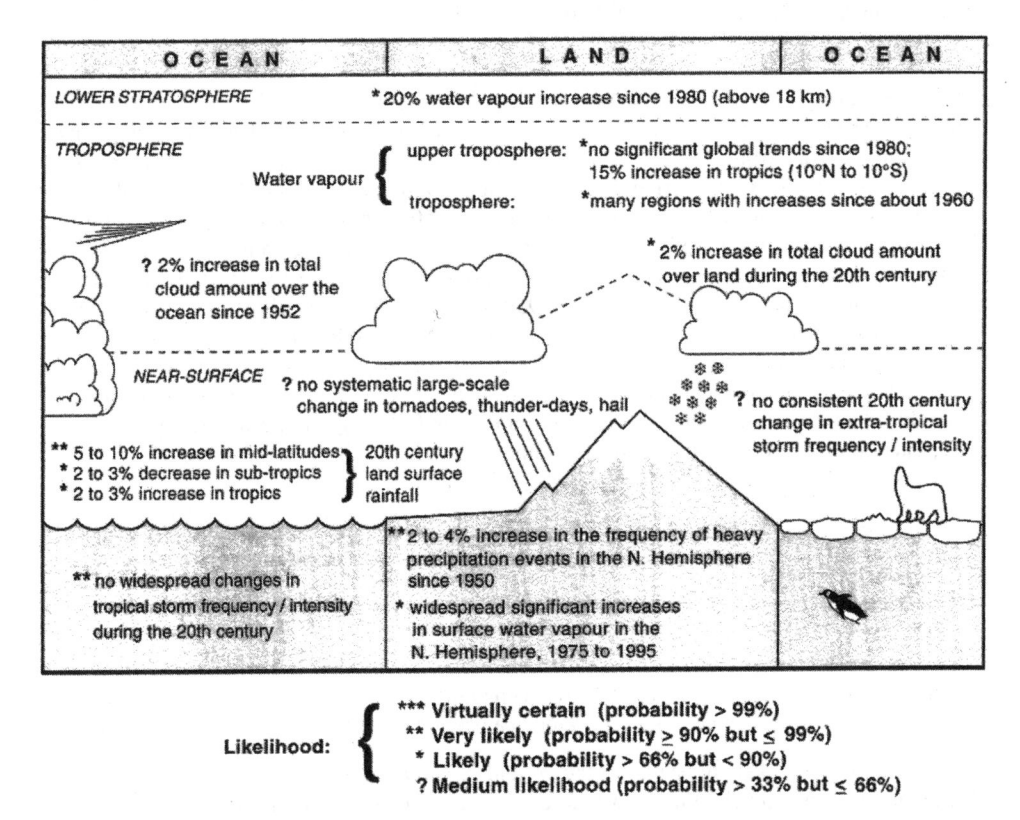

Figure 9. Schematic of observed variations of various hydrological and storm-related indicators (IPCC, 2001a).

- Runoff is most sensitive to climate changes in arid and semi-arid regions.
- Relative change in runoff exceeds the relative change in evapotranspiration only in regions where the *runoff ratio*, w, (ratio of long-term average runoff to long-term average precipitation) is less than 0.5.
- Overall it is expected that very large increases in average runoff in response to predicted warming, unless there is a compensating large increase in land evapotranspiration.
- Higher surface temperatures will increase global precipitation in the range of 3% to 11%, probably leading to even greater relative increases in runoff.

Loaiciga et al. (1996) reviewed the findings and limitations of predictions of hydrologic responses to global warming.

2.2.6 *Streamflow variability*

Karl & Riebsame (1989) used historical data from the US with a water balance model to study the sensitivity of streamflow to changes in temperature and precipitation. Their study concluded that a 1° to 2°C temperature change typically had little effect on streamflow. On the other hand, a relative change in precipitation magnified to a one to six-fold change in relative streamflows.

Schaake (1990) and Nemec & Schaake (1982) showed that on an average basis (e.g., using mean annual precipitation, evaporation, and runoff) that the *elasticity of runoff* with respect to precipitation is greater than one, and is larger related to the *long-term runoff ratio*, which is less in arid than in humid climates.

Because streamflow is climatically determined and consequently highly variable, it is instructive to look at historical data on streamflow variability in order to develop a context for detecting and predicting the effects of climate change on the hydrologic cycle. Dingman (2002) illustrates that there is a much higher relative variability of streamflow in arid regions (Upper and Lower Plains and the Southwest) than in humid regions (Northeast and Northwest). Also evident is that the time series of river discharge show considerable synchronism over the United States.

Long records of flow in large rivers such as the Nile River show evidence of the climate-related persistence. The Nile River has the longest streamflow record in the world with information available from 622 to 1520 and from 1700 to the present. Riehl & Meitin (1979) discussed three contrasting patterns in the record: (1) the period from 622 to about 950 had periods of high flows alternated with periods of low flow, with these cycles lasting from 50 to 90 years with moderate amplitude; (2) the period from 950 to 1225 had no major trends; and (3) the period from 1700 to present had alternating periods of high and low flow, having cycles of 100 to 180 years with much higher amplitudes than during the period from 622 to 950. These periods of variability appear to be linked to global climatic fluctuations. Eltahir (1996) found that the Nile River flows were influenced by the El Nino-Southern Oscillation (ENSO) cycle. Richey et al. (1989) studied the Amazon River record (1903–1985) and found no indication of climate or land-use change over that period; however, they did find a two to three year period of declining flow following the warm phase of the ENSO cycle. They found periods of high flow were coincident with the ENSO cold phase.

2.2.7 *Groundwater*

The effects of climate change on groundwater sustainability include Alley et al. (1999):

- Changes in groundwater recharge resulting from changes in average precipitation and temperature or in seasonal distribution of precipitation.
- More severe and longer lasting droughts.
- Changes in evapotranspiration resulting from changes in vegetation.
- Possible increased demands of groundwater as a backup source of water supply.

Surficial aquifers are likely to be part of the groundwater system that is most sensitive to climate. These aquifers supply much of the flow to streams, lakes, wetlands, and springs. Because groundwater systems tend to respond more slowly to short-term variability in climate conditions than the response of surface water systems, the assessment of groundwater resources and related model simulations are based on average conditions, such as annual recharge and/or average annual discharge to streams. The use of average conditions may underestimate the importance of droughts (Alley et al., 1999).

The impacts of climate change on (a) specific groundwater basins, (b) the general groundwater recharge characteristics, and (c) groundwater quality has received little attention in the literature. Vaccaro (1992) addressed the climate sensitivity of groundwater recharge, finding that a warmer climate (doubling CO_2) resulted in a relatively small sensitivity to recharge and depended on land use. Sandstrom (1995) studied a semi-arid basin in Africa and concluded that a 15% reduction in rainfall could lead to a 45% reduction in groundwater recharge. Sharma (1989) and Green et al. (1997) reported similar sensitivities of the effects of climate change on groundwater in Australia. Panagoulia & Dimou (1996) studied the effect of climate change on groundwater-streamflow interactions in a mountainous basin in central Greece. They realized large impacts in the spring and summer months as a result of temperature-induced changes in snowfall and snowmelt patterns. Oberdorfer (1996) looked at the impacts of climate change on groundwater discharge to the ocean using a simple water-balance model to study the effect of changes in recharge rates and sea-level on groundwater resources and flows in a California coastal watershed. The impacts of sea-level rise on groundwater will include increased intrusion of salt water into coastal aquifers.

2.3 *Water-resource-system effects*

Studies that have looked at the impact of climate change on water-resources management include: Lettenmaier et al. (1999), Kirshen & Fennessey (1993), Lettenmaier & Sheer (1991), Arnell & Reynard (1995), Nash & Gleick (1991), Nemec & Schaake (1982).

A broad assessment was performed by Lettenmaier et al. (1999) of the sensitivity to climate change of six major water systems in the United States (Columbia River basin, Missouri River basin, Savannah River basin, Apalachicola-Chattahooche-Flint (ACF) River basin, Boston water supply system and Tacoma water supply system). The investigators evaluated system reliability (percentage of time the system operates without failure), the system resiliency (ability of a system to recover from failure), and system vulnerability (average severity of failure). Some thoughts or conclusions from this study were:

- Most important factors determining climate sensitivity of system performance were changes in runoff, even when direct effects of climate change on water demands were affected.
- Sensitivities depended on the purposes for which water was needed and the priority given to those uses.
- Higher temperatures increased system use in many basins, but these increases tended to be modest. Effects of higher temperatures on system reliability were similar.
- Influence of long-term demand on system performance had a greater impact than climate change when long-term withdrawals are projected to substantially grow.

Regional assessments can focus on three major questions (IPCC, 1996b; Miles et al., 2000):

1. How sensitivity is the region to climate variability?
2. How adaptable is the region to climate variability and change?
3. How vulnerable is the region to climate variability and change?

Sensitivity is the degree to which a system will respond to a change in climatic conditions (IPCC, 1996b). *Adaptability* refers to the degree to which adjustments in systems' practices, processes, or structures to projected or actual changes of climate are possible (IPCC, 1996b). Vulnerability defines the extent to which climate change may damage or harm a system (IPCC, 1996b). *Vulnerability* depends not only on a system's sensitivity but also on its ability to adapt to new climatic conditions. Major (1998) discussed the role of risk management methods for climate change and water resources.

2.4 River basin/regional runoff effects

Numerous studies of the effect of climate change on various river basins have been reported in the literature in recent years. These include: Ayers et al. (1993), Cohon (1991), Georgakakos & Yao (2000a, b), Georgakakos et al. (1998), Hamlet & Lettenmaier (1999), Hay et al. (2000), Hotchkiss et al. (2000), Hurd et al. (1999), Kaczmarek et al. (1996), Kirshen & Fennessey (1993), Lettenmaier & Gan (1990), Leung & Wigmosta (1999), McCabe & Ayers (1989), Miles et al. (2000), Nash & Gleick (1991), Stone et al. (2001) and many others. Studies on two major river basins (the Missouri River Basin and the Columbia River Basin) in the United States are discussed in this section for the purposes of illustrating the types of studies and modeling efforts that can be performed.

2.4.1 Missouri River Basin

Stone et al. (2001) studied the impacts of climate change on the Missouri River Basin water yield. They considered the effect of doubling the atmospheric carbon dioxide on water yield using a Regional Climate Model (RegCM) and the Soil and Water Assessment (SWAT) model, which is based upon the following daily water balance equation (Equation 9).

$$SW_t = SW_{t-1} + P_t - Q_t - ET_t - SP_t - QR_t \qquad (9)$$

where SW_t is the soil water content for the current day, SW_{t-1} is the soil water content for the previous day, P is the precipitation, Q is surface runoff, ET is the evapotranspiration, SP is percolation or seepage, and QR is return flow. The SWAT model is a rainfall-runoff model that computes surface runoff using the SCS curve number method, includes variable storage coefficient channel routing, reservoir routing, a storage routing combined with crack-flow model for percolation, a snowmelt algorithm based upon mean air temperature and soil layer temperature, a lateral subsurface flow model based upon a kinematic storage model, groundwater flow model based upon storage, transmission losses based upon Lane's method, evaporation based on the Penman-Monteith, Hargreaves, or Priestly-Taylor methods, and sediment yield based upon the modified universal soil loss equation (MUSLE). Precipitation can be either historic or stochastic using a first-order Markov chain model.

The water yield changes were summarized using three-month seasonal averages, with the winter season beginning in January. The time period from 1965 to 1989 was used. Average water yields were calculated for each season for the baseline run and for a doubling of the CO_2. Results of this analysis are as follows (Stone et al., 2001):

- Precipitation and temperature both increase more in the northern part of the basin, thus resulting in the greatest changes in the predicted water yields.
- Water yields (modeled) for the spring, summer, and fall seasons showed more distinct spatial trends than water yields for the winter months. SWAT simulation results exhibited dramatic increases (70% and much more in some subbasins) in water yields across the northern and northwestern portions of the Missouri River Basin in the spring, summer, and fall.
- Southeastern portions of the basin displayed an overall decrease (most decreased by less than 20% but some decreased by more than 80%) in water yields during theses three seasons. During the spring and summer the decreased water yields in the southeastern portions are enough to lower the total water yield of the entire basin by 10% to 20%.
- Water balance summaries for the principal tributaries indicate the large increase of runoff in the northern subbasins and decreased runoff in the southern subbasins.
- Six main stem reservoirs are operated in the upper river basin and control 55% of the area upstream of the reservoirs with the storage capacity of three times the mean annual runoff from the area. Total main stem reservoir storage changed dramatically from the baseline conditions, doubling the average annual releases, enough additional water under doubling CO_2 conditions to provide 24 cm of additional water annually for all irrigated areas in the Missouri River Basin.

2.4.2 Columbia River Basin

Hamlet & Lettenmaier (1999) and Miles et al. (2000) studied the impact of climate change on the hydrology and water resources of the Columbia River Basin. The Columbia River System and the major dam locations are shown in Figure 10. The response of the Columbia River Basin to climate change is a complex interaction among climate, hydrology, and physical water system and water resources management practices. Models were linked together to simulate the response of the climate/hydrology/water resources system.

Figure 11 shows the overview of the suite of the three linked models used by Hamlet & Lettenmaier (1999) and Miles et al. (2000) to investigate the climate, hydrology and water resources system of the Columbia River Basin. On it, the grey shaded area defines the drainage basin upstream of the Dallas, Oregon. Triangles represent major storage reservoirs (size of icon represents relative storage capacity), and circles are major run-of-river projects.

Figure 12 illustrates the dominant impact pathway through which changes in regional climate are manifested in the basin. Some of the major findings from the Hamlet & Lettenmaier (1999)

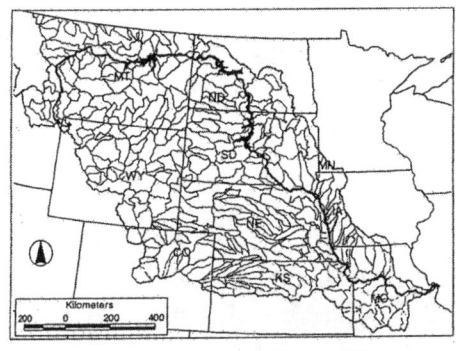

The Missouri River basin separated into the 310 USGS 8-digit subbasins

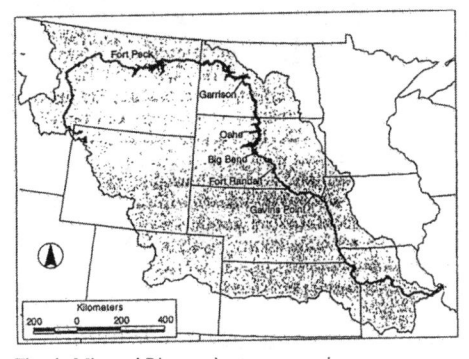

The six Missouri River main stem reservoirs

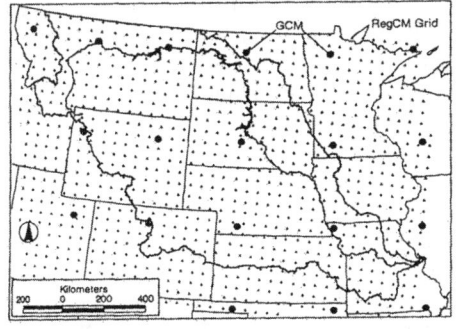

Missouri River basin GCM and RegCM grid spacing

Figure 10.   Missouri River Basin (Stone et al., 2001).

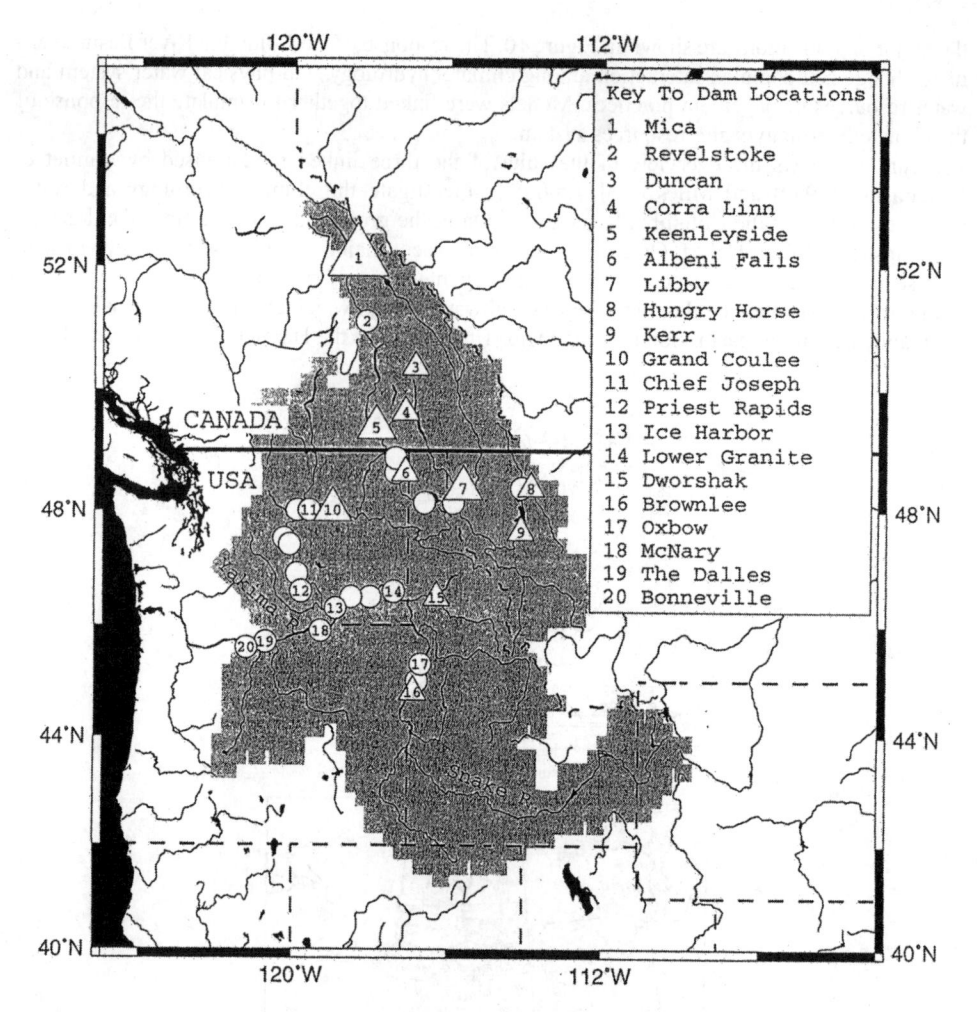

Figure 11. The Columbia River Basin and major dams (Hamlet & Lettenmaier, 1999).

are outlined below:

- Hydrologic response to climate change will tend to shift flow volumes from the summer to the winter.
- Management of the system to compensate for this effect would require that water be stored in the winter and released in the summer, which is basically the opposite to what is done now.
- Energy production, however, has its greatest economic value during the winter months. The result would be potential reductions in the required spring/summer flood storage and a shift in primary hydropower production to the summer.

2.4.3 *Summary of regional runoff effects found in the United States*
- *Arid and semi-arid western regions* – relatively modest changes in precipitation can have proportionately larges impacts on runoff and higher temperatures result in higher evaporation rates, reduced streamflows, and increased frequency of droughts.
- *Cold and cool temperate zones* – where a large portion of annual runoff comes from snowmelt – the major effect of warming is the change in timing of streamflow, both timing and intensity of the peaks. A declining proportion of the total precipitation occurs as snow with rising temperatures and the remaining snow melting sooner and faster in the spring resulting in more winter

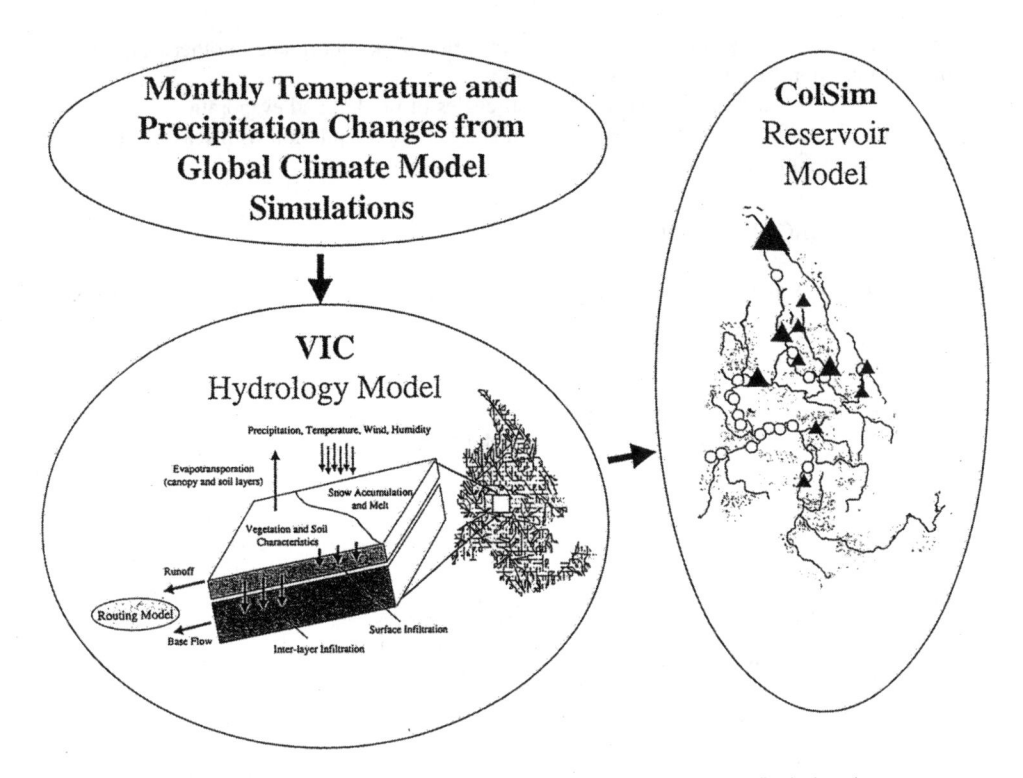

Figure 12. Overview of the suite of linked models used to investigate the climate/hydrology/water resources system (Hamlet & Lettenmaier, 1999).

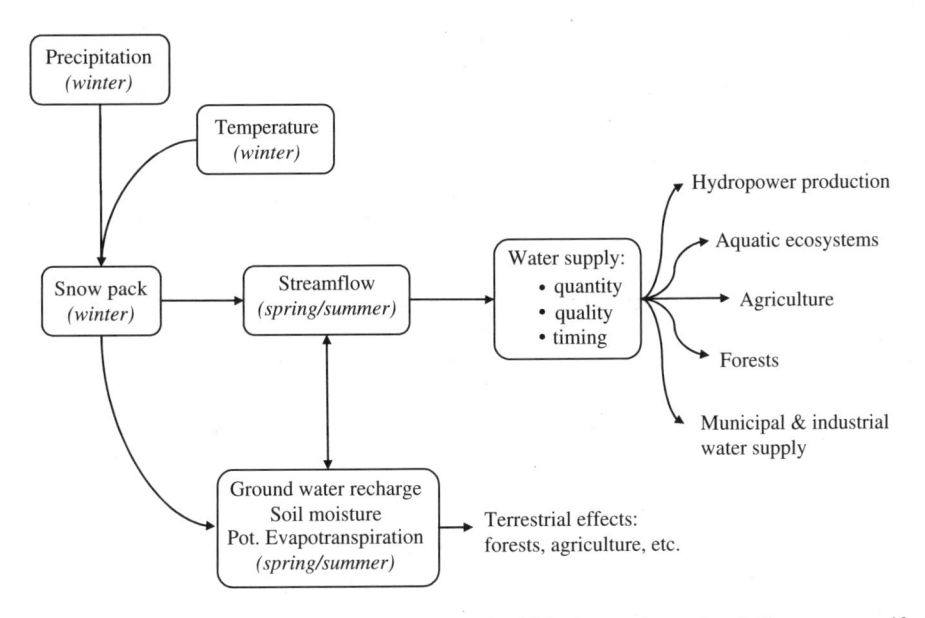

Figure 13. The dominant impact of pathway through which changes ion regional climate are manifested in the Pacific Northwest (Miles et al., 2000).

runoff. In some basins there may be increases in spring peak runoff, and in other basins runoff volumes may significantly shift to winter months.
- *Warmer climates* – in the south where seasonal cycles of rainfall and evaporation are predominant – runoff is affected much more significantly by total precipitation with seasonal cycles of rainfall and precipitation.

3 WATER MANAGEMENT OPTIONS

Table 2 summarizes some supply-side and demand-side adaptive options for various water-use sectors.

Table 2. Supply-side and demand-side adaptive options (IPCC, 2001b).

Supply side		Demand side	
Option	Comments	Option	Comments
Municipal water supply			
Increase reservoir capacity	Expensive; potential environmental impact	Incentives to use less (e.g., through pricing)	Possibly limited opportunity; needs institutional framework
Extract more from rivers or groundwater	Potential environmental impact	Legally enforceable water standards (e.g., for appliances)	Potential political impact; usually cost-inefficient
Alter system operating rules	Possibly limited opportunity	Increase use of grey water	Potentially expensive
Inter-basin transfer	Expensive potential	Reduce leakage	Potentially expensive to reduce to very low levels, especially in old systems
Desalination	Expensive; potential environmental impact	Development of non-water-based sanitation systems	Possibly too technically advanced for wide application
Irrigation			
Increase irrigation source capacity	Expensive; potential environmental impact	Increase irrigation-use efficiency	By technology or through increasing prices
		Increase drought-toleration	Genetic engineering is controversial
		Change crop patterns	Move to crops that need less or no irrigation
Industrial and power station cooling			
Increase source capacity	Expensive	Increase water-use efficiency and encourage energy efficiency	Possibly expensive to upgrade
Use of low-grade water	Increasingly used		
Hydropower generation			
Increase reservoir capacity	Expensive; potential environmental impact. May not be feasible	Increase efficiency of turbines; encourage energy efficiency	Possibly expensive to upgrade
Navigation			
Build weirs and locks	Expensive; potential environmental impact	Alter ship size and frequency	Smaller ships (more trips, thus increased costs and emissions)
Flood management			
Increase food protection (leeves, reservoirs)	Expensive; potential environmental impact	Improve flood warning and dissemination	Technical limitations in flash flood areas, and unknown effectiveness
Catchment source control to reduce peak discharges	Most effective for small floods	Curb floodplain development	Potential major political problems

4 EUROPEAN REGION

4.1 *Climate scenarios for Europe*

Some of the key features of climate scenarios (synthetic scenarios, palaeoclimatic analogs, and scenarios from GCM output) generated for Europe include (IPCC, 2001):

- Annual temperatures over Europe warm at a rate between 0.1° and 0.4°C per decade. Warming is greatest over Southern Europe (Spain, Italy, and Greece) and Northeastern Europe (Finland and Western Russia) and least along the Atlantic coastline of the continent (IPCC, 2001).
- During winter the continental interior of Eastern Europe and Western Russia warms more rapidly (0.15°–0.6°C per decade) than elsewhere.
- During summer warming has a strong south-to-north gradient, with Southern Europe warming at a rate of between 0.2° and 0.6°C per decade and Northern Europe warming between 0.08° and 0.3°C per decade.
- Winters classified as cold (occurring 1 year in 10 during 1961–1990) become much rarer by the 2020s and disappear by the 2080s.
- Hot summers become much more frequent.
- Agreement between models for these future temperature changes is greatest over Southern Europe in winter and in summer this regions shows the greatest level of disagreement between model simulations.
- All model simulations show warming in the future across all of Europe during all seasons.

The key features of precipitation include:

- Widespread increases of annual precipitation (between +1% and +2% per decade) in Northern Europe, smaller decreases (maximum −1% per decade) across Southern Europe, and small or ambiguous changes in central Europe (France, Germany, Hungary, Belarus).
- Winter and summer patterns of precipitation change have a marked contrast.
  - Most of Europe gets wetter in the winter (between +1% and +4% per decade).
  - Summers show a strong gradient of change between Northern Europe (wetting as much as +2% per decade) and Southern Europe (drying as much as −5% per decade).

4.2 *Changes in hydrologic cycle*

Calculations at the European scale under most climate change scenarios indicate the following (Arnell, 1999):

- Northern Europe would experience an increase in annual average streamflow.
- Southern Europe would experience a reduction in streamflow.
- In much of mid-latitude Europe runoff would decrease or increase by about 10% by the 2050s; however, the natural multi-decadal variability in runoff may be larger than the change resulting from climate change.
- In the Mediterranean regions, climate change is likely to exaggerate considerably the range in flows between winter and summer.
- In maritime Western Europe the range is likely to increase, but to a lesser extent.
- In more continental and upland areas, where the snowfall makes up a large portion of winter precipitation, a rise in temperature results in more precipitation as rainfall causing an increase in winter runoff and a decrease in spring snowmelt, significantly changing the timing of streamflow.
- Further east and at higher altitudes most of the precipitation continues to be snowfall, so that the distribution of flow throughout the year is altered very little.
- Across most of Europe low-flow frequency generally increase; however in some areas where the minimum occurs during winter, the absolute magnitude of low flows may increase because winter runoff increases (season of lowest flow shifts toward summer).
- For much of Europe the riverine flood risk generally would increase and in some areas the time of greatest risk would change from spring to winter.

Table 3. Key impacts on climate change on European water resources (IPCC, 2001b).

Public water supply	Reduction in reliability of direct river abstractions	Kaczmarek, et al. (1996)
	Change in reservoir reliability (dependent on seasonal change in flows)	Dvorak et al. (1997)
	Reduction in reliability of water distribution network	
Demand for public water supplies	Increasing domestic demand for washing and out-of-house use	Herrington (1996)
Water for irrigation	Increased demand	Kos (1993)
	Reduced availability of summer water	Alexandrov (1998)
	Reduced reliability of reservoir systems	
Power generation	Change in hydropower potential through the year	Grabs (1997)
	Altered potential for run-of-the-river power	Saelthun et al. (1998)
	Reduced availability of cooling water in summer	
Navigation	Change (reduction?) in navigation opportunities along major rivers	Grabs (1997)
Pollution risk and control	Increased risk of pollution as a result of altered sensitivity of river system	Mander & Kull (1998)
Flood risk	Increased risk of loss and damage	Grabs (1997)
	Increased urban flooding from overflow of storm drains	
Environmental impacts	Change in river and wetland habitats	

Table 4. Examples of adaptive techniques in the water sector (IPCC, 2001b).

Impact	Supply-side	Demand-side
Reduced water-supply potential	Change operating rules	Reduce demand through pricing, publicity, statutory requirements, etc.
	Increased interconnections between sources	
	New sources	
	Improved seasonal forecasting	
Increased stress on irrigation water	New sources (e.g., on-farm ponds storing winter runoff)	Increase irrigation efficiency
		Change cropping patterns
Increased flood risk	Increased flood protection	Accept higher risk of loss
		Reduced exposure by relocation
Reduced navigation opportunities	Increase flood protection	Smaller ships
	Increase dredging	
Reduced power	Install/increase water storage	Increase cooling water-use efficiency

4.3 Water management

Table 3 summarizes the key potential impacts on European water resources. Two general statements can be made concerning the water management:

– Impacts of climate change on European water resources and their management depend not only on the change in hydrology but also (perhaps more particularly) on characteristics of the water management system.
– The more stressed the system is under current conditions, the more sensitivity it will be to climate change.

4.4 Adaptation potential for water

The two approaches to adaptation in the water sector are the *supply side* (change the water supply) and the *demand side* (alter exposure to stress). Table 4 summarizes some examples of adaptive techniques in the water sector that may be appropriate for the effects of climate change. The supply-side techniques are most widely used at present and demand-side techniques are used less; however, demand-side techniques are the focus of increasing attention. The ideal adaptive strategy is one that can be altered in time as more information becomes available. A water management agency's adaptive response to climate change depends on procedures for considering strategies as well as adaptive capacity. *Long-term adaptations* refer to major structural changes to overcome changes caused by climate change. The three broad response strategies distinguished by Klein et al. (2000) for coastal regions are as follows:

– Reduce the risk of the event by decreasing its probability of occurrence.
– Reduce the risk of the event by limiting its potential effects.
– Increase society's ability to cope with the effects of the event.

5 UNCERTAINTIES

As pointed out by many investigators, the limitations of state of art climate models are the primary sources of uncertainty in the experiments that study the hydrologic and water resources impact of climate change. Future improvement to the climate models, hopefully resulting in more accurate regional predictions, will greatly improve the types of experiments to more accurately define the hydrologic and hydraulic impacts of climate change. Future precipitation and temperature are the primary drivers for determining future hydrologic response. Because of the uncertainties of the predictions of the future precipitation and temperatures, the uncertainties of the hydrologic responses of various river basins are uncertain, resulting in uncertainties of our future hydraulic resources.

REFERENCES

Alexandrov, V. 1998. A Strategy Evaluation of Irrigation Management of Maize Crop Under Climate Change in Bulgaria. *Proceedings of the Second International Conference on Climate and Water.* Espoo, Finland, 17–20 August 1998. Lemmela, R. and Helenius, N. (eds.). Vol. 3. 1998.

Alley, W. M.; Reilly, T. E. & O. L. Franke. 1999. *Sustainability of Ground-Water Resources.* U. S. Geological Survey Circular 1186, U. S. Geological Survey, Denver Colorado, 1999.

Arnell, N. W. & Reynard, N. S. 1995. Impact of Climate Change on River Flow Regimes in the United Kingdom. *Water Resources Research.* 1995.

Arnell, N. W. 1999. The Effect of Climate Change on Hydrological Regimes in Europe: A Continental Perspective. *Global Environmental Change.* 9: 5–23, 1999.

Arnold, J. G.; Srinivasan, R.; Muttiah, R. S. & Williams, J. R. 1998. Large Area Hydrologic Modeling and assessment, Part I: Model Development. *Journal of American Water Resources Association.* 34(1): 73–89, 1998.

Arnold, J. G.; Srinivasan, R.; Muttiah, R. S. & Allen, P. M. 1999. Continental Scale Simulation of the Hydrologic Balance. *Journal of American Water Resources Association.* 35: 1037–1051, 1999.

Ayers, M. A.; Wolock, D. M.; McCabe, G. J.; Hay, L. E. & Tasker, G. D. 1993. *Sensitivity of Water Resources in the Delaware River Basin to Climate Variability and Change.* U. S. Geological Survey Open-File Report 02–52, 1993.

Burkett, V. & Kusler, J. 2000. Climate Change: Potential Impacts and Interactions in Wetlands of the United States. *Journal of the American Water Resources Association.* 36: 313–320, 2000.

Chalecki, L. H. & Gleick, P. H. 1999. A comprehensive Bibliography of the Impacts of Climate Change and Variability on Water Resources of the United States. *Journal of the American Water Resources Association.* 35: 1657–1665, 1999.

Chow, V. T.; Maidment, D. R.; & Mays, L. W. 1988. *Applied Hydrology.* McGraw-Hill, New York, 1988.

Cohon, S. J. 1987. Influences of Past and Future Climates on the Great lakes Region of North America. *Water International.* 12: 163–169, 1987.

Cohon, S. J. 1991. Possible Impacts of Climatic Warming Scenarios on Water Resources in the Saskatchewan River Sub-Basin, Canada. *Climatic Change.* 19: 291–317, 1991.

Dickinson, R. E. 1989. Uncertainties of Estimates of Climate Change: A Review. *Climate Change.* 15: 5–14, 1989.

Dingman, S. L. 2002. *Physical Hydrology*. Second Edition, Prentice-Hall Inc., Upper Saddle River, New Jersey, 2002.

Dvorak, V.; Hladny, J. & Kasparek, L. Climate Change Hydrology and Water Resources Impact and Adaptation for Selected River Basins in the Czech Republic. *Climate Change*. 36, 93–106, 1997.

Eagleson, P. S. 1986. The Emergence of Global-Scale Hydrology. *Water Resources Research*. 22: 6S–14S, 1986.

Eltahir, E. A. B. 1996. El Nino and the Natural Variability of the Flow of the Nile River. *Water Resources Research*. 32: 131–137, 1996.

Falkenmark, M & Lindh, G. 1993. *Water and Economic Development, Water in Crisis: A Guide to the World's Fresh Water Resources*. Edited by P. H. Gleick, Oxford University Press, Inc., Oxford, 1993.

Firth, P. & Fisher, S. G. (eds). 1992. *Climate Change and Freshwater Ecosystems*. Springer-Verlag, New York, 1992.

Fisher, S. G. & Grimm, N. B. 1996. Ecological Effects of Global Climate Change on Freshwater Ecosystems with Emphasis on Streams and Rivers. *Water Resources Handbook*. Edited by L. W. Mays. McGraw-Hill, NY, 1996.

Georgakakos, A. & Yao, H. 2000a. Climate Change Impacts on Southeastern U. S. Basins, 1: Hydrology. *Water Resources Research*. U. S. Geological Survey Open-File Report 00-334, 2000a

Georgakakos, A. & Yao, H. 2000b. Climate Change Impacts on Southeastern U. S. Basins, 2: Management. *Water Resources Research*. U. S. Geological Survey Open-File Report 00-334, 2000b

Georgakakos, A.; Yao, H.; Mullusky, V & Georgakakos, K. 1998. Impacts of Climate Variability on the Operational Forecast and Management of the Upper Des Moines River Basin. *Water Resources Research*. 34: 799–821, 1998.

Gleick, P. H. (editor). 1993. *Water in Crisis: A Guide to the World's Fresh Water Resources*. Oxford University Press, Inc., Oxford, 1993.

Gleick, P. H. (lead author). 2000. *Water: The Potential Consequences of Climate Variability and Change for the Water Resources of the United States*. A Report of the National Water Assessment Group, For the U. S. Global Change Research Program, Pacific Institute for Studies in Development, Environment, and Security, Oakland, CA, September 2000.

Grabs, W. 1997. *Impact of Climate Change on Hydrological Regimes and Water Resources Management in the Rhine*. Cologne, Germany, 1997.

Green, T. R.; Bates, B. C.; Fleming, P. M.; Charles, S. P. & Taniguchi, M. 1997. Simulated Impacts of Climate Change on Groundwater Recharge in the Subtropics of Queensland, Australia. *Subsurface Hydrological Responses to Land Cover and Land Use Changes*. Kluwer Academic Publishers, Norwell, U. S. A., pp. 187–204, 1997.

Gregory, J. M.; Mitchell, J. F. B. & Brady, A. J. 1997. Summer Drought in Northern Midlatitudes in a Time-dependent CO_2 Climate Experiment. *Journal of Climate*. Vol. 10, pp. 662–686, 1997.

Hamlet, A. F. & Lettenmaier, D. P. 1999. Effects of Climate Change on Hydrology and Wate Resources Objectives in the Columbia River Basin. *Journal of the American Water Resources Association*. 35: 1597–1624, 1999.

Harris, R. N. & Chapman, D. S. 1997. Borehole Temperatures and a Baseline for 20th-Century Global Warming Estimates. *Science*. 275, 1618–1621, 1997.

Hay, L. E.; Wilby, R. L. & Leavesley, G. H. 2000. A Comparison of Delta Change and Downscaled GCM Scenarios for Three Mountainous Basins in the United States. *Journal of the American Water Resources Association*. 36: 387–398, 2000.

Herrington, P. 1996. *Climate Change and the Demand for Water*. Her Majesty's Stationary Office, London, United Kingdom, 1996.

Hostetler, S. W. & Small, E. E. 1999. Response of North American Freshwater Lakes to Simulated Future Climates. *Journal of the American Water Resources Association*. 35: 1625–1638, 1999.

Hotchkiss, R. H.; Jorgensen, S. F.; Stone, M. C. & Fontaines, T. A. 2000. Regulated River Modeling for Climate Impacts Assessments: The Missouri River. *Journal of the American Water Resources Association*. 36: 375–386, 2000.

Houghton, J. T.; Jenkins, G. J. & Ephraums, J. J. (editors). 1990. *Climate Change: The IPCC Scientific Assessment*. Cambridge University Press, Cambridge, 1990.

Hurd, B.; Leary, N.; Jones, R. & Smith, J. 1999. Relative Regional Vulnerability of Water Resources to Climate Change. *Journal of the American Water Resources Association*. 35:1399–1410, 1999.

IPCC. References available at www.ipcc.ch

IPCC. 1990. *Climate Change. The Intergovernmental Panel on Climate Change Scientific Assessment*. Houghton, J. T.; Jenkins, G. J. & Ephrauns, J. J. (editors), Cambridge University Press, Cambridge, United Kingdom, 1990.

IPCC. 1996a. Climate Change 1995: The Science of Climate Change. Intergovernmental Panel on Climate Change. Contribution of Working Group I to the *Second Assessment Report of the Intergovernmental Panel on Climate change*. Cambridge University Press, Cambridge, New York, 1996a.

IPCC. 1996b. Climate Change 1995: Impacts, Adaptations, and Mitigation of Climate Change: Scientific-Technical Analysis. Intergovernmental Panel on Climate Change. Contribution of Working Group II to the *Second Assessment Report of the Intergovernmental Panel on Climate Change*. Cambridge University Press, Cambridge, New York, 1996b.

IPCC. 1996c. Hydrology and Freshwater Ecology. In Climate Change 1995: Impacts, Adaptations, and Mitigation of Climate Change. Intergovernmental Panel on Climate Change. Contribution of Working Group II to the *Second Assessment Report of the Intergovernmental Panel on Climate Change*. Cambridge University Press, Cambridge, New York, 1996c.

IPCC. 1998. *Regional Impacts of Climate Change: An Assessment of Vulnerability*. Intergovernmental Panel on Climate Change. Cambridge University Press, Cambridge, New York, 1998.

IPCC. 2001a. *Climate Change 2001: The Scientific Basis*. Intergovernmental Panel on Climate Change. www.ipcc.ch, 2001a.

IPCC. 2001b. *Climate Change 2001: Impacts, Adaptation, and Vulnerability*. Intergovernmental Panel on Climate Change. www.ipcc.ch, 2001b.

IPCC. 2001c. *Climate Change 2001*. Intergovernmental Panel on Climate Change. www.ipcc.ch, 2001c.

Kaczmarek, Z.; Arnell, N. W. & Stakhiv, E. Z. 1996. Water Resources Management. *Intergovernmental Panel on Climate Change, Second Assessment Report*. Chap. 10, Cambridge University Press, 1996.

Kaczmarek, Z.; Napiorkowski, J. J. & Strzepek, K. M. 1996. Climate Change Impact on the Water Supply System in the Warta River Catchment. *Water Resources Development*. 12(2): 165–180, 1996.

Karl, T. R.; Knight, R. W. & Plummer, N. (nd). Trends in High Frequency Climate Variability in the Twentieth Century. *Nature*. 377: 217–220. 1995.

Karl, T. R. & Quayle, R. G. 1988. Climate Change in Fact and Theory: Are we Collecting the Facts? *Climate Change*. 13: 5–17, 1988.

Karl, T. R. & Riebsame, W. E. 1989. The Impact of Decadal Fluctuations in mean Precipitation and Temperature on Runoff: A Sensitivity Study over the United States. *Climate Change*. 15: 423–447, 1989.

Kaufmann, R. K. & Stern, D. I. 1997. Evidence for Human Influence on Climate from Hemispheric Temperature Relations. *Nature*. 388: 39–44, 1997.

Kiehl & Trenberth. 1997. Earth's Annual Global Mean Energy Budget. *Bulletin American Meteorlogical Society*. 78, 197–208, 1997.

Klein, R. J. T.; Aston, J.; Buckley, E. N.; Capobianco, M.; Mizutani, N.; Nicholls, R. J.; Nunn, P. D. & Ragoonaden, S. 2000. Coastal Adaptation Technologies. *IPCC Special Report on Methodological and Technological Issues in Technology Transfer*. Edited by Metz, et al. Cambridge University Press, Cambridge, U.K. and New York, NY, U.S.A., 2000.

Kirshen, P. H. & Fennessey, N. M. 1993. *Potential Impacts of Climate Change upon the Water Supply of the Boston Metropolitan Area*. U. S. Environmental Protection Agency, 1993.

Kos, Z. 1993. Sensitivity of Irrigation and Water Resources Systems to Climate Change. *Journal of Water Management*. 41(4–5):247–269, 1993.

Lettenmaier, D. P. & Gan, T. Y. 1990. Hydrologic Sensitivities of the Sacramento-San Joaquin River Basin, California, to Global Warming. *Water Resources Research*. 26: 69–86, 1990.

Lettenmaier, D. P. & Sheer, D. P. 1991. Climatic Sensitivity of California Water Resources. *Journal of Water Resources Planning and Management*. 117(1): 108–125, 1991.

Lettenmaier, D. P.; McCabe, G. & Stakhiv, E. Z. 1996. Global Climate Change: Effect on Hydrologic Cycle. *Water Resources Handbook*. Edited by L. W. Mays. McGraw-Hill, NY, 1996.

Lettenmaier, D. P.; Wood, A. W.; Palmer, R. N.; Wood, E. F. & Stakhiv, E. Z. 1999. Water Resources Implications of Global Warming: A U. S. Regional Perspective. *Climate Change*. 43(3); 537–579, 1999.

Leung, L. R. & Wigmosta, M. S. 1999. Potential climate Change Impacts on Mountain Watersheds in the Pacific Northwest. *Journal of the American Water Resources Association*. 36(6): 1463–1471, 1999.

Levine, J. S. 1992. Global Climate Change. *Climate Change and Freshwater Ecosystems*. Edited by Firth, P. and Fisher, S. Springer-Verlag, New York, 1992.

Loaiciga, H. A. et al. 1996. Global Warming and the Hydrologic Cycle. *Journal of Hydrology*. 174: 83–127, 1996.

Major, D. C. 1998. Climate Change and Water Resources: The Role of Risk Management Methods. *Water Resources Update*. 112: 47–50, 1998.

Mander, U. & Kull, A. 1998. Impacts of Climatic Fluctions and Land Use Change on Water Budget and Nutrient Runoff: The Porijogi. *Proceedings of the Second International Conference on Climate and Water*. Espoo, Finland, 17–20 August 1998. Lemmela, R. & Helenius N. (eds). Vol. 2, 884–896, 1998.

McCabe, G. J. & Ayers, M. A. 1989. Hydrologic Effects of Climate Change in the Delaware River Basin. *Water Resources Bulletin*. 25(6): 1231–1242, 1989.

Miles, E. L.; Snover, A. K.; Hamlet, A. F.; Callahan, B. & Fluharty, D. 2000. Pacific Northwest Regional Assessment: Impacts of the Climate Variability and Climate Change on the Water Resources of the Columbia River Basin. *Journal of the American Water Resources Association*. 36(2): 399–420, April 2000.

Mitchell, J. F. B. 1989. The "Greenhouse" Effect and Climate Change. *Reviews of Geophysics*. 27: 115–139, 1989.

Mitchell, J. F. B.; Johns, T. C.; Gregory, J. M. & Tett, S. F. B. 1995. Climate Response to Increasing Levels of Greenhouse Gases and Sulphate Aerosols. Nature. 376: 501–504, 1995.

Nash, L. L. & Gleick, P. H. 1991. Sensitivity of Streamflow in the Colorado, Basin to Climatic Changes. *Journal of Hydrology*. 125: 221–241, 1991.

Nemec, J. & Schaake, J. C. 1982. Sensitivity of Water Resource Systems to Climate Variation. *Journal of Hydrological Sciences*. 27: 327–343, 1982.

Oberdorfer, J. A. 1996. Numerical Modeling of Coastal Discharge. *Predicting the Effects of Climate Change, Groundwater Discharge in the Coastal Zone*. Edited by . Buddemeier, R. W. Proceedings of an International Symposium, Moscow, Russia, pp. 85–91, July 1996.

Pacific Institute. References on climate change: www.pacinst.org/CCBib.html

Panagoulia, D. & Dimou, G. 1996. Sensitivities of Groundwater-Streamflow Interactions to Global Climate Change. *Hydrological Sciences Journal*. 41, 781–796, 1996.

Richey, J. E.; Nobre, C. & Deser, C. 1989. Amazon River Discharge and Climate Variability: 1903 to 1985. *Science*. 246: 101–103, 1989.

Riehl, H. & Meitin, J. 1979. Discharge of the Nile River: A Barometer of Short-period Climatic Fluction. *Science*. 206: 1178–1179, 1979.

Rind, D. R.; Goldberg, R.; Hansen, J.; Rosensweig, C. & Ruedy, R. 1990. Potential Evapotranspiration and the Likelihood of Future Drought. *Journal of Geophysical Review*. Vol. 95, pp. 9983–10004, 1990.

Saelthun, N. R. et al. 1998. Climate Change Impacts on Runoff and Hydropower in the Nordic Countries. *TemaNord*. 552, 170, 1998.

Sandstrom, K. 1995. Modeling the Effects of Rainfall Variability on Groundwater Recharge in Semi-Arid Tanzania. *Nordic Hydrology*. 26, 313–320, 1995.

Santer, B. et al. 1996. A Search for Human Influences on the Thermal Structure of the Atmosphere. *Nature*. 382; 39–46, 1996.

Schaake, J. C. 1990. *From Climate to Flow, in Climate change and U. S. Water Resources*. Ed. by Wagner, P. E. John Wiley and Sons, Inc., New York, 1990.

Sharma, M. L. 1989. Impact of Climate Change on Groundwater Recharge. Proceedings of the *Conference on Climate and Water*. Volume I, Helsinki, Finland, Valton Painatuskeskus, pp. 511–520, September 1989.

Stone, M. C.; Hotchkiss, R. H.; Hubbard, C. M.; Fontaine, T. A.; Mearns, L. O. & Arnold, J. G. 2001. Impacts of Climate change on Missouri River Basin Water Yield. *Journal of the American Water Resources Association*. 37(5): 1119–1129, October 2001.

Tett, F. B.; Mitchell, J. F. B.; Parker, D. E. & Allen, M. R. 1996. Human Influence on the Atmospheric Vertical Structure: Detection and Observations. *Science*. 274: 1170–1173, 1996.

Vaccaro, J. J. 1992. Sensitivity of Groundwater Recharge Estimates to Climate Variability and Change, Columbia Plateau, Washington. *Journal of Geophysical Research*. 97 (D3): 2821–2833, 1992.

Vinnikov, K. Y.; Robock, A.; Speranskaya, N. A. & Schlosser, C. A. 1996. Scales of Temporal and Spatial Variability of Midlatitude Soil Moisture. *Journal of Geophysical Research*. Vol. 101, pp. 7163–7174, 1996.

Vorosmarty, C. V.; Green, P.; Salisbury, J. & Lammers, R. B. 2000. The Vulnerability of Global Water Resources: Major Impacts from Climate Change or Human Development? *Science*. 289: 284–288, 2000.

Vorosmarty, C. V.; Hinzman, L.; Peterson, B.; Bromwich, D.; Hamilton, L.; Morison, J.; Romanovsky, V.; Sturm, M. & Webb, R. 2002. Artic-CHAMP: A Program to Study Artic Hydrology and Its Role in Global Change. *EOS Transactions*. American Geophysical Union, 83(22), May 28, 2002.

Walsh, J. et al. 2001. *Enhancing NASA's Contributions to Polar Science: A Review of Polar Geophysical Data Sets*. Commission on Geosciences, Environment, and Resources, National Research Council, 124pp., National Academy of Press, Washington, D. C., 2001.

Wetherald, R. T. & Manabe, S. 1999. Detectability of Summer Dryness Caused by Greenhouse Warming. *Climate Change*. Vol. 43, No. 3, pp. 495–522, 1999.

Wigley, T. M. L. & Jones, P. D. (nd). Influence of Precipitation Changes and Direct CO_2 Effects on Streamflow. *Nature*. 314: 149–151. 1985.

Demand management of water in a regulatory environment

D. Howarth
National Water Demand Management Centre, Environment Agency of England and Wales, United Kingdom

ABSTRACT: With a relatively high rainfall, the United Kingdom (UK) is often perceived as water plentiful, but with a high population density the water availability per capita is no greater than many African and Middle Eastern countries. Faced with relatively low water availability and the possibility of rising demands placing ever more stress on the water environment, in the last five years demand management options have begun to be considered, for economic and environmental reasons, as a means of ensuring that supply and demand are kept in balance. The paper sets out the background to the water industry in England and Wales including the roles of the regulators: central Government, the economic regulator (Office of Water Services, OFWAT) and the environmental regulator (Environment Agency, EA). Recent droughts and the ensuing debate show how demand management came to be considered as a viable alternative to resource development. The paper describes how demand management is being implemented through the regulatory framework and how that framework has evolved to incorporate new and existing regulatory processes and tools: water resources planning, price setting, leakage targets, abstraction licensing, water efficiency plans, water fittings regulations and Integrated Pollution, Prevention and Control.

1 INTRODUCTION

1.1 Rainfall and water availability

Annual average rainfall in England and Wales is 897 mm varying from 604 mm in East Anglia to 1,312 mm in Wales. This level of rainfall would tend to give the impression that water is plentiful, however when the *annual internal renewable water resources* are expressed as per capita, England and Wales has a value of 1,334 m^3/(person annum), compared to the European average of 8,576 m^3/(person annum) (WRI, nd). The high population density is compounded by a rainfall gradient from the wetter highlands of the west and north to the dry south-east of England that runs in the opposite direction to a population density gradient. The situation is likely to get worse in future with economic expansion in the south-east due to the proximity of mainland Europe and climate change bringing warmer weather with declining rainfall. In some areas of England and Wales, most notably the south-east of England, there are significant areas of water stress with well-documented cases of environmentally damaged sites resulting from an historic over-licensing of water resources. Major investment programmes are taking place and will continue to take place to remedy over-abstracted sites.

1.2 Structure of the water industry

The water industry of England and Wales was privatised in 1989 and now comprises 10 water and sewerage companies and 13 water supply only companies supplying 15,236 Ml/day to 23.3 million properties. The companies range in size from Thames Water serving 3.3 million properties to Tendring Hundred Water supplying just 67,200 properties. The private water companies of

England and Wales and the public water utilities of Scotland and Northern Ireland are collectively represented by their trade organisation, Water UK.

The primary statutory duty of the water companies relates to the supply of water and in doing so they are answerable to their shareholders and the economic and environmental regulators. The turnover and profits of the regulated business come from supplying water.

1.3 Water use

The "public water supply", i.e., water provided by privatised and regulated water companies accounts for some 45% of all water abstracted as shown in Figure 1. Water used for power generation (32%) is generally returned to the river close to the point of abstraction without significant changes to water quantity or quality. Water abstracted directly for agricultural use accounts for just 1.1% of the total (0.8% is spray irrigation) although this is concentrated in time (summer) and space (south and east of England).

The public water supply, in total 14,991 Ml/day (2000–01), can be further broken down as shown in Figure 2.

1.4 The regulatory regime

Privatisation of the water industry took place in 1989, moving it away from public accountability (although the degree of that accountability was questionable) making regulation necessary. The 23 water companies are regulated by the Office of Water Services (OFWAT, the economic regulator), the Environment Agency (EA, the environmental regulator) and the Drinking Water

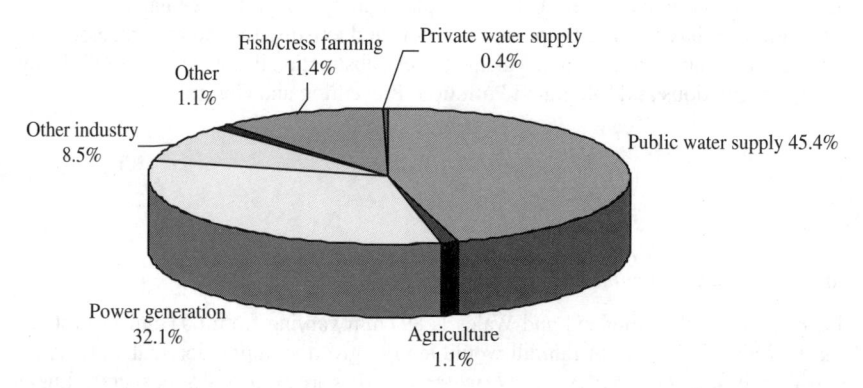

Figure 1. Water abstraction (by volume) (EA, 2001a).

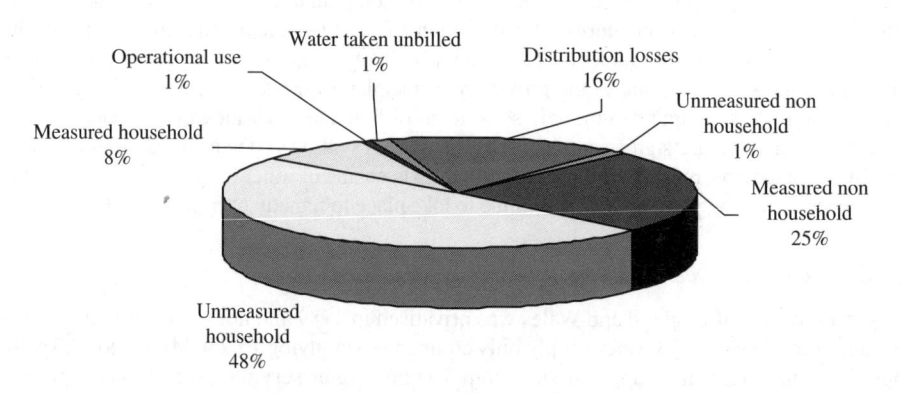

Figure 2. Public water supply (by volume) (OFWAT, nd-a).

Inspectorate (DWI, for drinking water quality). The Secretaries of State for the Environment, Food and Rural Affairs and for Wales have a wider role in developing policy and the legislative framework, often driven by compliance with EU Directives, as depicted in Figure 3.

Working within the framework established by Government, it has been OFWAT and the EA that have been most influential in introducing demand management concepts into water company water resources planning. Demand management is consistent with and an important part of both environmental and economic regulation.

OFWAT's duties are to ensure that:

- Water and sewerage functions are properly carried out through England and Wales.
- Water companies are able (in particular, by securing reasonable returns on their capital) to finance the proper carrying out of these functions.

The Director General of Water Services (of OFWAT) is required by the Water Industry Act 1991 to use his powers in a manner which he considers best calculated to promote economy and efficiency on the part of water and sewerage undertakers.

The EA is a non-departmental public body sponsored by the Department of Environment, Food and Rural Affairs (DEFRA). The Agency was created in 1996 to provide integrated environmental protection, management and enhancement and play a central role in putting the Government's environmental policies into practice.

The Agency has major responsibilities for the management and regulation of the water environment and for controlling industrial pollution and wastes. The Agency's main responsibilities in relation to water resources are:

- To take any necessary action to conserve, redistribute, augment and secure the proper use of water resources.
- To publish information about the demand for water and available resources.

The Agency fulfils these responsibilities principally by two means: firstly in its role as the abstraction licensing authority for all water abstractions in England and Wales and secondly by a water resource planning process with the water companies.

1.5 *Demand management*

Demand management has been defined as: *the implementation of policies or measures that serve to control or influence the consumption or waste of water* (UKWIR & EA, 1996). This wide-ranging definition is consistent with the use of the term in this paper.

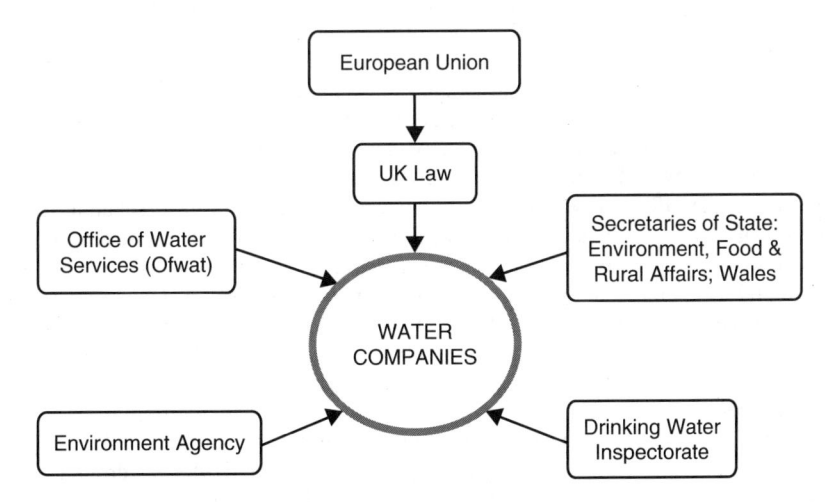

Figure 3. The regulation of the water industry.

2 RECENT HISTORY

2.1 The 1988–92 drought

Privatisation of the water industry in 1989 took place during a drought that in some parts of England lasted four years (1988–92). Southern and eastern England was short of approximately ten months rainfall. It was not a drought of extremes but the combination of four successive dry winters and intervening hot, dry summers was sufficient to seriously deplete water resources. In the south and south-east of England groundwater levels fell to their lowest level for 100 years. Many rivers also fell to very low levels, primarily due to low rainfall but also due to reduced contribution to river flows from groundwater. In 1990–91 some 20 million people were subject to a hosepipe ban and 6 million lived in areas where emergency and potentially damaging abstractions took place in order to maintain supplies. The privatisation of the water industry had not been popular and consequently during this drought media and public attention began to focus on water company failures, whether real or perceived. The disappearance of some rivers, primarily due to the drought but exacerbated by over abstraction was considered by many to be environmentally unacceptable. In June 1992 the National Rivers Authority (NRA, a predecessor body of the EA) issued the following statement (NRA, 1992):

> "Before any new sources are developed, it is essential that water companies make sure they are doing all they can to reduce leakage and to carry out effective demand management. The NRA supports selective domestic metering, with an appropriate tariff, in areas where water resources are stressed. Where it can be shown that proper attention is not being given to the control of leakage, or where appropriate consideration has not been given to the introduction of selective metering, the NRA will not grant licences for new sources".

This was significant in that a clear message was given to the water industry that further abstractions would not be granted until demonstrable effort had been applied to demand management.

2.2 The aftermath of the 1988–92 drought

In response to the drought in 1992 the Government published a consultation document *Using water wisely* (DE, 1992) which considered in detail a range of options for demand management. In 1992–93 OFWAT issued its first annual report *The cost of water delivered and sewage collected* (OFWAT, nd-b) which included water demand component data from all companies (including leakage). In 1994 the NRA published the country's first national water resources strategy since 1973, *Water – Nature's precious resource* (NRA, 1994) following a two-year consultation period. The document, with the stated aim of being an environmentally sustainable water resources development strategy was founded on the concepts of sustainable development, the precautionary principle and demand management. High, medium and low demand scenarios reflected differing growth rates and differing levels of demand management. In the low scenario the report concluded that no strategic resource developments would be required for thirty years.

At privatisation the NRA came into being as recognition of the need to separate regulatory activity from water company planning and operations. The NRA had demonstrated, during the drought, that it was going to take its duty to "secure proper use of water resources" seriously, and this was followed up by the inclusion of demand management in its national water resources strategy. The decision was also taken in 1993 to establish a National Centre for Water Demand Management with a mission *to provide a focus for information and expertise to ensure the acceptance of water conservation throughout society*. The original terms of reference for the Centre included providing expert and specialised support to its own water resources staff in addition to promoting the concepts externally. One of its first tasks was to publish in 1995 *Saving water: the NRA's approach to water conservation and demand management* (NRA, 1995). The 1988–92 drought, allied to focused regulatory roles had succeeded in placing demand management firmly on the water resources agenda.

2.3 The 1995 drought

Following the privatisation of the water industry in 1989 many water companies relaxed their efforts on leakage as they sought to reduce staffing levels. Leakage for the water companies of England and Wales reached a peak in 1994–95 of 5,112 Ml/day or 31% of distribution input (or "water into supply"). In 1995 a significant drought occurred that resulted in 39% of the population being subjected to hosepipe bans. Particular problems were experienced in Yorkshire where the use of 700 road tankers for a three month period were required in order to maintain supplies to the West Yorkshire districts of Bradford, Calderdale and Kirklees. Supply problems experienced during the 1995 drought were largely regarded as being due to mismanagement by the water companies. High levels of leakage and soaring executive salaries left the water companies with a major public image problem.

2.4 The aftermath of the 1995 drought

Following the 1995 drought the Government published *Water resources and supply: An agenda for action* (DE, 1996) which required the EA to revise its national and regional water resources strategies in consultation with water companies and to be fully involved with water companies' new resource development plans. The House of Commons Environment Committee conducted an inquiry into water conservation and supply (HCEC, 1996). Key recommendations included:

- Leakage reduction must be the water companies' top priority in relation to water conservation.
- Mandatory leakage targets should be set for water companies.
- Water companies should take responsibility for leaks on customer-owned supply pipes.
- The adoption of a six litre standard for all new toilets and the introduction of incentives to encourage people to replace their existing toilets.
- Ecolabelling schemes to be introduced urgently on water using household appliances.
- Demand management and leakage reduction should be tried first, and tried seriously.
- Establishing a national body which will campaign, through education and through the provision of grants and other incentives for the sustainable and efficient use of water in the home and in industry.
- Water companies to carry out micro-component analyses of water use in their company area and to place the results in the public domain.

The water companies already had a legal duty to develop and maintain an efficient and economical system of water supply. In 1996 an additional duty was placed on water companies to "promote the efficient use of water by its customers" (WIA,1991).

2.5 The water summit

Just three weeks after being elected in May 1997 the Labour Government produced a ten-point plan for a world class water industry, with eight of the points directly addressing demand management issues:

1. OFWAT to set tough mandatory leakage targets.
2. All water companies expected to provide free leak detection and repair service to household customers' supply pipes.
3. Water companies to have a statutory duty to conserve water in their operations.
4. Water companies to carry out with vigour, imagination and enthusiasm their duty to promote the efficient use of water by their customers.
5. Water companies to consider the role of the Environment task Force (a youth unemployment initiative) in improving the efficiency of water use.
6. New water regulations will include significantly tighter requirements for water efficiency.
7. A Government review of the system of charging for water, including future use of rateable value and metering policy.
8. Compensation for customers affected by drought related restrictions.
9. Companies to publish locally details of their performance in reducing leakage.
10. A review of the abstraction licensing system, with environmental protection.

Mandatory leakage targets have been set annually since 1997, which have resulted in leakage being reduced from 5,112 Ml/day in 1994–95 to 3,240 Ml/day in 2000–01, a fall of 35%, although it has risen to 3,410 Ml/day in 2001–02. New water metering legislation has been passed and new water regulations introduced. In addition the review of the abstraction licensing system will allow the EA to assess the sustainable level of abstraction for river catchments, and revoke all or parts of abstraction licences where the sustainable abstraction level has been exceeded and/or where water is being used inefficiently. The Government succeeded in moving water demand management higher up the political agenda where it has remained to date.

3 REGULATORY DRIVERS FOR DEMAND MANAGEMENT

The droughts of 1988–92 and 1995 have been significant in demonstrating the sometimes small margin that exists of supply over demand. Such droughts are not enough in themselves to make it self-evident that demand management has an important role to play in managing the supply demand balance. A typical policy response to a drought is often the advancement of the development of new water resources. The fact that no major water resource developments have been advanced in England and Wales is largely due to the roles of economic and environmental regulation.

3.1 *Economic regulation*

OFWAT has to ensure that water companies are properly financed to carry out their functions and to ensure that water company customers receive a good and cost-effective service. In relation to water resources there are four areas where their policies have been influential in relation to demand management.

3.1.1 *Price setting*
A five-year price setting process (the Periodic Review) determines the water prices that companies can charge customers. This is arrived at by consideration of the water companies' Asset Management Plans, of which supply-demand balance is a component. The Periodic Review process, operating on a five yearly cycle has been criticised for being too short term in its approach. Consideration is currently being given as to how longer term interests can be accounted for in the process. As the best value for customers is being sought, the Director General of Water Services will need to be convinced that any proposed resource developments are less expensive than demand management approaches. For the first time in 2004 OFWAT's supply-demand balance aspect of the Periodic Review will be streamlined as part of the Agency's water resources planning process.

3.1.2 *Charging for water*
Currently only 19% of households in England and Wales are metered, with most of these being installed since privatisation. Where meters have been installed studies have shown that demand is reduced by around 10% (NMTG, 1993). OFWAT has set out expectations that water companies should adopt metering as a method of charging in the longer term, on the basis that growth should be paid for in relation to use, although OFWAT also recognise the demand management benefits. However, with current government policy conferring the right to remain unmetered in existing households, most meters being installed are for optants (i.e., people choosing to pay by volume because their bill will be reduced at their current level of use). As a result, at the last price setting process OFWAT constrained the number of meters companies could install so as to limit the price rises for the remaining unmetered customers.

3.1.3 *Water efficiency*
Water companies have had a legal duty to promote water efficiency by their customers since 1996. It is the role of OFWAT to ensure that the companies comply with this duty. Water companies were asked to prepare a water efficiency plan in 1996 and updates have been requested since then.

OFWAT's expectations have been that water companies should be active in the following areas:

- Supply pipe repairs and replacements.
- Installation of toilet cistern devices.
- Household water audits (by water company or self-audit).
- Non-household water audits.

OFWAT has reported that water company water efficiency activity (excluding metering and leakage control) has resulted in a water saving of 789 Ml/day across England and Wales, representing 5.2% of distribution input (water into supply) at a cost of 91 p/m^3 (typical resource development costs are 35–70 p/m^3). Water company water efficiency performance is reported annually to OFWAT and summarised in *Leakage and the efficient use of water* (OFWAT, nd-a).

3.1.4 Leakage

Since the water summit of 1997, OFWAT has set water leakage targets and monitors water company performance against them. The targets were initially set pragmatically based on existing leakage level and resource surplus. Latterly the targets have been set on water companies own calculations of their "economic level of leakage", subject to their assessment being deemed robust by OFWAT. If water companies fail to meet their target they are required to report on progress every three to four months to OFWAT. If a water company continues to fail to achieve its target ultimately the company could face the loss of its operating licence. Leakage performance is also presented in OFWAT's *Leakage and Efficient Use of Water* (OFWAT, nd-a).

3.2 Environmental regulation

3.2.1 Abstraction licensing and discharge consenting

The EA administers around 50,000 abstraction licences and receives some 1,500 to 2,000 new applications each year. The majority of these licences are not owned by water companies, but by industrial and agricultural abstractors. The Agency can refuse licence applications if the need is not considered reasonable and is now in the process of developing Catchment Abstraction Management Strategies (CAMS) for all river catchments. It is still to be determined whether, as part of the CAMS process, the Agency will have the powers to convert all licences to time-limited status. The advantage of this would be that when a decision is made on whether the licence should be renewed at the end of the period, the Agency can take into consideration whether any environmental damage has occurred. The Agency typically adopts this approach for new licences but has not been able to apply such conditions retrospectively. The EA, through its R&D programme has developed a range of water use benchmarks for industries that abstract water directly from the environment, that can be used for assessing water use requirements in licence applications (EA, 1998). All industries that discharge wastewater into watercourses require a discharge consent from the EA to do so. Refusal of discharge consents because of possible contributions to flooding and worsening water quality has made industries and building owners opt for more sustainable solutions such as rainwater collection for toilet flushing.

3.2.2 Water resources planning

The EA also requires water companies to submit Water Resources Plans every five years that have to be agreed with the EA. In 2004 working collaboratively with OFWAT, the submissions will also form part of the price setting process. The EA also monitors annually water company outturn data against the original plan. In March 2001 the EA published its National and Regional Water Resources Strategies (EA, March 2001), which set the framework for water resources decision-making for the next twenty-five years. The strategies include demand forecasts based on future socio-economic scenarios that offer considerable potential for further reductions by water conservation and other demand management measures.

3.2.3 Integrated Pollution Prevention and Control (IPPC)

IPPC is a regulatory system that employs an integrated approach to control the environmental impacts of certain activities, by determining controls for industry to protect the environment through a single permitting process. The IPPC regulations have been made under the IPPC Act 1999 that implemented European Commission Directive 96/61. The Directive covers only specified industries, but some of those use large volumes of water. IPPC Guidance for the Food and Drink, Paper and Pulp and Pigs and Poultry industries set out benchmarks and best practice water use for compliance with the Directive. Full implementation will be across 30 industry sectors and 6,000 installations on a phased basis to 2007.

3.2.4 The National Water Demand Management Centre

The EA established a National Water Demand Management Centre in 1993 as a focus for expertise in water demand management matters, with a mission statement to *provide a focus for information and expertise to ensure the acceptance of water conservation throughout society*. The Centre has been very proactive and succeeded in ensuring that demand management is and remains high on the water resources agenda. The Centre involves itself in advising, promoting and improving technical, economic, social and environmental understanding of demand management. The Centre exists to provide advice to internal and external colleagues on demand management matters as well as the promotion of demand management concepts. Credible promotion is underpinned by a high level of understanding some of which is gained from its own research programme. To single out just two of the Centre's publications, the *Demand Management Bulletin* (EA-NWDMC, 1993–2002) is read throughout the world, and the "Fact Cards" (EA, 2001b), explaining what water conservation devices are available, how they work, potential savings and list of suppliers, have proved extremely popular.

3.3 Government initiatives

3.3.1 Sustainable development strategy

In 1999 the Government updated its a sustainable development strategy (DETR, 1999a) and defined four objectives intended to be met simultaneously:

- Social progress that recognises the needs of everyone.
- Effective protection of the environment.
- Prudent use of natural resources.
- Maintenance of high and stable levels of economic growth and employment.

Water demand management is wholly consistent with the second and third objectives, offers opportunities for employment (fourth objective) and could arguably be regarded as a form of social progress (first objective).

3.3.2 The Water Supply (Water Fittings) Regulations 1999

Limiting the water use of new appliances can make significant contributions to demand management in the longer term as people change their appliances. This has been successful in the USA where the 1992 Energy Act set maximum use standards for toilets (6 litres), taps (faucets) and showers (both 9 l/min maximum flow). Due to England and Wales not having a history of water-wasting fixtures the new Water Supply (Water Fittings) Regulations of 1999 (Department of the Environment, Transport and the Regions) will not have such a dramatic effect. The new Regulations are shown in Table 1, and compared to the previous water byelaws.

Only with toilets has significant progress been made, by reducing the single flush volume and allowing the re-introduction of dual flush toilets. Whilst the maximum allowable water use of clothes washers and dishwashers has declined, pressure to harmonise with the rest of Europe has resulted in the adoption of a standard which very few machines would fail. Power showers that deliver flow rates of 40–50 l/min can be purchased in the UK and are growing in popularity. No regulation exists because of the perceived problems of enforcement. Enforcement is at the point of installation rather than at manufacture or sale although this is currently under review.

Table 1. Appliances and water regulations.

Appliance	Water Byelaws, pre 1999	Water Supply (Water Fittings) Regulations 1999
Toilets	7.5 l/flush	6 l and re-introduction of dual flush
Clothes washers	180 l/cycle	120 l/cycle
Dish washers	7 l/place setting	4.5 l/place setting
Showers	None	None

3.3.3 *Communication – "Are you doing your bit?"*
In 2000 the Government supported an advertising campaign to influence the public to change their behaviour for the sake of the environment. A high profile campaign of television and newspaper advertisements involving media celebrities included aspects of water conservation. The Government also supports an annual water month in July that includes water conservation messages.

3.3.4 *Industrial, commercial, institutional and agricultural*
Envirowise is a Government sponsored programme to improve the competitiveness and environmental performance of British industry, being funded from DEFRA and the Department of Trade and Industry. Envirowise produce free guides and case studies for a range of industries to encourage other industries to adopt good practice. Reducing water use is frequently identified as an area where significant cost saving can be made. Envirowise also organise and sponsor seminars to disseminate good practice and encourage the setting up of waste minimisation clubs where industries can share the costs of buying in expertise. In 2000 the Government's procurement arm, The Office of Government Commerce, embarked on the Watermark project, which has the objective of setting good practice water use benchmarks for public sector buildings. DEFRA has also produced reporting guidelines for company reporting on water (DETR, 2001), and a guidance document (MAFF, 1999) for farmers to help them examine and where appropriate improve current irrigation practices.

4 DEMAND MANAGEMENT POLICIES

The regulatory approach has created the conditions for demand management polices and practices but it should be noted that this is largely by mandating activity rather than encouragement by incentives. The policies being followed by the water companies and others are as follows:

4.1 *Leakage control*

Leakage control policy and practice is an area where the UK water industry probably leads the world. In 1994 the industry published the *Managing Leakage* (UKWIR, 1994) series of reports covering leakage measurement, comparative performance measures, pressure management, using night flow data and the economics of leakage control. The reports also introduced the concept of a component based approach to leakage which has since revolutionised leakage management in the UK. The component based approach recognises that at any given time leakage consists of bursts (that can be detected) and background leakage (that cannot be located by conventional methods). The relative proportions of burst and background leakage influence the policies that need to be adopted. For example, a company whose leakage is primarily background leakage needs to consider mains replacement and/or pressure management rather than spending on manpower to find leaks.

The majority of water companies have sectorised their networks into District Meter Areas (hydraulically discrete areas of between approximately 1,000–3,000 properties). From analysis of night flows the company can estimate the level of leakage and determine where detection efforts should be concentrated. Recent developments include acoustic noise loggers placed strategically on the network listening for the sound of a leak. A patroller unit, housed in a vehicle receives signals from the logger that indicates a leak exists. In this way the network can be monitored by simply driving around it. Some water companies are experimenting with siting acoustic loggers permanently on the network whereas others are using them in a reactive mode. Such developments are likely to reduce the costs of leakage control making it more economic to operate at increasingly lower levels of leakage.

Pressure management has been proven to be a very cost-effective approach to reducing leakage and twelve companies claim to have reduced the pressure to over half of their customer base. The longer-term solution to leakage is clearly mains replacement, however it is an expensive option. The water companies are currently renewing their networks at the rate of around 1.5% per year. For a company to renew at a greater rate requires them to convince OFWAT that it is cost-effective to do so.

4.2 Household metering

Household metering remains a contentious issue. Since 1999 compulsory metering of households has been prohibited but metering is allowed under the following circumstances:

- Free option scheme – water companies must provide a free meter to any customer who requests one. The customer has a "right to revert" up to one year after the installation.
- New homes – metering is the normal charging method.
- Households that use water for "non-essential" purposes, e.g., garden sprinkler.
- Water stressed areas – a water company can appeal to the Secretary of State for designation, if granted the company can then meter all households compulsorily (there have been no applications to date).
- At change of occupancy (since customers only have a right to remain unmeasured only in their *present* home).

The difference in the proportions of metered households across the water companies is largely due to the varying degree with which the above policies are pursued, with some companies showing more enthusiasm than others.

Data in the public domain are limited that demonstrate the effect on demand of moving from unmeasured to measured charging. The National Metering Trials (1988–92) (NMTG, 1993) concluded that the introduction of a volumetric charge reduced average demand by 10% and peaks by 30%. In the last price setting process (1999) OFWAT allowed for a 5% reduction in demand by meter optants (those electing to be metered, rather than it being imposed), unless the company could demonstrate otherwise. Very little is known about the price elasticity of water supplied to households and although Government, OFWAT and the EA wish to see companies experiment with tariffs (e.g., rising block, seasonal) few companies have shown themselves willing to do so.

4.3 Water efficiency

In 1996, in furtherance of their new water efficiency duty, OFWAT required water companies to prepare Water Efficiency Plans. The plans were updated in 2001. On an annual basis the water companies report to OFWAT on their water efficiency activity and a summary is included in OFWAT's *Leakage and the efficient use of water report* (OFWAT, nd-a), from which Table 2 has been reproduced. It is clear from the table that the principal activities are the issuing of cistern devices (which reduce the toilet flush by displacing a volume of water) and self-audit packs (which enable the householder to better understand their water use and the scope for savings). Both of these activities are low cost in that they can be mailed out to the customer. It is important to recognise that low cost activities are not necessarily cost-effective. The water companies are required to report both the costs and water saved of these activities. At present lack of data on water efficiency options results in water companies not seriously considering such options in their water resources planning. Concern remains that water companies are motivated by carrying out the minimum level of activity that will satisfy OFWAT rather than a genuine desire to save water. Gathering robust data and

Table 2. Water industry progress in promoting the efficient use of water 1996–2001.

	1996–97	1997–98	1998–99	1999–00	2000–01	Total
Supply pipe repairs						
Total repaired	36,500	76,240	77,024	73,586	66,951	330,301
Free	19,128	67,199	67,707	62,693	54,602	271,329
Charged	17,372	9,041	9,317	10,893	12,349	58,972
Supply pipe replacements						
Total replaced	1,126	9,366	11,643	12,766	7,556	42,457
Free	0	3,248	5,393	6,311	3,871	18,823
Charged	1,126	6,118	6,250	6,455	3,685	23,634
Cistern devices						
Number distributed	366,297	2,770,715	1,419,987	1,417,388	497,216	6,471,603
Installed by customer	n/c	n/c	n/c	518,303	137,432	655,735
Installed by company	n/c	n/c	n/c	4,384	541	4,925
Other household water saving devices installed (water butts and spray guns)	16,100	48,287	85,388	104,753	46,293	300,821
Household water audits						
Carried out by company	865	12,467	14,120	11,739	28,077	67,268
Self audit packs issued	1,000	1,251,860	2,009,486	1,551,809	3,737,285	8,551,440
Non-household water audits						
Carrried out by company	1,094	5,479	10,276	8,764	3,316	28,929
Self audit packs issued	n/c	n/c	n/c	28,463	15,557	44,670
Water regulation inspections	n/c	n/c	n/c	17,671	36,864	54,535
Total savings/costs						
Total savings	n/c	312	201	157	119	789
Achieved/assumed (Ml/d)	n/c	106,153	96,757	32,264	26,453	261,528
Total cost of savings (£1,000)	n/c	93	132	56	61	91

n/c = not collected.
For reference: 23.3 m households in England and Wales.

making them widely available remains a considerable challenge for the water industry. The industry, to its credit, has devised best practice guidelines for the assessment of costs and savings from water efficiency programmes.

The author believes that the following options are promising for England and Wales and merit further investigation:

- Household water audits (carried out by company or third party on company's behalf).
- Industrial, commercial and institutional water audits.
- Converting existing single flush toilets to dual flush operation.
- Fitting controllers to urinals (or waterless retrofit).
- Appliance exchange programmes (clothes and dishwashers).
- Rainwater harvesting for new properties.

While the duty remains one of promoting water efficiency and doubts remain about cost-effectiveness it is difficult to see how progress will be made in investigating these and other options.

4.4 *Public awareness programmes/communication*

Even with the best technological solutions possible it is generally recognised that for success in managing water demand it is vital to obtain the participation and support of the general public. This is difficult for water utilities in England and Wales that have traditionally not had a close relationship with their customers. Privatisation of the industry has made the public suspicious of water company motives and less accepting of their failings.

In 1996 one water company stated: "We are keen to encourage the voluntary adoption by customers of more efficient washing machines, low flush WCs and other water saving devices. But a more pro-active approach, as is adopted in parts of America, is probably too intrusive for our customers" (EA, 1997).

Most water companies have issued public summaries of their water efficiency plans and communicated the importance of saving water with billing information. In addition many carry out the following activities:

- Mobile visitor centres.
- Water saving plays in schools.
- High profile sales of rainwater butts.
- Developing curricula material for schools.
- Gardening leaflets.
- Web based information.

The challenge is to move water conservation away from being a drought response mechanism to one that is geared towards the longer term. It is not easy to convey a message of "saving water" when the country is suffering severe flooding (as it did in November 2000).

Communication strategies are not the sole preserve of water companies. The Non Governmental Organisations (NGOs) such as Royal Society Protection of Birds, the National Trust and the Council for the Protection of Rural England also communicate the need to conserve water for the sake of the environment.

5 DEMAND 1992–93 TO 2001–02

The success of the regulatory regime as an instrument for managing demand can be approximated by analysing the changes in demand components, from 1992–93 the first year of reliable data. It is worth stating that the availability of these data and the fact that they are published annually for each water company is a direct consequence of the regulatory regime. Published figures, audited by the economic regulator, have been very powerful in stimulating debate about water company policies and performance.

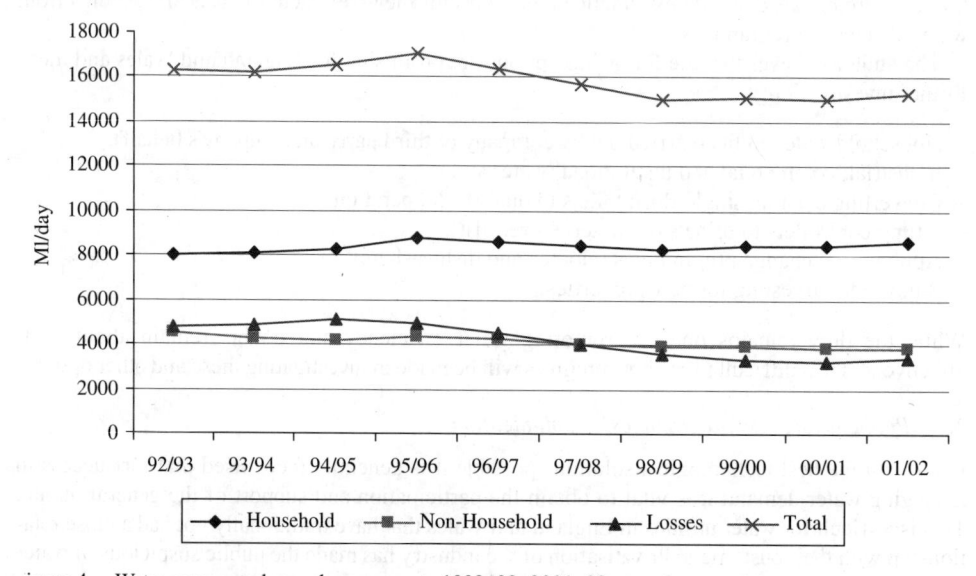

Figure 4. Water company demand components 1992–93, 2001–02.

Summarising the trends:

- Distribution input ("water into supply") has declined by 2,036 Ml/day from 1995–96 to 2000–01, but an increase of 245 Ml/day was recorded in 2001–02.
- The principal reason for this is the reduction in water loss (1,737 Ml/day over the same period).
- Non-household use has declined (due to closure of high water using industry as well as adoption of water conservation).
- Household use (and unmeasured per capita consumption) has been relatively stable for the last four years although unmeasured household per capita consumption has risen from 140 l/(h day) in 1992–93 to 152 l/(h day) in 2000–01.

This analysis would suggest that demand in England and Wales has stabilised and so it could be concluded that the regulatory approach to demand management and the climate that has been created has been successful in stabilising demand. However water company water resources plan submissions to the EA in 1999 included rising demand forecasts.

6 THE CHALLENGES AHEAD

It would appear that the regulatory framework in England and Wales has played a major role in managing demand. The inexorable rise in demand, commonly portrayed in demand forecasts has not happened. Water supplied by water companies is currently at its lowest level since the 70s. Water company demand forecasts submitted to the EA in 1999 predicted stable leakage and non-household demand but an overall growth driven by a rise in household consumption. The predictions were that this growth would be driven by falling occupancy rates and increased water use for power showers and outdoor use which presumably (since insufficient detail was provided) would be greater than the demand reduction due to increasingly water efficient household appliances (toilet, clothes washer, dish washer).

Over the next decade five challenges will need to be met if England and Wales is to reach the goal of sustainable water consumption. These challenges are distinct but are related to each other as shown in Figure 5, and all existing within the context of the EU Water Framework Directive (WFD).

6.1 *How much demand management is needed?*

Demand management is not an end in itself. The EA, in its licensing procedures is required to balance the needs of abstractors with the needs of the environment. As part of its CAMS process the Agency

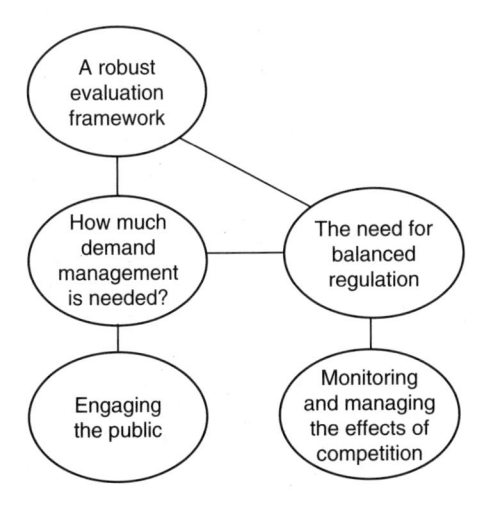

Figure 5. Relationship between challenges.

will be determining for each river catchment, its sustainable level of abstraction. Where current levels of abstraction are in excess of this abstractors will be required to reduce their demands. The sustainable level of abstraction will be determined by modelling (Resource Assessment Methodology) and public consultation. This approach is consistent with the requirements of the WFD. Once this level is reached there should be no requirement to reduce demand further unless it is in the direct economic interests of the water supplier or water user to do so. In time the definition of what constitutes a "sustainable level of abstraction" may change so it is important that it is regularly reviewed.

6.2 A robust evaluation framework to assess supply side/demand side and related information needs

In 1996 the EA and UK Water Industry Research (UKWIR) published the *Economics of Demand Management* (UKWIR & EA, 1996). This represented an agreed approach to assessing supply and demand options in the same economic framework. This guidance, used for the 1999 Periodic Review and for Water Resource Plan submissions has been recently modified and updated "The Economics of Balancing Supply and Demand" in time for the Periodic Review 2004. The guidance encompasses a least cost approach where options are chosen to bridge predicted supply deficits on the basis of their unit costs in pence/m^3. The method is not controversial but the numbers used in the analyses can be, particularly environmental valuation, where a low valuation can often result in an environmentally damaging resource development appearing to be the least expensive option.

Another problem is that, other than leakage control, relatively few demand management options have been properly evaluated and so basing a long term strategy on under researched demand management options is seen as a risky strategy. Monitoring and evaluating water savings is expensive and to date there has been an expectation that water companies should fund such projects from their revenue budget with no specific allowance in price setting. Most have been reluctant to do so, other than on a very small scale. Discussions are currently taking place over whether allowances for water efficiency projects could be made in the next Periodic Review. A well-developed evaluation framework will only work successfully with data to populate it.

6.3 Ensuring that regulation delivers

This paper has demonstrated that the strong regulatory framework for water in England and Wales has ensured that demand management has played, and will continue to play an important role in managing the supply-demand balance both for economic and environmental reasons. However, the majority of these regulatory tools are related to the setting of a standard and then expecting the water companies to meet it, with penalties imposed should they fail. There a few incentives whereby rewards are offered for water companies and other abstractors who actively pursue demand management policies.

At the last Price Review (1999) there was some uncertainty over whether demand management options would be allowed for in price limits. The effect of this uncertainty resulted in water companies predominantly choosing new resource options (where a rate of return on their investment is guaranteed) to maintain their supply-demand balance. The situation was not helped by the lack of data making it difficult for robust analyses of the costs and benefits of demand management options to be submitted as part of the Review. OFWAT has indicated that it will be clearly communicated that in the next Price Review demand management options will be allowed for in price limits, thereby ensuring equal treatment with supply side options.

However, if there was a desire to see more demand management the regulatory regime could be amended to facilitate this. Metering and charging by volume is seen as a demand management tool but as the number of metered users increases the relationship between profits and sales becomes more closely aligned. De-coupling profits from sales volumes, as has happened in the US energy industry (Weizsacker et al., 1998), perhaps allied to the water company keeping at least a share of the savings resulting from its own demand management activity may offer a way forward.

There is also scope for improving the Water Supply (Water Fittings) Regulations in future years. A maximum flow standard for showers could usefully halt the growth in high flow (power) showers. Toilet, clothes washers and dishwasher volumes could be further reduced in future.

6.4 *Engaging the public*

Regulation, acting alone is limited in what it can achieve. It can mandate water companies to keep their leakage levels low, it can mandate the size of toilets and it can ensure that householders who wish to use garden sprinklers must pay for that use by volume. We have also seen that it can ensure water companies promote water efficiency by their customers and that they consider it their water resources plans. However, for appliances to be purchased that are more efficient than specified by regulations and for the public to make behavioural changes (for example shorter showers) they need to understand the importance of the issue and be engaged in the solution. There is a lot of evidence in England and Wales that the public think that it is the role of Government to solve environmental problems. People do not believe their individual actions can make a difference; they cannot appreciate the collective effect of millions of small actions. A culture change needs to be initiated towards more public involvement in water resources planning and management where the starting point has to be the public understanding the need to conserve water. Such an awareness programme requires central and local government, water companies, regulators, NGOs all working together to deliver consistent messages. The relationship between company and customer will need to change markedly.

The WFD requires member states to encourage active involvement of all interested parties in its implementation, in particular of all interested parties in the production, review and updating of river basin management plans. This requirement could provide the catalyst for greater public involvement.

6.5 *Monitoring and managing the effects of competition*

The Government has stated (DEFRA, 2002) that the forthcoming Water Bill will include proposals for new entrants (to the market) to be licensed to compete with the existing water companies by using their own source of water (accessing the existing water company's network to transport the water to a customer, i.e., "common carriage") or by buying water from the statutory undertaker to supply a customer (i.e., providing retail services only). The Government is adopting a cautious approach by proposing that competition should be subject to industrial consumers using over 50 Ml/year, although this could be reduced in future if the Government considers that this would be in the consumer interest. Competition is likely to affect demand in two ways. Firstly the price will fall leading to a rise in demand (price elasticity has been estimated to be approximately -0.3% in the non-household sector) (Thackray, 1999). But in a competitive market it has been suggested that one way that new entrants would compete for customers would be by offering water conservation expertise as part of the arrangement. The Government believes that competition will make industrial consumers more aware of costs. It remains to be seen what the overall effect on demand will be.

7 THE FUTURE ROLE OF THE NATIONAL WATER DEMAND MANAGEMENT CENTRE

This paper has shown that initiatives to manage demand have been taken for a variety of reasons: to obtain best value for the water customer, to protect the environment, to make British industry more competitive, to save money in the public sector, to improve public image, to meet targets and to achieve sustainable development. Even in the same water company the activity is not always well co-ordinated with, for example, the regulatory department determining metering policy, the operations department being responsible for leakage control and the public relations department promoting water conservation. The National Water Demand Management Centre, by virtue of the fact that its focus is water demand management, is able to see how all these disparate programmes fit together, enabling it to see issues developing, research needs and to provide expert advice on the technical, social, economic aspects of demand management.

The National Water Demand Management Centre has been in existence since 1992. When demand management policies are fully integrated into water company water resources plans, all abstractors are using water in an efficient manner and the public are engaged in the decision-making process and water conservation has become second nature, then the Centre will no longer be needed. This would not appear to be likely in the foreseeable future.

The views expressed in this paper are those of the author and not necessarily those of the Environment Agency.

REFERENCES

DE. 1992. *Using water wisely*. Department of the Environment.
DE. 1996. *Water resources and supply: An agenda for action.* Department of the Environment.
DEFRA. 2002. *Extending opportunities for competition in the water industry in England and Wales.* Department of the Environment, Food and Rural Affairs July 2002.
DETR. 1999. *A better quality of life: A strategy for sustainable development for the UK.* Department of the Environment, Transport and the Regions.
DETR. 2001. *Environmental reporting guidelines for company reporting on water.* Department of the Environment, Transport and the Regions December 2001.
EA. 1997. *Saving water taking action.* Environment Agency.
EA. 1998. *Optimum use of water for industry and agriculture dependent on direct abstraction.* R&D Technical Report W157. Environment Agency.
EA. 2001a. *Water resources for the future: A strategy for England and Wales.* March 2001. Environment Agency.
EA. 2001b. *Water conservation fact cards.* Environment Agency.
EA-NWDMC. 1993–2002. *Demand Management Bulletin.* Issues 1–55. Environment Agency – National Water Demand Management Centre.
HCEC. 1996. *An inquiry into water conservation and supply.* House of Commons Environment Committee.
MAFF. 1999. *Irrigation water assessment and management plan.* Ministry of Agriculture, Fisheries and Food. February 1999.
NMTG. 1993. *Water metering trials – final report.* National Metering Trials Group. Water Services Association, Water Companies' Association, Office of Water Services, WRc and Department of the Environment.
NRA. 1992. *Board statement.* National Rivers Authority.
NRA. 1994. *Water – Nature's precious resource.* National Rivers Authority.
NRA. 1995. *Saving water: the NRA's approach to water conservation and demand management.* National Rivers Authority.
OFWAT (nd-a). *Leakage and the efficient use of water 2000–01 Report.* Office of Water Services.
OFWAT (nd-b). *The cost of water delivered and sewage collected 1991–92.* Office of Water Services.
Thackray, J. 1999. Incentives for large users: Are we getting them right? *Demand Management Bulletin.* No. 33. February 1999.
UKWIR. 1994. *Managing Leakage.* UK Water Industry Research. WRc publications.
UKWIR & EA. 1996. *The economics of demand management.* UKWIR and Environment Agency.
WIA. 1991. *Water Industry Act.* Section 93A amended by the Environment Act 1995.
Weizsacker, E. Lovins, A. & Lovins, L. 1998. *Factor four doubling wealth, halving resource use.* Earthscan Publications.
WRI (nd). website http://www.wri.org. World Resources Institute.

Sustainable water management in agriculture

E. Fereres & D. Connor
Instituto de Agricultura Sostenible, Universidad de Córdoa, Spain

ABSTRACT: There is general agreement that population increase, coupled with economic growth and mounting awareness of the need for nature preservation, are now subjecting existing freshwater resources to considerable pressures. Looking at current and future trends of agricultural production worldwide, it is clear that irrigation is essential to meet current food demand and that irrigation will have to increase in the future. The use of large amounts of water by crops is dictated by the evaporative demand of the environment and is tightly associated with biomass production and yield. Increases in potential biomass production per unit water transpired during the next 10–20 years will be small despite promises from biotechnology. Breeding for yield potential will also provide a small contribution in the near future, so the real opportunity to increase the productivity of water resides in closing the current large gap between actual and potential yields. To increase the effectiveness of irrigation in a sustained fashion, without detrimental impacts on the environment, a number of measures need to be considered, particularly: improving water management at farm and district levels and reforming institutions to allow for user participation.

1 INTRODUCTION: FEATURES OF THE HYDROLOGIC CYCLE IN AGRICULTURAL SYSTEMS

The urban, industrial, agricultural, and environmental sectors all use water diverted from surface and groundwater sources and storage facilities. Agriculture is the primary user of diverted water; two thirds of all diversions are used for irrigation worldwide. Rainfall in excess of that infiltrating the soil runs off to watercourses and eventually, to the oceans. On land, water evaporates from plant and soil surfaces, driven by solar radiation. Evaporated water condenses to fall as rain elsewhere, closing the hydrologic cycle.

In any watershed, incoming water in the form of rain or irrigation must be balanced by water lost as evaporation from soils (E) and transpiration (T) from plants (termed evapotranspiration, ET), runoff, deep percolation and stored in the soil. It is possible to quantify a water balance at many scales, from an individual field, to a farm, a hydrologic basin or a region, up to the global scale.

Water that evaporates from a watershed is considered a loss or consumption, while water running off can be recovered downstream and may not be lost to the system. Thus, there are consumptive and non-consumptive uses of water. Water applied as irrigation may be used consumptively in the ET process, while the network and runoff losses may be recovered downstream and used by others. Water used within a basin is not always consumed in that basin and can be used several times before it leaves the basin. Thus, water conservation efforts may or may not lead to net water savings depending on whether the water saved is part of the recoverable or the unrecoverable losses.

Every time water is used, its solute load increases, as does also the chance of it picking up contaminants; the result is a deterioration of water quality. Such deterioration has many diverse impaction everything from human health to ecosystem services, and it directly affects the availability of water supply.

As discussed below, crop plants loose large amounts of water through ET, keeping less than 1% of what they transport from the soil to the atmosphere during their life cycle. Crop consumptive use is met by soil water stored from rainfall and/or irrigation. When soil water supply is insufficient to meet the evaporative demand, crops undergo water stress and their production levels are usually reduced. Irrigation is aimed at avoiding water stress during periods of insufficient rainfall, even though often the irrigation supply is insufficient to fully meet the crop demand.

Effective water management in both rain fed and irrigated agriculture has very similar goals at the field and farm scales: maximizing the use of stored soil water and minimizing losses to runoff and percolation. There is no reason to isolate rain fed from irrigated agriculture, since in many agricultural systems a continuum exists from rain fed to limited to full irrigation supply.

Irrigated agriculture is often seen as a water source for alternative uses, given the large proportion of the diverted water that is used for irrigation, and it could play an important role in mitigating water scarcity if integrative, coordinated approaches to water management are pursued at the basin and regional levels. Lack of control of runoff and percolation losses in agriculture may lead to a major source of non-point pollution that negatively affects the environment. While some losses may be unavoidable, effective water management is the primary tool to minimize pollution from agricultural systems.

From the standpoint of food production, there is significant uncertainty in our capability to meet world food demand for the next two decades and even less confidence in our ability to achieve the goal of eradicating or reducing the hunger that presently affects almost 800 million people worldwide. It is in light of these uncertainties that the use of water in agriculture becomes a matter of primary importance. This work focuses on water management in irrigated agriculture, because food production will be increasingly dependent on irrigation in the future. Irrigation is a multifaceted, complex activity with deep cultural roots and with many ramifications in rural societies; thus, this chapter delineates the complexities associated with the practice of irrigation and its improvement and evaluates the opportunities and constraints that exist to conserve water in irrigated agriculture in the future.

2   THE CHALLENGE OF PRODUCING SUFFICIENT FOOD

The most remarkable and most ignored fact of world population growth in recent decades is the concomitant increase in food production that has made such growth possible. Several population studies indicate that by the year 2025, world population, currently at six billion, will approach eight billion. Will there be sufficient food for all? The challenge that we face is enormous, and too often ignored by affluent societies.

To evaluate the prospects of meeting this challenge, it is instructive to study how the last doubling of world population – from three to six billion in just three decades – has been fed. Increasing agricultural production requires more planted area and/or improvement in production per unit area. Since the 60s, the area devoted to crop production worldwide has hardly changed from a figure of about 1.5 billion ha. Thus, the spectacular growth in food production of recent decades has been brought about almost exclusively by improvements in agricultural productivity. Science and technology developed new crop varieties that could take full advantage of available climate and improved management methods, particularly of increased soil fertilization with nitrogen. Crop yield increases of the major staples have been truly remarkable both at the potential (theoretical maximum for a given location), and actual (average) levels. Today, potential grain yields of wheat or rice vary between 10–15 t/ha while those of maize may exceed 20 t/ha. While increases in yield potential have been important, closing the gap between actual and potential yield has had even more influence in the improvement of agricultural productivity.

In areas of favorable soils and climate and with access to capital (human and economic) to employ new technologies, the gap between actual and potential yield has shrunk to the point that record yields obtained by some farmers in recent years are essentially at potential levels (Evans & Fischer, 1999). On the contrary, the gap between actual and potential yields in areas less endowed for agricultural production (physically and socially) continues to be substantial.

Among the environmental factors adversely contributing to the reduction in crop yield, insufficient water is probably the most important worldwide. Water-limited crop yields are often a small fraction of what can be produced when water supply is sufficient. It is not surprising, therefore, that since the beginning of agriculture, man has attempted to remove the constraints of water deficits, either by using water conservation strategies and tactics in rain fed agriculture (Loomis & Connor, 1992) or by developing additional water supplies to irrigate crops. Although irrigation has been practiced since ancient times, it has been during the last four decades when it has really contributed to feeding the world. Although the present 260 Mha of irrigated land represents only 17% of the world cropped area, it contributes more than one third of world food production. Thus, irrigation has played a very important role in the intensification of agricultural production that has been so important in successfully matching the population expansion of recent decades.

Looking at food production in the future, it appears that irrigation will have to play a role even more important than it does today. Several studies have forecasted world food supply and demand 20 to 25 years from now. The 2020 vision of IFPRI (IFPRI, 1999) predicts that demand for cereals will increase by 40% above present levels, and possibly more if there are improvements in diet. To meet such demand, IFPRI (1999) predicted a modest 6%, increase in cultivated area but a substantial increase in productivity. The same analysis predicted that either irrigated areas must increase by about 40 Mha (over the present 260 Mha) or that the areas currently irrigated would have to increase their productivity by over 20% in 20 years. Given that most of the expansion in irrigated lands is required in the developing countries (IFPRI, 1999) and considering the physical and economic limitations to irrigation expansion, together with the losses of irrigated lands to urbanization and salinization, it is inevitable that continuous improvement of productivity in irrigated agriculture is the major avenue for meeting the increased food demand in the future. There are, however, severe constraints for further productivity (i.e., yield) gains in many irrigated areas that are discussed below. The large productivity gains observed in the 60s and 70s have given way to more modest increases in the 80s and 90s (Table 1) to the point that many authors have speculated that world crop yields are approaching a plateau. There are a number of reasons for the slowdown in productivity growth shown in Table 1, but dramatic changes in productivity from present trends are improbable in the short run, despite the promises made by the biotechnology advocates.

Recent trends in irrigation management in many world areas include a shift towards intensification of crop production in irrigated systems. Such intensification has also played a major role in increasing crop productivity but has often associated increases in water use. The rice-wheat system of the Indo Gangetic Plains is a good example (Timsina & Connor, 2001). Irrigation, together with short season, improved cultivars of both crops, allow them to be grown in sequence in areas that were previously limited to a single rice or wheat crop per year. In the most favorable parts of the rice-wheat system, there is also the possibility to grow a third, of legume crop, after wheat and before rice. In Bangladesh where this and many other intensive systems are practiced to meet a burgeoning population, average cropping intensity is now 2.5 (two and half crops per year).

Table 1. Annual growth (%/year) in cereal crop area, total production and yield, production per unit area, 1967–1994 (IFPRI, 1999).

All cereals	Peak green revolution 1967–1982			Post – green revolution 1982–1994		
	Area	Production	Yield	Area	Production	Yield
Developed countries	0.23	1.92	1.69	−1.27	0.01	1.30
Developing countries	0.48	3.36	2.87	0.46	2.34	1.87
World	0.37	2.61	2.24	−0.24	1.27	1.51

The off-season production of vegetable crops under plastic in Almería (Spain) started with one to two crops per year in the autumn to early spring period (Castilla & Fereres, 1990) but is now expanding to cropping intensively for 8–10 months a year, with July and August as the only months when the majority of the greenhouses are renewed. Such expansion has increased considerably the agricultural water requirements of the area, as the evaporative demand of late spring and early summer is much higher than that of the winter season, despite the use of white paint over plastic greenhouses after April (Fernández et al., 2001). Modernization of irrigation schemes also provide incentives for intensification: one example in the Ebro Valley of Spain (Zapata & Ederra, 2002) shows that after land consolidation there was an increase in the area devoted to horticultural crops, and even though efficiency of irrigation increased from 57% to 75%, the water requirements were estimated to be about 5% higher than before the modernization. Nevertheless, river diversions were reduced by 30% by the efficiency improvement, even though consumptive use was actually increased (Zapata & Ederra, 2002).

The examples cited above suggest that as we increase crop productivity by intensification, it will not be reasonable to expect, at the same time, that agricultural water use will be reduced. Rather, one would expect that more intensive systems would have greater irrigation demands but that the crop productivity gains would largely exceed the increase in water consumption, leading to higher water productivity.

3 THE PROCESS OF IRRIGATION: NATURE AND CONSTRAINTS

Irrigation is by far the largest consumer of developed fresh water on a global basis. Seckler et al. (1998) estimated that irrigated agriculture currently uses about 72% of per capita water diversions while industry and urban diversions amount to 19% and 7%, respectively. Such a high level of water use by agriculture raises two frequent key questions: Why does irrigation demand so much water, and could irrigation demand be reduced? Much of this paper focuses on answering those two questions.

The analysis of the processes involved in determining irrigation water use is divided into two parts; first considered are the processes starting at the level of the crop up to the individual farm. Afterwards, an integrative analysis at the district and watershed levels is presented.

3.1 Irrigation: from the crop to the farm

Crop plants require a continuous supply of water to replace the water evaporated (transpired) from their aerial organs (mainly leaves). Such demand arises because the leaves are exposed to strong evaporative demand (strong fluxes of solar and thermal radiation and warm, dry air) while the internal plant surfaces are nearly saturated with water. For carbon dioxide (the substrate for photosynthesis) to enter the leaves, the microscopic leaf pores (stomata) must be open. But when the pores are open, water vapor freely escapes from the interior of the leaves. To sustain the water flow needed to prevent tissue dehydration, plants have evolved elaborate water gathering and transport systems. Water flows from the soil into the root system and is transported via conductive xylem vessels to the leaves where it replaces that evaporated to the atmosphere. Thus, from a purely physical viewpoint, plants are water transport systems from a source, the soil, to a sink, the atmosphere. Such systems are capable of transporting large amounts of water, equivalent, in a typical summer day, to depths of 6 to 8 mm over the land. This is generally several times the plant weight. However, a small imbalance in the transport process – that is hardly detectable – may occur in response to alterations in water supply or demand, thus creating a plant water deficit. Such very mild water deficits are often detrimental to yield, even though large amounts of water are being transported by the plants at the same time (Hsiao et al., 1976).

3.1.1 Transpiration and crop yield
The seasonal evaporation losses from crop and soil surfaces are quite large relative to the amount of yield produced. It is not uncommon to require from 100 to more than 1000 kg of water per kg of harvested crop yield. As good conditions for high productivity (high solar radiation and favorable

temperatures) are also conducive to high rates of evaporation, high yields and high water consumption are unavoidably linked. It is useful to define here indicators of the productivity of water in agriculture, which is the ratio between production and water used. There are several definitions of water productivity; water use efficiency (WUE) is defined as the ratio between production and evapotranspiration (ET), the sum of evaporation from the soil (E) and transpiration (T) from crop surfaces. For a maize crop that produces 24 t/ha of above-ground biomass or dry matter and 12 t/ha of grain yield and has a ET of 660 mm, WUE is 36.4 kg/(ha mm) for dry matter and 18.2 kg/(ha mm) for grain yield, respectively. Another measure of water productivity is transpiration efficiency (TE), defined as the ratio between production and transpiration. In the above example, assuming that E is 30% of ET, TE would be 51.9 kg/(ha mm) for dry matter and 26.0 kg/(ha mm) for grain yield, respectively.

As a first approximation, dry matter production and ET of individual crops are linearly related. This fact has led to the development and use of water production functions (e.g., Vaux & Pruitt, 1983) to predict yield as a function of water use. The empirical nature of these relationships has limited their transferability from place to place, even though they are often treated as if they were universal truths. Among most engineers and economists that use water production functions, there is little awareness of limitations of the functions and of the processes that actually determine water-limited crop productivity under field conditions.

3.1.2 *The water balance of a field*

Crop water demand is met in most agricultural areas of the world by rainfall. Irrigation is applied in areas where natural rainfall is insufficient to meet the crop water requirements during the growing season. The disposition of water over an irrigated field during and after irrigation is displayed in Figure 1. Water is lost in four ways: (1) direct evaporation from the soil surface; (2) evaporation from plant surfaces, (transpiration); (3) surface run-off; and (4) deep percolation below the crop root zone.

A water balance equation may be derived from Figure 1 stating that the water entering an irrigated field either changes the root zone water content or must leave the field.

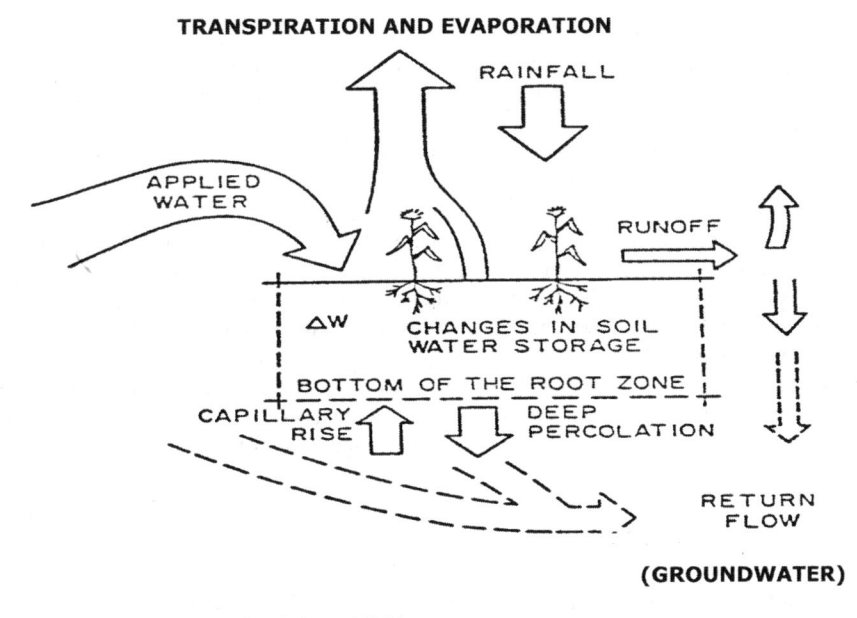

Figure 1. The water balance of an irrigated field.

Mathematically,

$$A + P = ET + R + D + W \tag{1}$$

where

A = applied irrigation water
P = precipitation
ET = evaporation + transpiration
R = run-off
D = drainage below the root zone (deep percolation; it may be negative if capillary rise occurs)
W = changes in soil water content in the crop root zone

Equation 1 states the water balance of a field from as seen by a hydrologist. From the farmer's viewpoint, the irrigation water requirement (I_R) is defined as:

$$I_R = ET - P_e + I_L \tag{2}$$

Where P_e is effective precipitation and I_L, are the irrigation losses resulting from the combination of R and D, which are largely unavoidable during application of water to land. Effective precipitation is defined as the fraction of the seasonal rainfall that contributes to the crop water requirements.

All irrigation waters contain salts. Evaporation from the soil and transpiration remove pure water and leave the salts behind. Those salts become increasingly concentrated with time and unless leached from the root zone by deep percolation crop productivity cannot be sustained under irrigated conditions due to the increased salinity. The fraction of deep percolation needed to displace the salts that accumulate in irrigation below the root zone is called the leaching requirement and must be included as part of the irrigation water requirement. This need for salinity control is of particular importance in areas where annual rainfall is insufficient to achieve a salt balance through leaching of the soil profile.

It is often considered that irrigation losses at the field scale should include not only surface run-off and deep percolation, but also spray evaporation losses in the case of sprinkler irrigation. Contrary to popular belief, much of the spray evaporation reduces evaporative demand and thus should not be considered a loss. In any event, spray evaporation is generally a small fraction of a sprinkler application and can be ignored under most conditions.

3.1.3 *Irrigation efficiency and uniformity*

When considering the partitioning of water applied to a field into the various fluxes indicated in Figure 1, it is important to quantify the proportion of irrigation water that is consumptively used by the crop and the proportion that is nonconsumptive use. Water used consumptively either leaves the basin or becomes inaccessible for further use; for instance, water lost in evaporation from the soil surface and transpiration from the foliage is considered consumed, while runoff and deep percolation are nonconsumptive uses of water. Irrigationists use the term *irrigation efficiency* to describe the fraction of applied water that is consumed by a crop. An additional concern relates to the spatial application of water within the field. Uniform distribution is desirable to avoid under-irrigation of some parts and overirrigation of others. Several parameters to quantify distribution uniformity have been proposed over the years based on measures of minimum and average depths of application of water over the field.

Confusion has frequently arisen over the terminology and definitions of irrigation efficiency and uniformity leading to recent efforts at clarifying the processes involved and the terminology used (Burt et al., 1997; Jensen, 1996). The current definition of irrigation efficiency used by water engineers is that of Burt et al. (1997):

$$IE = \frac{\text{Volume of irrigation benefical used}}{\text{Volume of irrigation water applied} - \Delta\ \text{storage}} (100\%) \tag{3}$$

where Δ *storage* is the change in stored water in the root zone of the field considered, and water is used beneficially in processes that contribute to the objectives pursued under good management.

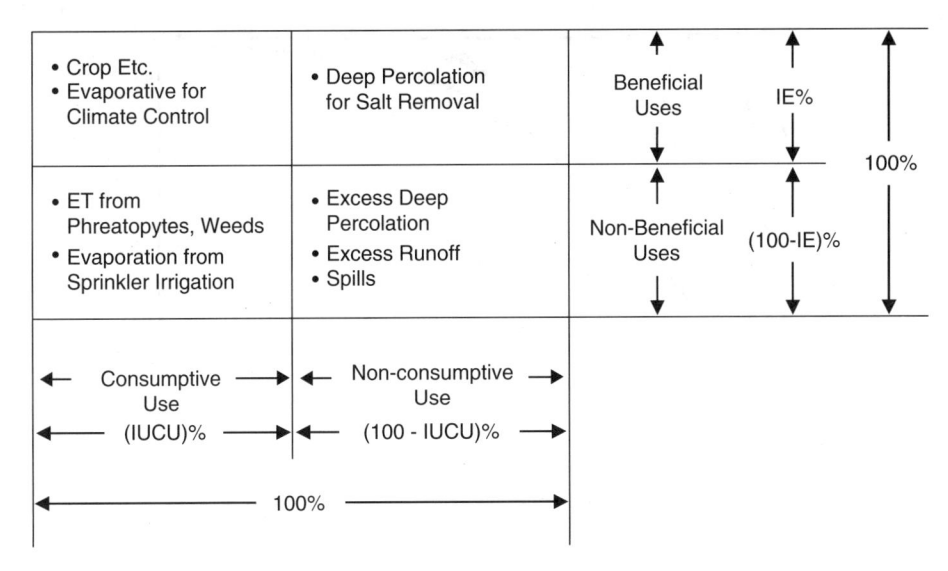

Crop Etc. • Evaporative for Climate Control	• Deep Percolation for Salt Removal	Beneficial Uses	IE%	
• ET from Phreatopytes, Weeds • Evaporation from Sprinkler Irrigation	• Excess Deep Percolation • Excess Runoff • Spills	Non-Beneficial Uses	(100-IE)%	100%
← Consumptive Use (IUCU)% →	← Non-consumptive Use (100 - IUCU)% →			
← 100% →				

Figure 2. The division between consumptive and non-consumptive uses is distinct from the division between beneficial and non-beneficial uses (IE = Irrigation Efficiency; IUCU = Irrigation Consumptive Use Coefficient defined as the percentage of the total water supply that is used consumptively).

Figure 2 taken from Burt et al. (1997) shows that there are differences between the water fractions that are beneficially and/or consumptively used in a field or a farm.

The other parameter indicative of the quality of irrigation is the distribution uniformity (DU), usually defined as shown by Equation 4.

$$DU = \frac{\text{Average depth in the quarter area that receives the lowest depth}}{\text{Average depth of water over the whole area}} (100\%) \quad (4)$$

The Christiansen Uniformity Coefficient (CUC), commonly used in sprinkler irrigation, is another measure of DU. It is defined as shown by Equation 5.

$$CUC = \left(1 - \frac{\sum |X - X_a|}{\sum X}\right)(100\%) \quad (5)$$

Where X is the depth of water in equally spaced catch cans in a grid, X_a is the average depth of water in all cans and the sign Σ indicates that all measured depths and deviations from the mean are summed. Performance evaluation of irrigation at the field or farm level is based in the quantitative assessment on the indicators described above.

3.1.4 *Relations between irrigation and crop yield*
While the relationship between transpiration and crop yield is linear, that between applied irrigation water and crop yield is usually curvilinear because the magnitude of deep percolation losses and evaporation from the soil surface tend to increase with the depth of irrigation water. The example shown in Figure 3 illustrates the relations between applied irrigation, distribution uniformity and yield. The two curves depict the relations between applied irrigation water and yield of maize for two levels of distribution uniformity (quantified as 70% and 90% CUC) for a typical situation in a semiarid area where 40% of crop ET is supplied by rainfall or soil water storage (Fereres et al., 1993). In the case of the irrigation depth required to achieve maximum yield at 70% CUC, applied irrigation is about twice the net irrigation requirement, leading to a very low irrigation efficiency of about 40%.

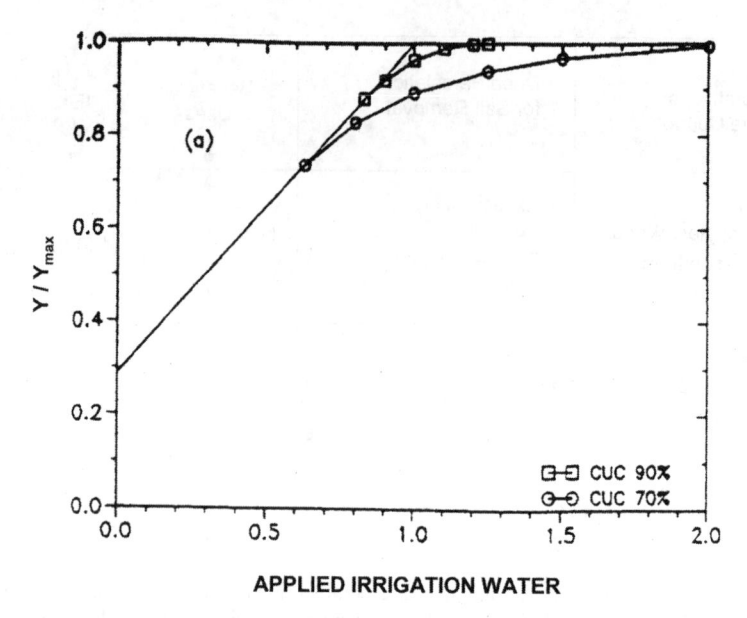

Figure 3. Relationship between applied irrigation water (relative to the net irrigation water requirements, NIR, to meet the maximum ET demand; values over 1.0 represent excess water applied which is needed to meet the NIR in the areas that receive less water, depending on the CUC value) and relative yield (Y/Y_{max}) of maize at Córdoba, Spain, under two levels of irrigation uniformity (Fereres et al., 1993).

The curvilinear nature of these relationships allows for economic analyses to identify the economically optimum irrigation depth. Such analyses, however, require simulation models that must be validated for each site to provide realistic estimates (Fereres et al., 1993).

4 IRRIGATION: FROM THE FARM TO THE WATERSHED

Significant improvement in the understanding of interactions between on-farm irrigation and basin hydrology has been achieved in recent decades. This has led to increased recognition that irrigation losses from the farm are not necessarily losses from the basin. The reason is that losses from runoff and deep percolation of one farm enter the general drainage network of the basin and may be recovered as part of the water supply of another farm downstream. Thus, irrigation losses are at least partly recoverable within a basin, therefore if the objective is water conservation, strategies for improving the efficiency of water use in irrigation must focus first on the nature of the potential water savings. Measures that reduce irrecoverable water losses represent a net reduction in the consumptive use in the basin. On the other hand, techniques applied to decrease recoverable losses will not reduce basin water demands and may even alter the water supply of downstream users. To illustrate the importance of these differences it will suffice here to describe results of a study on irrigation water use in the Western USA published about 20 years ago (Interagency Task Force Report, 1979). The study evaluated a number of water conservation and improved irrigation efficiency measures, on and off-farm, in terms of the potential reduction in irrigation water demands. While the estimated reduction of diversions for irrigation exceeded 30 million acre-feet, most of that water was reused and the reductions in irrecoverable losses, representing the estimated net water savings (of additional supplies) were only between two and five million acre-feet.

It is surprising that, even though the idea and the practice of water reuse are obvious, they are not always included in the analyses of water supply for irrigation at the watershed or regional levels. It is often assumed that once water is withdrawn, it is lost to further use.

The evaluation of water use of the various sectors could be very misleading in many cases if based on diversions without consideration of water reuse. For instance, the estimate that 72% of water use goes to irrigation (Seckler, 1993) probably overestimates irrigation water use because much water diverted for irrigation ends up being reused for other purposes within a basin. Seckler et al. (1998) pointed out that most international data sets on the water supply of countries ignore water-recycling effects. For instance, the national hydrologic studies in Spain use a single figure of 20% of diversions as the average reuse in irrigation for the whole country. The truth is that tracing and quantifying recoverable water within and between basins is a difficult task and that the quantity and quality of hydrologic data in most cases are not adequate for that purpose. Therefore, renewed efforts are needed to obtain reliable information for quantifying reuse in irrigation at basin and regional levels. This is essential to determine the net water savings that may be affected by conservation measures.

5 RECYCLING IRRIGATION WATER: IMPACTS ON THE ENVIRONMENT

The positive aspects of water reuse must be balanced against the deterioration of water quality and the effects that such degradation, which occurs during reuse, have on the environment. Figure 4 illustrates the same water balance of Figure 1 but with the addition of the processes that contribute to the degradation of the water quality in return flows from irrigated lands. Sediments, nitrate ions, pesticides and other materials carried from croplands by water are all detrimental to other water users, and threaten the sustainability of irrigated agriculture and other connected ecosystems. The control of water pollution is of paramount importance to extend the use of a given water diversion as far as possible.

Conservation efforts to increase the efficiency of water use in irrigation must pay attention to non-point source pollution as well as to the distinction between conserving recoverable and irrecoverable water losses in irrigation. There have been too many cases of irrigation-related environmental problems that could have been avoided or minimized had due attention been paid to conserving water.

Tanji (1990) reviewed a number of cases where irrigation return flows had detrimental impacts on the environment. The case of the Kesterson ponds in the San Joaquin Valley of California is notable,

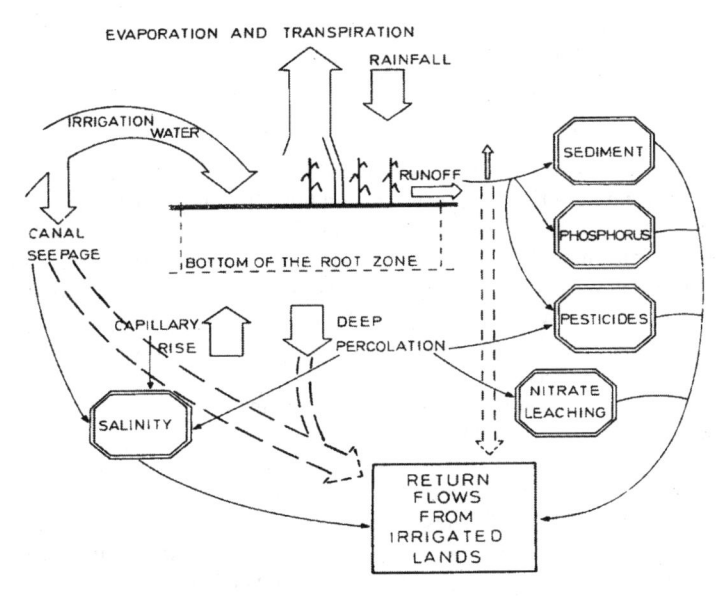

Figure 4. The water balance of an irrigated field showing the processes that contributes to the degradation of irrigation return flows.

where, the drainage flows from irrigated lands carried selenium, which poisoned fauna using the ponds. The response of regulatory agencies was directed at eliminating the drainage outflow from the area contaminated with Se, about 17,000 ha. By mid 1986, all drains from that area (which receives less than 200 mm of annual rainfall) had been plugged, and since then irrigation has been carried out without the evacuation of salts. To maintain the salt content of root zone below the tolerance thresholds for various crops, farmers in that area must take measures if they wish their irrigated agriculture to remain permanent. Precision water management and scheduling (applying the amount needed at the right time) and the use of irrigation systems with very high uniformity are some of the practices that farmers in that area have taken to manage the salts in such critical situation. Nevertheless, it should be understood that irrigation cannot be sustainable in many areas of the world, unless adequate drainage is provided and safe disposal of the salt load is achieved. Gardner (1993) and van Shilfgaarde (1994) have clearly stated the changes and risks that irrigated agriculture is facing that threaten its sustainability. The US National Research Council (1996) explored the future of irrigation in the USA and concluded that there will not be dramatic changes in the future. That study and others highlighted some of the benefits of irrigation to society, often widely ignored, and suggested future directions for irrigated agriculture in North America. It would be desirable for other countries to undertake similar analyses considering their site-specific economic and cultural conditions.

6 REDUCING AGRICULTURAL WATER USE: MYTHS AND REALITIES

Because irrigated agriculture is the largest user of developed water, it is important in looking at the future, to assess the potential that exists for reducing irrigation water demand worldwide.

If the primary goal of irrigation is food production, increasing crop yield per unit irrigation water is an important objective. To increase the productivity of irrigation water it is necessary to:

– Increase biomass production per unit water transpired.
– Increase harvestable yields per unit water transpired.
– Reduce evaporation from soil.
– Reduce deep percolation and run-off losses to sinks.

We now consider briefly the potential that each of these four options have for increase by combining advances, not only from science and technology but also from the improvement of water management and from institutional reforms.

6.1 *Improving the water productivity of crops. The potential of biotechnology*

Since the beginning of this century, data has been collected on the amount of biomass produced per unit of water transpired, defined above as transpiration efficiency, (TE). All scientific evidence points to the principle that for a given crop species, TE depends primarily on the evaporative demand of the location where the crop is grown (Tanner & Sinclair, 1983). There are differences in TE among species; in warm climates, those with a C-4 photosynthetic pathway (such as maize) having higher TE values than wheat or rice, which have C-3 pathways. The effects may be reversed in cold climates. Until now, attempts to alter the photosynthetic pathways of crop plants have failed, including recent attempts using genetic engineering.

The one factor that can be modified is the evaporative demand (defined as reference ET or ET_o) by growing the crops at times when ET_o is low (i.e., winter). Under low ET_o, biomass production per unit water loss increases relative to the situations of high ET_o. There have been some gains in TE already by shifting summer crops to spring or fall, but there is a limit to improving TE by changing the growing season as in many climates, because low ET_o is associated with low solar radiation and low temperatures, which slow down or can even prevent the growth of summer crops. Breeding for tolerance to low temperature is therefore an important objective for the improvement of TE.

Growing off-season crops can increase TE by at least two-fold in some cases. A comparison of WUE of tomatoes grown outdoors in the summer of the Central Valley of California (Pruitt et al.,

1984) versus winter tomatoes grown in greenhouses at a similar latitude in Almería, Spain (Castilla & Fereres, 1990) reveals values from 15 kg/(ha mm) in the summer to about 55 kg/(ha mm) for the winter greenhouse crop. This is a large improvement in WUE, achieved by displacing the season and by the use of plastic, unheated greenhouses. Obviously, there are economic and marketing constraints limiting greenhouse expansion in irrigated agriculture.

While TE is a conservative attribute of a crop, the amount of harvestable yield per unit water transpired (WUE_y) could be increased. Indeed WUE_y has been raised dramatically by plant breeding during the Green Revolution. Is it possible to raise WUE much further? That would require increasing the harvest index (the proportion of biomass that is harvested as yield), which is already approaching theoretical limits in many improved varieties.

In spite of the promises from biotechnology, it is unlikely that drastic improvements in raising potential yield or harvest index of the major crops will be realized in the next 10 to 20 years. This is because crop yield depends on many genes and on their interactions with the environment. In addition, plants have already been subject to billions (in the case of photosynthesis) to thousands (in the case of harvestable yields) of years of selection. At present, biotechnology has addressed successfully crop improvement objectives by working with traits that depend on one or a few genes, selecting for some defect modifications and for resistance to pests, herbicides, and diseases. Breeding for drought resistance using current and future technologies as they become available, however, can produce no more than limited, incremental improvements of crop yield under limited water supply in the next one to two decades.

The major opportunities for improving WP, and that demand priority effort, reside within the management component. Inadequate management is the primary cause of the low WP that today characterizes rain fed and irrigated systems. There are also challenges in the biophysical area related to maximizing WP under limited or deficit irrigation. It is very likely that many irrigated areas will not have full supplies in the future and will be forced to use limited supplies in an optimal fashion. Actually, some large, existing irrigated areas exist were under-designed for political reasons and will continue to suffer chronic restrictions in water supply. Research at optimizing a limited amount of water has been carried out in the past, but the new tools of spatial analysis and simulation modelling have much to add to the development of effective tools for advising irrigators in optimal scheduling methods. One other major challenge in rain fed but also, irrigated systems, is the need to maximize the potential contribution of soil-water storage. Water conservation measures that increase the fraction of rainfall that is used as transpiration need to be developed and tailored to each major system. While many of the opportunities discussed above have been partly researched, a major gap exists in most agricultural systems between what is known to increase WP and what is actually applied. This is primarily because there has been little or no involvement of social scientists in the research and extension efforts.

6.2 *Reducing evaporation from the soil*

When soils are wet, evaporation (E) rates are normally high and related to the solar radiation incident on the surface. However, as soon as the soil surface dries in one or two days, E rates decrease markedly due to limitations to soil water transfer to the surface. The amount of radiant energy incident to the soil surface is the key factor governing loss by E. Thus, E rates may be high when crop cover is small, as during the initial crop growth stages, and when soil is frequently wetted by rain or irrigation. On the contrary, in situations where the crop canopy fully shades the ground, E is low, perhaps 10% of ET or less. Averaged over a season, E may be a substantial component of ET, in particular in rain fed crops where growth is sparse and the soil surface is exposed for much of the season. However, the rapid growth rates under irrigation reduce seasonal E values for field crops to a largely unavoidable proportion of 20%–35% of ET.

One of the perceived benefits of drip irrigation is the reduction in soil evaporation relative to other irrigation methods that wet the whole ground surface. It is true that important water savings, via reduced E, can be accomplished using drip irrigation in young orchards, but the method is not feasible for irrigation of the major field crops. Even if it were economical to use drip, E savings in row crops would be small or negligible, as shown in a study that compared drip and furrow irrigation in processing tomatoes (Pruitt et al., 1984). A recent assessment of the potential E savings by changing from surface to subsurface drip in an olive orchard, based on Bonachela et al. (1999), gave an estimate of less than 7% of the seasonal depth of irrigation. Such savings could be important in situations of water scarcity but cannot justify the change in irrigation systems in most cases.

6.3 Reduction of irrigation system losses

Science and technology have provided over the years a wide range of irrigation method and techniques that can improve irrigation effectiveness and reduce losses to runoff and deep percolation. Recent developments on surface irrigation optimization are providing guidelines for design of surface systems where losses are kept to a minimum. If soil variability and topography prevent the use of surface systems, a variety of sprinkler and drip systems are available that offer high potential efficiency and uniformity. Thus, a wide variety of technical solutions are available to solve most irrigation problems at the individual field scale. But, are water losses at the farm and district levels being reduced?

The problems at the system level are of a different nature. To optimize the design and operation of existing surface irrigation systems may require changing the water delivery procedures and flows, land consolidation and changes in the design and operation of delivery networks. Such changes not only require capital investments but agreements among various users as well. The introduction of pressurized systems (sprinkler and drip) in collective irrigation networks also require other changes in physical infrastructure (e.g., reservoirs). Thus, there must at least be economic incentives and access to capital in order to effect the changes needed to increase the potential efficiency and uniformity at the system level.

Before introducing measures to reduce irrigation system losses, knowledge of the potential net water savings is required. Here, there have been important scientific and technological limitations to the quantification of water sources and sinks at the district and basin levels. New technological developments based on the use of modeling, remote sensing and GIS are allowing the assessment of irrigation return flows and the impacts of various on-farm conservation measures on the net water savings and the environmental impacts at the basin level.

7 THE ROLE OF MANAGEMENT AND INSTITUTIONS

We have seen that technologies are available that could increase irrigation effectiveness worldwide. To realize such potential, both improved water management and institutional reforms are required.

Optimization of on-farm water management may be achieved by using irrigation scheduling (IS) techniques that determine the date and amount of irrigation (Fereres, 1996). For the most part, IS techniques have already been developed, although new refinements are possible in the future (Fereres et al., 1999; Goldhamer et al., 1999). Such techniques have not been widely adopted by farmers in the developed countries due to lack of economic incentives and/or appropriate regulations. There are indications that increased adoption of IS technologies in the more advanced irrigated areas is taking place (Fereres, 1996). In the developing countries, however, where irrigation networks have often been designed to meet only a fraction of the crop demand, the proposal of using IS techniques is not realistic. In those cases, the starting point there would be to collect basic soil and climatic data that can be used to establish seasonal requirements for planning purposes at the district or watershed levels.

Institutional reforms are advocated for more effective management of water resources in agriculture, including the introduction of market forces to deal with water as an economic good (e.g., Briscoe, 1997). In many world areas, there are two prerequisites to treating water as an economic good so that market forces can determine the allocation of water to agriculture. One is the guarantee of water deliveries to farmers and the other is guaranteed and defined water rights. Both conditions are often lacking in developing countries where irrigation entitlements are often unclear. In the developed countries, however, the introduction of market forces in the management of water in irrigated agriculture is a desirable development for the increase of irrigation effectiveness, particularly in situations of water scarcity.

Seckler et al. (1998) have analyzed world irrigation demand for the year 2025. Although the study can only be a rough approximation to the real world because the quality of the data is questionable, it is a refreshing attempt to obtain a global view of this important problem. One conclusion of the study is that about half of the projected increase in demand from 1990 to 2025 could be provided by increased irrigation effectiveness via conservation.

8 CONCLUSION

In conclusion, given the increased water demands worldwide by other users that have priority over irrigated agriculture and because significant improvements in the biological efficiency of the crop water use process are not possible, at least in the short run, and, it is evident that conservation of irrigation water is the most important new source of water for agriculture now and in the future. As we have seen in this paper, technologies already exist to conserve irrigation water; however, technologies alone will not provide permanent solutions to the many problems of irrigated agriculture. The path towards a more sustainable irrigated agriculture requires multidisciplinary approaches to problem solving and user participation at all levels.

REFERENCES

Bonachela, S., Orgaz, F., Villalobos, F.J. & Fereres, E. 1999. Measurement and simulation of evaporation from soil in olive orchards. *Irrigation Science*. 18(4):205–211.

Briscoe, J. 1997. Managing Water as an Economic Good. pp. 339–361. *Water: Economics, Management and Demand*. Kay, M., Franks, T. & Smith, L. (eds). Chapman and Hall, London.CAST. 1996. Future of Irrigated Agriculture.Task Force Report 1997. CAST, Ames, IA, USA.

Burt, C.M., Clemmens, A.J., Strelkoff, T.S., Solomon, K.H., Bliesner, R.D., Howell, T.A. & Eisenhauer, D.E. 1997. Irrigation performance measures: efficiency and uniformity. *J. Irrig. Drain. Div.*, ASCE. 123(6):423–442.

Castilla, N. & Fereres, E. 1990. *The Climate and Water Requirements of Tomatoes in Unheated Plastic Greenhouses*. Agric. Mediterran. (Pisa) 120:1–7.

Evans, L.T. & Fischer, R.A. 1999. Yield Potential: Its Definition, Measurement, and Significance. *Crop Sci.* 39:1544–1551.

Fereres, E., Orgaz, F. & Villalobos, F.J. 1993. Water use efficiency in sustainable Agricultural systems. *International Crop Science I*. Buxton, D.R., Shibles, R., Forsberg, R.A., Blad, B.L., Asay, K.H., Paulsen, G.M. & Wilson, R.F. (eds). Crop Science Society of America Inc. pp. 83-89. Madison, Wisconsin, USA.

Fereres, E. 1996. Irrigation Scheduling and Its Impact on the 21st Century. Keynote in *International Conference on Evapotranspiration and Irrigation Scheduling*. San Antonio, Texas. ASAE. pp. 547–553.

Fereres, E., Goldhamer, D., Cohen, M., Girona, J. & Mata, M. 1999. Continuous Trunk Diameter Recording is a Sensitive Indicator for Irrigation of Peach Trees. *California Agriculture*. 53(4):21–25.

Fernández, M.D., Orgaz, F., Fereres, E., López, J.C., Céspedes, A., Pérez, J., Bonachela, S. & Gallardo, M. 2001. *Programación del riego de cultivos hortícolas bajo invernadero en el sudeste español*. CAJAMAR, Almería, España. 78 pp.

Gardner, W.R. 1993. The future of irrigated agriculture. pp. 97–99. *International Crop Science I*. Buxton, D.R., Shibles, R., Forsberg, R.A., Blad, B.L., Asay, K.H., Paulsen, G.M. & Wilson, R.F. (eds). CSSA, Madison, WI. USA.

Goldhamer, D., Fereres, E., Cohen, M., Mata, M. & Girona, J. 1999. Sensitivity of continuous and discrete plant and soil water status monitoring in peach trees subjected to deficit irrigation. *Journal of American Society for Horticultural Science*. 124(4):437–444.

Hsiao, T.C., Fereres, E., Acevedo, E. & Henderson, D.W. 1976. Water Stress and Dynamics of Growth and Yield of Crop Plants.pp. 281–306. *Water and Plant Life. Ecological Studies*. Springer-Verlag, New York.

IFPRI. 1999. *World Food Prospects: Critical Issues for the Early Twenty First Century*. International Food Policy Research Institute. Washington D.C. 49 pp.

Interagency Task Force Report. 1979. *Irrigation water use and management*. USDI, USDA and EPA, Washington, DC. 133 pp.

Jensen, M.E. 1996. Irrigated agriculture at the crossroads. pp. 17–33. *Sustainability of Irrigated Agriculture*. Pereira, L.S., Feddes, R.A., Gilley, J.R. & Lesaffre, B. (eds). NATO Serie E: Applied Sciences.

Loomis, R.S. & Connor, D.J. 1992. *Crop Ecology. Productivity and management in agricultural systems*. pp. 538. Cambridge Univ. Press, Cambridge, UK.

Pruitt, W.E., Fereres, E., Henderson, D.W. & Hagan, R.M. 1984. Evapotranspiration Losses of Tomatoes under Drip and Furrow Irrigation. *California Agriculture*. 38 (5–6):10–11.

Seckler, D. 1993. Designing water resources strategies for the twenty-first century. *Water Resources and Irrigation Division*. Discussion Paper 16. Winrock International, Arlington, Virginia, USA.

Seckler, D., Molden, D. & Barker, R. 1998. *World Water Demand and Supply, 1990 to 2025: Scenarios and Issues*. International Water management Institute. Research Report 19. Colombo, Sri Lanka, 40 p.

Tanji, K.K. 1990. Drainage and Return Flows in Relation to Irrigation Management. Irrigation of Agricultural Crops. *Agronomy*. 30:1057–1087. Am. Soc. Agron. Madison Wis. USA.

Tanner, C.B. & Sinclair, T.R. 1983. Efficient Water Use in crop Production: Rearch or Re-Search?. pp. 1–27. *Limitations to efficient Water Use in Crop Production*. Dinauer, R.C., Harp, C. & Buxton, D.R. (eds). Madison, WI. USA.

Timsina, J. & Connor, D.J. 2001. *Productivity and management of rice-wheat cropping systems: issues and challenges*. Field Crops Research 69:93–132.

US National Research Council. 1996. *A New Era for Irrigation*. Narional Academy Press, Washington D.C. 203 pp.

van Shilfgaarde, J. 1994. Irrigation – a blessing or a curse. *Agric. Water Manage*. 25:203–219.

Vaux, H.J. & Pruitt, W.O. 1983. *Crop Water Production Functions. Advances in Irrigation*. Academic Press. New York. USA. 2:61–97.

Zapata, N. & Ederra, I. 2002. Changes induced by a modernization project in an irrigation district. pp. 151–152. *VII Congress of the European Society for Agronomy*. Villalobos, F. & Testi, L. (eds). Córdoba, Spain.

Sustainable water management in cities[1]

B. Chocat
Unité de Recherche en Génie Civil, Institut National des Sciences Appliquées de Lyon, France

ABSTRACT: Provision of safe drinking water, flood protection, drainage and sanitation rank highly among the needs of all societies. Various means have been used to provide these essential services. By now, most of cities of the developed world rely on "all by networks" systems. These systems are now proving not to be effective or efficient in the developed world. Furthermore, they are very expensive. Hence, it does not seem logical to use these systems in developing countries where the most essential services are not yet broadly provided.

Four different scenarios shows that, without a clear and strong political and social involvement, the sustainability of water management in most of the cities is at least questionable.

Nevertheless, holistic approaches, could offer a way out. This will imply to find new ways of dealing with water in the cities. Even the required changes will be inefficient without more flexible institutional arrangements and increased water awareness among all stakeholder groups.

1 INTRODUCTION

"Water is a basic human need and a key component of development, it is a fundamental resource for food production as well as for enhancing social well-being and providing for economic growth. It is also the lifeblood of the environment. Already today, it is a scarce resource in large parts of the world. It is estimated that about one-sixth of the world population lacks access to safe drinking water and one third lacks sanitation. If present trends continue, two out of every three people on Earth will live in countries considered to be "water stressed" (Kofi Annan, Millenium Report, paragraph 274). According to World Bank projections, by 2025 40% of the global population are likely to face some form of water shortage, with one in five suffering severe shortages. Global climate change could further exacerbate the problem".

This paragraph has been extracted of the report *"G8 Initiative on Conflict and Development"*. It shows that the importance of water is now understood at the highest political level.

Cities cover much less than 1% of the earth area. Yet it is the place where most of the people live. It is the place where both the problems and the needs are the most important. The concentration of a great quantity of population and activities on a small area involve the need of a great amount of good quality water. But the concentration of a great quantity of population and activities on a small area also implies an increase in runoff and flood hazard, in pollutant emissions and in ecosystem deterioration. For that reason, it is probably in cities that the need for a sustainable management of water is the most important.

First we will present a short resume of the history of urban water management and explain the origin of the usual "everything by networks" solution. In the second paragraph we will evaluate the efficiency of this system, and explain why developing countries cannot deal with it. The third

---

[1] Largely based on Chocat et al. (2002)

paragraph will be dedicated to a kind of game: trying to imagine what the future could be. Finally, in the last paragraph we will try to develop the hopes and the limits of sustainable approaches.

2   URBAN WATER MANAGEMENT: A SHORT HISTORY

Provision of safe drinking water, flood protection, drainage and sanitation rank highly among the needs of all societies. Since early civilization, various means have been used to provide these essential services. Some of the earliest urban water systems were built about 5,000 years ago in the time of the Mesopotamian Empire (Wolfe, 2000). The Roman Empire is very often quoted as an example for an efficient management of water in the city. In that time, water needs were very important in Rome, more than 500 l/(day capita). Such needs obviously justified a strong organization and an efficient system.

Sanitation practices deteriorated after the decline of the Roman Empire, and, in most of European towns, surface drains and streets were used as the main means of conveyance and disposal of all kinds of wastewaters. Aqueducts also deteriorated and the quantity of water used in the city drastically diminished to some tenth of liters per habitant and day. Maneglier (1991) speaks of the "dry city" to qualify this typical town of the middle age in Europe.

Yet, in that time, urban water management could be qualified, in a certain way, as sustainable. Stormwater was collected and stored in cisterns, and constituted an important resource, especially in southern Europe or in fortified cities. Wastewaters fed urban creeks with organic elements which were necessary for the industry of paper, leather or cotton cloth. The harvesting of faeces was essential for the production of organic fertilizers. The infiltration of urine into urban soils formed saltpeter that was used to make gunpowder.

Even more, some cities developed integrated systems for managing the urban cycle of water. For example, in Venice, city squares were used to collect stormwater. Under each square, an underground water storage reservoir was dug, filled with sand, and made watertight with clay walls and bottom, to prevent saltwater intrusion. Fresh water filtered through, and stored in the sand was collected in wells. During this era, neither horses nor pigeons were allowed on the permeable pavement covering these squares!

Yet, during the 17th and the 18th Centuries, urban streams became so noxious that they had to be covered over and turned into sewers. Numerous epidemics of typhoid and cholera in Europe and the United States, particularly between the 1830s and 1870s, prompted city governments to find other solutions for dealing with water supply and sewage disposal (Wolfe, 2000).

At that time, three arguments proved to be decisive in the choice of the "everything by networks" solution we still enjoy today in most of developed countries:

- A pseudo-scientific one, based on a strange analogy between the human body and the urban settlement: if the lack of blood circulation in a limb made it sick, the same principle would apply to the lack of water circulation in a part of the city. So, the continuous flow of water appeared to be necessary, and tanks and cisterns became two examples of "pestilential stagnation".
- A political argument, particularly strong in France after the Revolution, which was based on the notion of equality: all citizens must be equal and treated equally by state and municipal administrators. A network for water supply or for wastewater disposal, common to all the citizens, constituted the best solution to provide equality.
- A technical argument: during Jules Verne's century, the idea that science and technology were without limit was the most generally admitted. For that reason, solutions based on huge technical systems were particularly appealing.

The construction of these systems also benefited from the fact that, in that time of colonialism, European countries could take advantage of the wealth of the whole world. Furthermore, the new born capitalism was able to mobilize the necessary funds, even with a long return on investment, and school development allowed to teach the essential hygiene rules.

These advances helped cope with flooding in towns and improved the health of their citizens. 150 years later, it could be considered that, in developed countries, most of cities are served by comprehensive and more or less efficient collective water systems. Yet all problems are not solved and it is interesting to assess the real state of the managing of water related problems.

3 CURRENT STATE: IS THE SITUATION SATISFACTORY?

3.1 *Current state*

To measure the level of satisfaction of the service given by the current state, it is obviously necessary to establish a clear distinction between developing countries and developed countries (Niemczynovicz, 1999).

In the industrialized world, tap water is available everywhere, its quality is severely checked and its price is acceptable by the great majority of citizens. Stormwater and domestic and industrial wastewater are conveniently removed effectively and efficiently. Even if some problems begin to appear (ageing of most of the systems, increasing pollution of many water bodies, etc.; Herz, 1998; USEPA, 1994), or if some recurrent ones resist or worsen (flooding hazards), the situation could be considered as satisfactory.

The situation is very different in developing countries for almost two-thirds of the world's population (Stephenson, 2001). Cities have no such inherited infrastructure, and many of their habitants struggle with recurrent flooding and the daily need to find convenient water to drink and to carry out the most fundamental personal ablutions. In 1997, more than 1.1 billion people in low and middle income countries lacked access to safe water supplies, and far more were without adequate sanitation (World Bank, 2002).

Lack of clean water and sanitation makes people sick. Water-related diseases kill an estimated three million Africans each year, while also digging a yawning gulf in productivity.

Everywhere in the world, but more hopelessly in developing countries, pollution caused by humans is contaminating lakes, rivers and underground water, reducing quality, limiting regeneration of freshwater ecosystems, and complicating treatment of water for domestic use. Industry, agriculture and inadequate sewage treatment are intensifying pollution, making reclamation of some water costly or impossible, thereby diminishing an essential resource.

As access to safe, clean water and effective sanitation is vital to improving human health and well-being, it is difficult to understand why, after more than one century of effort, the situation is so bad. A quick review of recent efforts by the international community helps understanding of present situation and shows that the problem is difficult to solve.

3.2 *The efforts in the recent past*[2]

On November 10th, 1980, the General Assembly of the United Nations proclaimed 1981-1990 as the International Drinking Water Supply and Sanitation Decade. The primary goal was full access to water supply and sanitation for all people in developing countries by 1990. At the Decade's end, universal access had clearly not been achieved.

In 1989, the OECD Council, drawing on lessons already learned from the Decade, urged members to review their institutional framework in the water sector. It asked them to pay special attention to rates, agreements, water rights, and unrecorded demand, in order to improve their policies for integrated resource management.

The International Water and Environment Conference held in Dublin in 1992, presented new approaches to freshwater resources, emphasizing integrated management of basins and protection of ecosystems with full attention to the environment, both economic and social.

In 1993, the World Bank published a document defining new objectives, and the FAO set the parameters for a program of action on water and sustainable development.

---

[2] Lake & Souaré (1997)

Yet in 2002, more than 2 billions of human being do not enjoy an acceptable access to drinking water or to sanitation, and the objective of the Johannesburg conference is only to reduce this number by 50%.

3.3 *Problems and challenges*

Several remaining challenges can be identified. The first challenge is population growth. The worst case is the African one where the increase rate is 3% a year, with nearly no sign of imminent reduction. This challenge is further compounded by rapid urbanization. Populations in cities such as Nairobi, Dar Es Salaam, Lagos and Kinshasa increased sevenfold from 1950 to 1980. In Latin America comparable cities grew half as fast, which still means that their population has been multiplied by 3 or 4 in 30 years. Population growth and urbanization are at the heart of many of developing countries problems.

A simple example can explain the difficulty of the problem. In a city as Lyon in France, with the level of investment that is currently accepted, it would take 100 years to build the existing water supply and sewerage systems. That means that, each year, the expenditure allows to serve one more per cent of the population. If the population increases 5% a year, then, the part of the population who is not served, increases about 4%.

Lyon is a very rich city compared with most of African, Asian or South American cities. So, it is easy to understand why, in spite of the important international help, the situation is deteriorating year after year.

The second challenge deals with water shortages. Water shortages are a growing problem in many parts of the world. To meet the needs of people and the environment, renewable freshwater resources of 1,700 cubic meters annually per capita are considered adequate. Below that, supplies are short: a figure between 1,700 and 1,000 means water stress, while less than 1,000 represents water scarcity. 10% of this water is needed for direct urban needs.

Globally, in 1950 only 12 countries with 20 million people faced water shortages of some sort. By 1990, it was 26 countries and 300 million people. Today, 166 million people in 18 countries suffer from severe water scarcity, while another 270 million in 11 additional countries are considered "water stressed". By 2025, affected populations will increase from 436 million to about 3 billion people, or about 40% of the world's population, most of them in poorest countries (World Bank, 2002).

Such numbers do not take into account variations in place and time (dry seasons, rapid runoff, geographic features, usage patterns in different areas). In reality, actual shortages are worse than national averages would indicate, and an "adequate" figure cannot justify complacency.

How can then the needs of the majority of the world's population be recognized and dealt with effectively? Are the techniques and paradigms, on which water supply as well as wastewater disposal were based in the past, still relevant today?

They are not, according to the experts of Habitat (ESC, 1996), who believe that *"the systems used in the developed world are not the most effective, efficient or indeed very logical"*, and they are certainly not sustainable (Argue, 1995; Marsalek & Chocat, 2001). Thus we need to find new ways of dealing with water in the cities and the time has come to develop new paradigms (Bacon, 1997; Malmqvist, 1999). For example, communities in developed countries will no longer be able to derogate responsibility for wastewater "disposal" and part of the new paradigm must include the recognition that "waste" water is in fact "resource" water (Otterpohl et al., 1997; Rauch et al., 1998).

Before trying to give some indications on the way that can be used, it is interesting to project some current trends and to see what can become of water management if we do not take care.

4 POSSIBLE FUTURES

There is a proverb which is often quoted, even though its source is rather unclear: *"prediction is a difficult art, especially when it deals with the future"*. Yet, it is possible to take the risk and try to imagine several possible futures. The four scenarios that are presented here are based on very different ideas. In the first one, the green scenario, we imagine that "political ecology" wins the game; it is a "politically based" scenario. The second one can be qualified as being "technology based": it describes what could arrive if engineers were fully in charge. In the third one, the only issue is the economical rule; it is the "financially based" scenario. The last one is probably the most realistic. It supposes that economical, technical and political driving forces act together, but with no cooperation, as it is the case today. These scenarios, as extreme as they are, give some trends on what future could be. They show that a sustainable solution cannot be found without an equilibrium between technical, political and economical issues.

4.1 The green scenario

The first possibility could be called the "green scenario". In that scenario, defenders of "small is beautiful" solutions have won the game. Such an approach has been formulated by Lawrence et al. (1999) as "total urban water cycle based management". It encompasses:

- Reuse of reclaimed wastewater (for pollution prevention, sub-potable water supply).
- Integrated stormwater, groundwater, water supply and wastewater based management (water supply, flow management, water and landscape provision, substitute of sub-potable water sources, and protection of downstream areas against urban impacts).
- Water conservation based approaches (efficient water use, reduced water demand for landscape irrigation, and substitute industrial processes with reduced water demand).

4.1.1 Main characteristics of the green scenario

Ecologists are in charge. They have imposed a so-called "balanced integrated approach" attempting to combine both "soft" and "hard" technology in achieving sustainability of water management. The key words are: sustainability, environmental concern, back to nature, "green cities", engineering "married" with social sciences, recycling of, both, wastewater and waste in small local cycles (house, lot, neighborhood), water harvesting...

In such a scenario, responsibility is decentralized. Long-term master plans for large-scale infrastructure disappear.

4.1.2 Risks and problems of the green scenario

This scenario is dangerous for at least three major reasons:

- The first one is that we are not sure how harmless the promoted solutions really are. Their potential impacts may be of cumulative and long-term nature, thus taking long time to manifest themselves. It is not clear whether nature alone can cope with these impacts. Some pollutants are persistent, and once they are in the ground or attached to bottom sediment in ponds and streams, they will stay there, accumulate and, under unfavorable bio-chemical conditions, enter the food chain or are released into the environment. This risk is exacerbated by the fact that nobody can control the quality of the installation since most of them are on private properties. The rainwater people harvest and drink can be polluted and dangerous without anybody being aware of it is before it is too late.
- The second risk follows from the fact that green solutions could be understood by local authorities as a convenient way to free themselves from the costly obligation to maintain their water infrastructure. On-site stormwater or wastewater facilities would be built and operated under the private property owner's responsibility. This transfer of responsibility and funding to the end-users might result in no or poor maintenance and, consequently, numerous small failures.
- The last, but not least, problem is that the way to proceed from the existing centralized systems to future decentralized systems is not obvious. The existing urban water systems are enormous infrastructures that took decades to build. Decentralized solutions also require large investments, and their wide-scale implementation would stretch over similar time spans. Thus, two functioning systems would have to be financed over a long period of time. During this period, fewer and fewer people would contribute to pay maintaining the old system that would obviously become

increasingly obsolete. When finally the centralized systems would become useless and could be turned off, the resulting write-off of (financial) capital would be unprecedented in times of peace.

In conclusion, the green scenario is not necessarily sustainable just because it appears more ecological (in a political sense). Its sustainability needs serious studies, particularly with respect to the long-term behavior of decentralized facilities and the feasibility of managing these facilities in the long run (ASCE, 1998). There is a clear risk underlined by Mikkelsen (2001) that we create a garbage dump in each garden and re-create a situation that Europe struggled with to overcome in the 60s when many illicit dump sites were finally closed down. The wide-spread distribution of small pollutant deposits could be a good solution to preserve the quality of the "nature", but it also might create health hazards that are unknown today.

4.1.3 Reasons why this scenario could happen

Probably, the main driving force towards the green scenario is the fact that it is politically and economically appealing (at least, in a short run) for those public utilities that struggle financially.

Its objectives are undisputed, both internationally and locally (e.g., Brundtland Commission, local Agenda 21). It receives "green political support", particularly in rich countries where people have bad consciousness over exploitation of nature and want "to do something good", and it is defended by a lot of enthusiastic "experts". From the sociological point of view, it appeals to well educated well-to-do part of the society, often living in up-scale developments or eco-villages.

4.2 The technocratic scenario

The technocratic scenario differs much from the green scenario. Publicly employed engineers are in charge and decide upon both, the technical approaches and the way the systems are managed. This scenario represents the classical approach, before privatization became an international issue.

4.2.1 Main characteristics of the technocratic scenario

The engineers in charge appreciate that society has delegated an important task to them. By applying the proven technology, coupled with redundancy and large safety factors, they make sure that the system does not fail. The solutions are not necessarily cheap, but they are robust and impressive from a technological point of view. Thus, this scenario is basically conservative, even if ambitious engineers might get a chance to apply advanced technologies to water systems: automation, control, robotics, real-time operation, third generation of communication, use of biotechnology for water quality control, new field measurement and laboratory equipment.

This system will be designed with sufficient numbers of fail-safe devices and fall-back alternatives, so that operational risks are kept small. Cost-benefit is one of the considerations, but not the most important one, and if in doubt, the technologically better solution is adopted, even if it is more expensive. This approach focuses strictly on good technology, i.e., solutions that an engineer is proud of. Water systems rely almost entirely on piped systems. The development of new water saving devices in household and industry and water recycling solutions in water scarce regions is managed by large companies and strictly supervised by central authorities (minimize recycling to minimize risks).

Long-term planning ("Grand Master Plan") is the rule to ensure best technical quality ("the pride of the engineer"). Maintenance and renewal of central systems with large size and with redundant components is the top priority. Responsibility for system operation and maintenance is centralized and mostly public. Water service remains to be a monopoly with no competition over services or approaches.

4.2.2 Risks and problems of the technocratic scenario

We certainly have the technology to solve urban water management problems in a "technocratic" way. We can distribute healthy fresh water in all the buildings. We can build huge sewer networks and use modern technologies, such as real time control, to protect cities against flooding, with some acceptable level of risk. We can build and manage end-of-pipe treatment plants capable of purifying dry or wet-weather effluents with a very high level of performance. We know how to rehabilitate streams in order to produce clean and valuable places for the public. But:

– The first problem is financial. Such solutions are expensive, and if implemented in top engineering quality, they are very expensive. There might not be sufficient political support and then this scenario gradually slides towards the "privatization" scenario (see below). Public utilities might need to join large companies, or risk being taken over by them, because private companies do not require political support for raising capital and making investments. In developed countries, it is likely that funding could be found for this scenario. People accept to pay more than $200 a month for phone, TV and Internet connections. It is reasonable to believe that people will accept to pay more for clean water, which is much more essential to their life, especially if they think that they have no other choice.

  However, due to the lack of funds, lack of engineering expertise, or operation and maintenance capacities, the technocratic solution is totally impossible to implement in developing countries. If this scenario would be the only way to provide urban water services, we would exclude 4/5 of the world population from access to clean water and from a minimum protection against flooding and public health hazards.

– The second problem is that nature is not easy to harness. Whatever level of protection is chosen against flooding, one has to accept that a rainfall greater than the design storm can occur. In that case, consequences could be very dramatic. In other words, technological perfectionism will avoid all but the most extreme problems, leaving people mentally unprepared for such rare, but terrible catastrophes.

4.2.3 *Reasons why this scenario could happen*

The technocratic scenario offers intellectual challenge and research opportunities. The engineering community can argue that "robust" solutions offer guaranteed performance and meet health and environmental protection goals, so the engineers in charge will be politically supported as long as the system functionality is guaranteed, service is provided, and money is not openly wasted.

The approach fits well into our present developed world, and profits slowly but surely from fast development in related fields. There seems to be a synergy potential (microbiology, gene manipulation, computers, nanotechnologies, etc.) that might be used in the future. Why should we apply "stone age" methods in urban water systems when modern technology, carefully and conservatively applied, solves the problem? In fact, is it not appealing to politicians, government and decision makers to show that we are in line with technological progress – the latest, the most powerful, the newest control devices – to ensure an equally good service to all citizens?

The scenario assumes that it is unrealistic to expect essential changes in individual and corporate behavior with respect to the environmental protection, or at least that such expectations are not needed. Thus, urban water management is relatively easy, since it operates largely independently of the political context.

Lastly, but not the least, the scenario assumes (and even requires) that most urban dwellers are not really interested in the details of urban water management, and politicians exercise low level of control as long as there is no trouble.

4.3 *The privatization scenario*

The privatization scenario might be a consequence of the situation, in which the credibility of the technocratic solution was undermined by repeated failures, obvious waste of money, or chronic under-funding. An urban water system can only be privatized once, but it is very tempting for politicians: privatize water infrastructure, sell the assets and use the money "to do something good" for the people. This idea becomes particularly attractive where aging infrastructure requires large investments and public agencies wish to avoid the need to raise such funds (by raising taxes) in public debate and with elections coming up (Barraqué, 1995).

Obviously, the leading idea of the privatization scenario is not to find the most beautiful solution to a technical problem, but to make money, i.e., provide a system with sufficient technical quality to deliver the service, but no more than that. In the worst and not unlikely situation, the water company might define the level of service itself (because it holds all the expertise), pursuing its own objectives and using a proper marketing approach.

4.3.1 Main characteristics of the privatization scenario

Investors and economists are in charge, and the buzzword is cost-efficiency. This is achieved by control of labor costs and limited investments into infrastructure. When funding shortfalls loom (i.e., economic losses), the private company will insist on increased service fees and/or re-negotiation of the original contract (it might even be formulated with this eventuality in mind). Sustainability of the approach is an issue, but only as long as the economical sustainability of the company is not at stake. As economic issues play the major role, the functionality of the system as well as ecological issues are just standards that have to be complied with at minimum effort. This scenario is governed by two key words: "efficiency and effectiveness" on one hand and "the true value of water" on the other.

The technological solutions will consist of "reasonably maintained" centralized systems with "decentralized pockets" (where profitable). Such a system is easier to manage, and the service charges are easier to invoice. Due to the physical nature of the system very few, but huge private companies have total regional monopolies. Competition only happens once, i.e., when the getting of the initial contract is at stake. Without extremely competent and politically strong regulation this kind of approach is bound to result in increasing service fees. How far environmental concerns will be taken seriously remains to be seen; regulatory pressure might push the issue, but carefully worded contracts will be a necessity in any case.

4.3.2 Risks and problems of the privatization scenario

The main risk is that the price for water service will become unreasonable, especially for developing countries. Market concentration could be enormous: only few companies could manage the water resources on the planet. Water could assume the same role as energy today: being no longer a natural resource (like the air we breathe) but a tradable commodity such as crude oil (Bryce, 2001).

The dilemma of water service companies is that regulators want to keep the fees down while investors require revenues and profits, which may be reinvested in other fields totally unrelated to water. A solution to the dilemma could be to offer "new" but deteriorated services to cut costs (e.g., bottled water in exchange of lower quality tap water).

Private companies might walk away from operating contracts, accompanied by legal disputes when encountering significant financial difficulties. They even might get bankrupt and stop operating over night.

Another risk is that companies could develop short-term strategies guaranteeing returns on investment, but disregarding long-term sustainability issues.

Employees might become frustrated by being denied the opportunity of doing their job properly and not being acknowledge for doing good job. Mergers and take-overs, accompanied by re-organization, will frequently lead to changes of staff, or changes of tasks that the staff perform. The experienced engineer, knowing his system in-and-out for decades, will become a relic of the past.

4.3.3 Reasons why this scenario could happen

Four major driving forces can make this scenario happen:

– Selling public water infrastructure creates large one-time income that can be used for many "good purposes".
– Everybody needs urban water services: given a "well-formulated" contract, water service can be a profitable business.
– Actual (or perceived) failures of the technocratic approach will support the opinion that "private is better than public".
– If privatization is one of several options, private industry will attract the best engineers and offer them opportunities and resources to develop their creativity and their innovation capacities. Thus, the competing public utilities will encounter a steady brain drain.

For private companies it is tempting to gain a business monopoly. At the same time, politicians have an opportunity to shed responsibility for under-funded, ageing water infrastructure and the need to raise capital, deal with municipal employee unions, public complaints about poor service, increases in fees, etc.

Nowadays, almost everywhere in the world (European Union, South America, even in some countries in Africa) we see a clear promotion of privatization presented as the (only) strategy to "make systems work". Shareholder value principle and the current development in many countries promote this development. It benefits from liberal political support of the argument that customer costs will become lower because of the greater economic efficiency of the private sector. Looking at current trends, this scenario might be the most likely to become our future reality.

Last but not the least, this scenario is somehow auto-engaging. Once it has reached a certain level of development and acceptance, it is difficult to change the strategy. The master of water resources and water utilities ultimately becomes the master of the game.

4.4 *The business-as-usual scenario*

The forth scenario is probably the most likely development, simply because is represents the current status of urban water management; we will continue as we have done for the past 30 years, without any clear strategy or political vision.

4.4.1 *Main characteristics of this scenario*

Nobody is in charge. This scenario is poorly defined, because the current level of services/approaches widely varies between countries and between jurisdictions. Practically everywhere, it is recognized as (or called) "not sustainable", yet it is the most common.

Its main characteristics are:

– Mixture of the green, technocratic and privatization scenarios.
– Lack of clearly formulated objectives and aspirations.
– No rational discussion about acceptable risks and opportunities.
– "Yes, large investments are necessary – but not now".
– "It is only family silver, we can sell it and do something good with it".
– Expertise disappears towards other sectors.
– Small investment into R&D, and hesitation to apply innovative approaches.

Stumbling between the technocratic tradition and green ideas, the pathway is unclear and the goals are fuzzy, and developments remain localized. However, the most pressing problems are attacked, just because it is necessary. Master plans are tried to be implemented, but with frequent political interventions after incidents and accidents. In case of problems, "quick-and-dirty" solutions are implemented, exercising activism and giving way to political pressure. All but the largest utilities experience a brain drain, and their lack of competence results in simplistic solutions of the type "one-size-fits-all". The overall result is a gradually decreasing quality of systems and services, and operation in a reactive mode.

On the other hand, it is the scenario that incorporates most approaches to management of water in cities that one can think of. In a Darwinist style, through this competition of approaches and survival of the fittest, i.e., through mutation, recombination and selection, an optimal approach to water management might evolve. This eventuality would require a number of prerequisites, including proper information and data, non-dogmatic discussions, sufficient public interest and the availability of enough time (i.e., slow urbanization) for such a process. In fact, there are hardly any countries where all these prerequisites exist.

4.4.2 *Risks and problems of this scenario*

Refusing to make a choice is obviously a dangerous idea, resulting in accumulation of disadvantages of the three first scenarios, without being able to understand why things are getting worse.

In the developed world this situation is further disguised by a rather low risk of acute (catastrophic) failure, because water systems usually work reasonably well and service deterioration is so slow that it goes unnoticed. However, larger problems are already being experienced in regions suffering from water scarcity.

Other problems emerge from poorly defined performance objectives and lack of clear-cut priorities. Practical solutions require frequent changes between approaches that are neither consistent nor continuing, and therefore unsustainable. Operations are under-funded and have to rely on general budget allocations and subsidies from senior levels of government, while the costs are high due to the necessity to manage different systems that are tried out in parallel. Finding money is more and more difficult, particularly when social and environmental concerns about water services decrease and professional marketing is lacking. Public urban water utilities find themselves squeezed between quarrelling pressure groups, that do not even acknowledge basic achievements, and the private industry that is eagerly waiting to take over. Or is there any other product in the world whose "brand name" variety costs 1000 times more than its "no name" equivalent (i.e., bottled versus tap water)?

Risks and acute problems of this scenario might not be so severe for the developed world, but they are unbearable for developing countries, where neither inherited infrastructure, nor money, nor experienced engineers are available. Again, the outlook for the developing world is particularly dull, and with business-as-usual the gap will increase quickly and endlessly between rich and poor countries.

4.4.3 *Reasons why this scenario could happen*

The most important reason is the resistance against change, visible as the inertia of technical, administrative, and political systems. Stated positively, in the absence of acute problems (catastrophes), the incentive for change is low. In fact, in most jurisdictions urban water services operate without dramatic failures and thus, attract limited attention of the public and politicians.

4.5 *Conclusion*

Is the future of urban water necessarily so gloomy? Not at least from a theoretical point of view, solutions do exist. There are even some good reasons for "hydro-optimism". Increasing water scarcity due to an increasing world population and equitable provision for all is raising the importance of water, not only as a fundamental human right, but also as a cornerstone for economic, health and food security Asmal (2001). Business being the main driving force of our civilization, business reasons should become good reasons to help the system change.

To achieve this evolution, we need to change our level of comprehension and to use a holistic approach. That could be the hope of sustainable approaches. This hope is developed in the next paragraph. Its materialization will need a strong political and social involvement (World Water Council, 2002).

5 SUSTAINABLE MANAGEMENT OF WATER IN CITIES: HOPE OR DEADLOCK?

5.1 *Sustainable management of water in cities: what does it mean?*

5.1.1 *What is sustainable development?*

Many definitions of sustainability can be found in the literature. It is not useful, and even probably not possible, to make an inventory of all these definitions. Yet it is interesting to give the most well-accepted ones.

The first definition, probably the well-known one, is set out in the Brundtland Report in 1987 (ESC, 1996):

"Sustainable development is development that meets the needs of the present without compromising the ability of future generations to meet their own needs".

Another interesting definition, given by the World Conservation Union, UN Environment Programme and World Wide Fund for Nature in 1991 (ESC, 1996) could be regarded as complementary:

"Sustainable development means improving the quality of life while living within the carrying capacity of supporting ecosystems".

Both of these definitions insist on the fact that sustainable development is a much broader concept than environmental protection. It must deal with future generations and with the long-term health and integrity of the environment. It implies concerns:

– For the quality of life.
– For equity between people in the present (including the prevention of poverty).
– For equity between generations (people in the future deserve an environment which is at least as good as the one we currently enjoy).
– For the social and ethical dimensions of human welfare.

The main issue is that further development should only take place as long as it is within the carrying capacity of natural systems.

5.1.2 *Cities and sustainability?*
Clearly, previous definitions show that addressing the sustainable development agenda provides new challenges for urban policy integration within holistic frameworks. Yet these definitions are difficult to use in a practical way.

The following more practical and local interpretation of sustainable development, provided by the International Council for Local Environmental Initiatives in 1994 (ESC, 1996), is helpful as we seek to apply the concept in urban areas:

"Sustainable development is development that delivers basic environmental, social and economic services to all residents of a community without threatening the viability of the natural, built and social systems upon which the delivery of these services depends".

Cities are both a threat to the natural environment and an important resource in their own right. It is also the place where most of people live, that means the place where human needs are the most important.

Cities affect the global system through, for example, energy and resource use, waste and polluting emissions. They affect regional systems though river catchments and flows, patterns of land use and stresses on surrounding rural areas which are subject to pollution, development and recreational pressures.

The challenge of urban sustainability is to solve both the problems experienced within the cities themselves (the focus of action in the past) and the problems caused by cities.

5.1.3 *What is sustainable water management?*
Following the previous definitions, it could be said that sustainable water management is management that meets current needs without compromising the ability of future generations to meet their own needs, both for water supplies and for a healthy aquatic environment (Schilling & Mantoglou, 1999).

The main objective of a sustainable management is to ensure a high quality of life by making the community's management of natural and man-made resources sustainable.

Water is the natural resource that forms the basis of all life. For that reason water is central to sustainable resource management. Good local water management must includes safe water supply and disposal, conservation of water resources, managing water as an economic resource and preserving it as a cultural asset. Effective water management is directly linked to the issues of poverty alleviation, social equity, and environmental sustainability, and thus to democratization and the transparency of decision-making.

5.1.4 What is sustainable water management in cities?

We can use the definition adopted in the MITRA project called "Sustainable urban water management": The basic requirements imposed on such a system are:

"Without harming the environment, urban water and wastewater systems should:

- Provide water for a variety of uses to households, factories, offices, schools and so on.
- Remove wastewater from users in order to prevent unhygienic conditions.
- Remove stormwater from streets, roofs and other surfaces in order to avoid damage from flooding" (Swedish EPA, cited by Malmquist, 1999).

5.2 Sustainable water management in cities: how can it be achieved?

First, Water Management must be understood as a part of the City Management: Sustainable Management of Water in the city is obviously impossible to achieve if it is not included in a Sustainable City Management (Stockholm International Water Institute, 2001).

That means that it is necessary to redefine the relationships between urban planners and the engineers in charge of water management (Harremoes, 1996).

Another key point is that Sustainable Development will only happen if it is explicitly planned for. Market forces or other unconscious and undirected phenomena will not solve alone the problems of sustainability. Agenda 21 specifies a thorough process of considering a wide range of issues together, making explicit decisions about priorities, and creating long term frameworks of control, incentives and motivation, combined with specified targets in order to achieve stated aims.

On a practical way, the process of sustainable urban management requires a range of tools addressing environmental, social and economic concerns in order to provide the necessary basis for integration:

- Collaboration and partnership.
- Policy integration.
- Market mechanisms.
- Information management.
- Measuring and monitoring.

Each tool should be considered as an element within an integrated system of sustainable urban management. It must also be clear that the vision of a coherent world-wide water policy is utopian at the moment and will be so for decades, despite initiatives such as the proposal for a world-wide Water Innovation Fund (Serageldin, 2001) and an International Water Centre (Swaminathan, 2001). Industrialized nations do have fully functional systems, whereas in developing and emerging countries this is not the case. Developed world technologies cannot be installed in developing countries in the future due to their high cost. Therefore, as any changes in these alternative system types will have to start from different baselines, a single unified world-wide strategy is unlikely to be achievable. Instead "adapted solutions" have to be sought that take into account the characteristics of the individual situation.

5.2.1 The case of developed countries

The optimum scenario in the industrialized world is probably some combination of the beneficial properties of the scenarios discussed earlier. However, key aspects of existing systems will remain in existence in the near future – at least in the densely populated areas. These systems have to be adapted with "green" solutions wherever appropriate and with strong incentives to adopt novel solutions.

There is agreement that both, storm and sanitary "wastewaters" provide a resource opportunity within the Integrated Water Resource Management framework (Ellis, 1995; Krebs & Larsen, 1997; Larsen & Gujer, 1997; Chocat et al., 2001; Marsalek & Chocat, 2001). Thus, our 20th Century paradigm of "waste" water needs to be modified to "opportunity" water.

The cornerstone of a realistic future vision is therefore decentralized wastewater treatment and utilization at the local level wherever this is practicable. This requires both technological development and greater individual and community responsibility. It may be achieved by a combination of source controls (provided by the industry) and technological development (biotechnology research). By applying efficient infiltration systems for the disposal of an important part of both the household effluent and the urban runoff, a sustainable solution is feasible, provided that groundwater quality is not compromised and groundwater aquifers can accept these extra inputs. In many cases, aquifers may then constitute a key resource for water supply. The need for integration of these efforts with other measures, i.e., total urban water cycle management, is obvious. *While it is essential to plan comprehensively, greater success could be achieved through discrete, manageable, and sequenced development.*

Where feasible, all opportunities for resource recovery should be taken, with water re-use and recycling, and utilization of nutrients from human wastes. This approach would:

- Create a greater awareness for water in the urban environment, which is largely unknown today, except in certain eco-villages (Hedberg, 1999).
- Ease a major economic and infrastructure problem of current water supply provision. The existing large pipe infrastructure systems would remain for the foreseeable future, as these would only be used for the transport of a diminishing amount of rain runoff (or household effluent of the same water quality). Hence the pressure to maintain or even expand the serviceability of these large and expensive systems would be significantly reduced.

The previous paragraph introduces the idea that scientific and technological changes, even if they are important, are probably not the most important. Some other points are also to be emphasized.

First, applying higher technology to decentralized solutions would shift a significant part of the cost (and responsibility) from the public to the individual. Yet, it is unlikely that this new system would function without an inspection and enforcement system. Why should this field be different from other fields of human behavior, in which we need laws, regulations, policing and penal systems?

The globalization of economic systems is also seen as a potential impediment with consequential institutional arrangements constraining innovation and the need to introduce more appropriate "local" solutions.

For such an approach to be viable, it would be necessary to change the current institutional systems, in which the wastewater utility (i.e., the asset owner) is valued according to the infrastructure assets it owns, and the revenue income is based on volumes and pollutants handled. Both "hard" infrastructure and handled wastewater volumes would clearly be reduced under such a scenario. As a consequence, many countries would need to fundamentally review the current institutional systems to ensure that the utilities were not under-valued due to the loss in assets.

Other barriers include professional reluctance to change from the "way-it-has-always-been-done", the QWERTY problem (Geldof, 1999). Evidence of "mono-thinking" and false beliefs is apparent even for water professionals (Biswas, 2001). Thus, new approaches to educating professionals will be needed, and new ways of providing real knowledge about water/wastewater systems will be required for new system "operators" (consumers) at the local level. This will be needed to achieve the "radical shift in thinking" required (Stockholm International Water Institute, 2001). The importance of ensuring that water professionals are competent, multidisciplinary and up-to-date must also be emphasized. There is even a need identified by Geldof (1999) that some experts in the water area should "de-learn" so that they would have a broader vision than they have now, as to what would be acceptable and appropriate solutions.

5.2.2 *The case of developing countries*

Optimum approaches to future urban drainage for developing countries will be slightly different from what is envisaged for developed countries, inasmuch as the fusion between existing and new technologies is not likely to be as big an issue.

This is both a virtue (no restriction due to existing systems) and a burden (as a functioning asset system does not exist). Megacities in developing countries face essentially the same problem with respect to urban water management as densely populated urban areas in the First World: land use. Many communities have to cope with dense overcrowding and no sanitation, and the associated increase in imperviousness, which inhibits the natural water cycle and requires an appropriate disposal of rain runoff. Typically, the developing countries with sanitation and drainage problems also have problems with inadequate water supplies. Note that increases in water supply will further influence local water imbalances.

Here the key principle must be the utilization of all "wastes" wherever feasible. It is likely that in many cases this may require high technology solutions and the fusion of tradition (pipe systems where necessary), with "green and ecological technologies" (infiltration, ponds, natural waterways, water harvesting) and modern technology (treatment and operation facilities where affordable). The principal consideration will be economic, and it is difficult to see how adequate resources can be made available to achieve what is required within the current globalized economic systems.

Where possible, the approaches adopted must take into account the need for easily repairable technological implementation, increased employment opportunities for local inhabitants, and the provision of basic education. Inherent flexibility and adaptability (key criteria for sustainability) must be built into water/wastewater systems due to the rate of change of circumstances in developing country communities. Greater "consumer" involvement and responsibility for water and wastewater services has been shown to work effectively at the community level in developing countries (Stephenson, 2001), but where sophisticated technologies have to be used, this may be problematic.

The last barrier is money, and it is probably the most difficult to overcome. The consensus that emerged from the World Water Forum, held at The Hague in March 2000, was that the annual investment required by 2025 to meet the world's needs for water for irrigation, industry, water supply and sanitation, and environmental management will increase to $180 billion, from the current $70–80 billion (World Bank, 2002).

We, in rich and developed countries, are we ready to assume this bill?

REFERENCES

ASCE (American Society of Civil Engineers). 1998. *Sustainability criteria for water resource systems.* WRPM Division and UNESCO International Hydrological Pogramme IV, Project M-4.3, Task Committee on Sustainability Criteria.

Argue, J.R. 1995. Towards a universal stormwater management practice for arid zone residential developments. *Water Science and Technology.* 32(1), 1–6.

Asmal, K. 2001. Water is a catalyst for peace. *Water Science and Technology.* 43(4), 24–30.

Bacon, M. 1997. Integrated planning for sustainable urban water resources. In: Sustaining Urban Water Resources in the 21st Century, Rowney A.C., Stahre P. and Roesner L.A. (eds.), *Proc. Engineering Foundation Conference*, Malmo (Sweden), ASCE, New York, pp. 15–29.

Barraqué, B. 1995. *Les politiques de l'eau en Europe.* Ed. La Découverte; Paris (France); 1995; 303p.

Biswas, A.K. 2001. Missing and neglected links in water management. *Water Science and Technology.* 43(4), 45–50.

Bryce, S. 2001. *The Privatisation of Water, Nexus Magazine*, 8(3) web: www.squirrel.com.au.

Chocat, B., Krebs, P., Marsalek, J., Rauch, W. & Schilling, W. 2001. Urban drainage redefined: from stormwater removal to integrated management. *Water Science and Technology.* 43(5), 61–68.

Chocat, B., Ashley, R., Ball, J., Marsalek, J., Rauch, W., Schilling, W. & Urbonas, B. 2002. Urban Drainage – Out-of-sight-out-of-mind? Position paper of the *IWA/IAHR Joint Committee On Urban Drainage;* to be published in 2003.

Ellis, J.B. 1995. Integrated approaches for achieving sustainable development of urban storm drainage. *Water Science and Technology.* 32(1), 1–6.

ESC (European Sustainable Cities). 1996. *Report by the expert group on the urban environment.* European commission directorate general XXI – environment, Nuclear safety and Civil Protection, WEB http://citiesnet.uwe.ac.uk. 1996.

Geldof, G. 1999. QWERTIES in integrated urban water management. *8th Int. Conference on Urban Drainage*, Sydney (Australia). Aug-Sept. Ed. Joliffe I.B., Ball J.E., ISBN 0 85825 718 1.

Harremoes, P. 1996. Dilemmas in Ethics: Towards a Sustainable Society. *Ambio*, 25(6), 390–395.

Hedberg, T. 1999. Attitudes to traditional and alternative sustainable sanitary systems. *Water Science and Technology.* 39(5).

Herz, R.K. 1998. Exploring rehabilitation needs and strategies for water distribution networks. *J. Water SRT-Aqua.* 47(6), 275–283.

Krebs, P. & Larsen, T.A. 1997. Guiding the development of urban drainage systems by sustainability criteria. *Water Science and Technology.* 35(9), 89–98.

Lake, E.B. & Souaré, E M. 1997. Water and Development in Africa. *International Development Information Centre, express n°09*, September 1997.

Larsen, T.A. & Gujer, W. 1997. The concept of sustainable urban water management. *Water Science and Technology.* 35(9), 3–10.

Lawrence, A.I., Ellis, J.B., Marsalek, J., Urbonas, B. & Phillips, B.C. 1999. Total urban water cycle based management. In: I.B. Joliffe and J.E. Ball (eds.), *Proc. of the 8th Int. Conf. on Urban Storm Drainage*, Sydney (Australia), Aug. 30– Sept. 3, 1999, vol 3, pp. 1142–1149.

Malmqvist, P.A. 1999. *Sustainable Urban Water Management.* Vatten, 55(1), 7–17.

Maneglier, H. 1991. *Histoire de l'eau, du mythe à la pollution.* Ed. F. Bourin, Paris (France), 1991.

Marsalek, J. & Chocat, B. 2001. International report on stormwater management (SWM). *1rst IWA congress.* Berlin. 16pp.

Mikkelsen, P. 2001. BMPs in Urban Stormwater Management in Denmark and Sweden. Linking Stormwater BMP designs and performance to receiving water impacts mitigation. *UEF conference.* Snowmass, Colorado (USA), to be published 15pp.

Niemczynovicz, J. 1999. Urban hydrology and water management – present and future challenges. *Journal of Urban Water.* 1(1), 1–14.

Otterpohl, R., Grottker, M. & Lange, J. 1997. Sustainable water and waste management in urban areas. *Water Science and Technology.* 35(9), 121–133.

Rauch, W., Aalderink, H., Krebs, P., Schilling, W. & Vanrolleghem, P. 1998. Requirements for integrated wastewater models driven by receiving water objectives. *Water Science and Technology.* 38(11), 97–104.

Serageldin, I. 2001. From Vision to action after the Second world water forum. *Water Science and Technology.* 43(4), 31–34.

Schilling, W. & Mantoglou, A. 1999. Sustainable water management in an urban context, In: Cabrera, E. (ed.). *Drought management planning in water supply systems.* Kluwer, Dordrecht, ISBN 0-7923-5294-7, pp. 193–215.

Stephenson, D. 2001. Problems of developing countries. *Frontiers in Urban Water Management: Deadlock or hope?* Ed. Maksimovic, C., Tejada-Guibert, J.A. IWA. ISBN 1 900222 76 0.

Stockholm International Water Institute 2001. Water Security for the 21st Century – Innovative Approaches. *Water Science and Technology.* 43(4).

Swaminathan, M.S. 2001. Ecology and equity: Key determinants of sustainable water security. *Water Science and Technology.* 43(4), 35–44.

USEPA 1994. National water quality inventory. *Report EPA 841-F-94-002.* US EPA, Washington, DC, USA.

Wolfe, P. 2000. History of wastewater. *World of Water 2000*, Supplement to PennWell Magazine, 24–36.

World Bank. 2002. Bridging Troubled Waters: A World Bank strategy. Precis. *World bank operations evaluation department*, n° 221, spring 2002.

World Water Council. 2002. Second draft of the "world water vision report", can be downloaded: http://www.worldwatercouncil.org/.

The future of desalination as a water source

J. Lora, M. Sancho & E. Soriano
Departamento de Ingeniería Química y Nuclear
Universidad Politécnica de Valencia, Spain

ABSTRACT: Desalination has already made a great contribution to the quality of life in the most arid regions of the world. Without desalination, many of these regions would have remained lifeless. Within the current situation characterized by an increasing global demand of water, an unequal distribution of fresh water and the continuous increase of the population, the technology of desalination is providing water suitable to human consumption, even in some countries where contamination reduces the quality of natural water resources. Thus, as a mean to enhance fresh water supplies, desalination contributes significantly to global sustainability. Desalination plants are increasing in number rapidly and producing more water than ever before, but this type of industry remains being expensive for much of the world's population. This problem is today being addressed through active research and development, devising novel and more efficient desalination configurations.

1 INTRODUCTION

1.1 *A great problem: the shortage of water*

The Agenda 21 and particularly its Freshwater chapter, as well as the Earth Summit celebrated in Johannesburg emphasize that water is a key factor for sustainable development. But water is a scarce resource indispensable for the survival of all life forms, and its quantity is limited, being one of the humanity's more serious problems nowadays.

The shortage of water refers to the limited resources of water to satisfy the current or projected demand in a specific location. The supplies of water can be limited in quantity or quality.

1.1.1 *The capacity*

The water demand is largely affected by the quick increase of world population, which reached 5,800 millions in 1996 and that will probably increase up to 6,100 millions in the year 2005, 8,000 millions in 2020 and 10,000 millions in 2050, according to estimations of United Nations. This nonstop increase and the change in the habits of the population's consumption have generated some unprecedented demands, given the limited available water resources. The UN estimates that the situation of water shortage will worsen considerably during the next 30 years if water management does not improve. To be exact, the UN estimates that on year 2025 the problem will affect around 35% of the world population.

The importance of the qualitative limitations is illustrated by the fact that approximately 97% of the water of the Earth is contained in the oceans and the seas with a high content of salts (35,000 mg/l). This fact means that the above vast resource is virtually useless for the human consumption without an appropriate treatment. Within the remaining 3% of surface water, 2% is present as ice in the polar caps and the glaciers, 0.3% is present in the atmosphere and only 0.1% in the rivers and lakes. The rest of surface water (0.6%) corresponds to underground aquifers, remaining almost half of the underground water in depths below 800 m.

Therefore it is not strange that water problems exist in the world although the 80% of the surface of the Earth are occupied by water. We have just said that in most of the cases the quality of the water does not allow its direct use by the population, and also, it is necessary to keep in mind that water and rain are not evenly distributed on the globe.

The nine countries of the world with more availability of fresh water resources are reported in the next Table 1.

In comparison with these countries, the whole of the European Union receives 1,171 km^3/year where Spain represents the 10%. In the opposed point, the countries with fewer resources are included in Table 2.

As a consequence of this unequal distribution of the water resources, many areas of the planet are subject to serious and recurrent droughts nowadays. A clear example of this is happening in Spain in the last years, where water is distributed in a very unequal way, concentrating 70% of the total resources on the peninsular north third. The insular territories with a great tourist appeal, together with most of the areas of the southeast coast suffer an unremitting deficit of water. This fact is seriously threatening the development potential in these zones because they concentrate at the same time the most innovative industry, the tourism and the specialized agriculture. In the periods of persistent drought, as it has happened very recently, the agriculture suffers catastrophic losses and population water demands cannot even be guaranteed.

Tables 3, 4 and 5 show the negative hydrologic balances for the horizons of the years 2002 and 2012 for different autonomous communities of the Spanish southeast coast. The data come from the studies carried out to the elaboration of the current National Hydrologic Plan (Sancho et al., 2002).

It is considered that each Spanish person consumes a hundred thousand liters of water on average per year, of which only about 700 liters (2 l/day) are drunk, and between 60,000 and 120,000 l/year (160–320 l/day) are consumed for other uses. This quantity is not the most important one because

Table 1. Countries with more availability of fresh water resources (Shiklomanov, 1998).

Country	km^3/year
Brazil	6,220
Russia	4,059
USA	3,760
Canada	3,290
China	2,800
Indonesia	2,530
India	1,200
Colombia	1,200
Peru	1,100

Table 2. Countries of the world with lower availability of fresh water resources.

Country	km^3/year
Kuwait	0.002
Malt	0.015
Gaza	0.046
United Arab Emirates	0.500
Libya	0.600
Singapore	0.600
Jordan	0.680
Israel	0.750
Cyprus	0.900

it represents approximately 12% of the total water consumed in Spain. On the other hand the agriculture consumes 80%, and the remaining 8% goes for industrial uses. Our water consumption is very high, being only by the United States, Canada and Russia, and so it can be deduced that we should save water with the shared effort of consuming the necessary fresh water quantity.

1.1.2 *The quality*

The water of a clean river, drinking water or, in other words, fresh water contains about 200 mg/l of salts; brackish waters about 5,000 mg/l and seawater about 35,000 mg/l. Most of industrial uses

Table 3. Forecast of the hydrologic balance in the Community of Andalucía.

	2002	2012
Resources (hm^3/year)		
Superficial	698	698
Undergrounds	228	228
Total	926	926
Demand (hm^3/year)		
Urban	317 (23.1 %)	374 (21.2 %)
Industrial	63 (4.6 %)	71 (4.0 %)
Agricultural	992 (72.3 %)	1319 (74.8 %)
Total	1,372	1,764
Deficit (hm^3/year)	446	838

Table 4. Forecast of the hydrologic balance in the Community of Murcia.

	2002	2012
Resources (hm^3/year)		
Superficial	352	352
Undergrounds	85	85
Total	437	437
Demand (hm^3/year)		
Urban	53 (8.0 %)	60 (8.6 %)
Industrial	12 (1.8 %)	13 (1.8 %)
Agricultural	601 (90.2 %)	628 (89.6 %)
Total	666	701
Deficit (hm^3/year)	229	264

Table 5. Forecast of the hydrologic balance in the Comunidad Valenciana.

	2002	2012
Resources (hm^3/year)		
Superficial	284	284
Undergrounds	382	382
Total	666	666
Demand (hm^3/year)		
Urban	139 (12.2 %)	160 (12.2 %)
Industrial	36 (3.2 %)	43 (3.3 %)
Agricultural	964 (84.6 %)	1102 (84.5 %)
Total	1139	1305
Deficit (hm^3/year)	473	639

require water that does not surpass 200 mg/l, and the limit is 1,000 mg/l when the water is used by agriculture.

Therefore, the water of a clean river can be used for all these purposes, except for some special cases in the industries (cooling water circuits, cleaning of electronic components, boilers of vapor, medical and pharmaceutical uses, etc.). On the other hand, the direct use of seawater for the three more important applications of fresh water would involve the death of people, the extermination of crops and the quick deterioration of industrial equipments.

There are several aspects related to water quality. On the one hand, in many arid regions ground waters contain high saline concentrations as a result of natural processes. On the other hand, the human action, due to the shortage of planning and to the application of irresponsible activities, has seriously contaminated (and keeps doing it) the available water supplies (surface and groundwater), causing an additional shortage for the direct use of this element.

The main problems that affect the quality of water are the municipal wastewaters improperly treated and the discharge of highly polluted industrial wastewaters due to the lack of appropriate controls.

The shortage of water is extended to the aquifers, where it takes place a reduction of water and a salination process due to the entry of seawater. The high quality ground water is being consumed much more quickly than the time needed to recover it. As a consequence, it is needed to drill more and more deep wells, as it is the case of many irrigable lands in a large part of the Mediterranean coast and insular territories of Spain. On the other hand, when the phreatic level is close to the sea, solutes concentration in water increases continually. This aquifer salination is especially serious in certain areas of the coast and the islands.

In view of this problematic situation different options for water management can be proposed: supply enhancement, demand management and institutional options (Lund, 2002). For example, increases in the storage capacity and transport can improve the yield of water distribution. Demand management tends to be applied at a local level, but it can also be applied at a regional scale by means of plumbing codes, regional water utility practices and standards, water pricing, and various forms of rationing. Wastewater reuse can also be considered to reduce net water demands and solve some wastewater disposal problems. Finally, different forms of agreements can be reached independently of the negotiations and discussions among the different responsible institutions for the administration of water: contracts, joint operations agreements, memoranda of understanding, formal or informal policies, joint powers authorities, merger of agencies and legal mandates from higher levels of governments.

Within the possibilities of water demand management the technology of the desalination is emerging, in the last years, as an alternative solution to the water supply problems. The characteristics of this technology as well as their future perspectives will be analyzed in the following chapters.

1.2 *Alternative solution: desalination*

Sun is the source of renewable energy and oceans are the great alternative source of water. Just as the sun is an alternative source of energy to meet future demands, oceans, rivers and lakes are alternative water resources. However, production of fresh water from these sources requires a specific infrastructure and also an important quantity of energy. The greatest natural desalination process, which occurs on the earth, is the hydrologic cycle. This natural machine acts as a continuous evaporation and pumping system. The sun supplies heat energy and this, together with gravity force, keeps water moving from the earth surface to the atmosphere by means of evaporation and transpiration, from the atmosphere to the earth as condensation and precipitation, and finally between points on the earth by flowing on rivers and groundwater.

Just as it happens in a cycle, this system of fresh water has neither beginning nor ending. From the point of view of the mass conservation law neither loss nor profit of water exists. However, the quantity of available water for the users may fluctuate. In fact, small changes on the behavior of the hydrologic cycle can cause unequal distributions of the natural fresh water on the earth and serious floods and periods of drought.

There are many references of philosophers like Aristotles, about obtaining fresh water from the oceans, but the first industrial plant was built in 1872, using solar distillation in a wooden structure covered with glass, in The Salines near Antofagasta (Chile) and was working up to 1908 with a capacity of 22.5 m^3/day. Since then, different technologies have been developed trying to imitate the great process of evaporation of the hydrological cycle. In these separation techniques, thermal systems or selective separation devices like semipermeable membranes have substituted the driven force given by the sun.

The beginning of commercial scale of the different desalination technologies has not taken place simultaneously, and along the last years the necessities of each technology have been adapted, by trying to respond in a most efficient way. In a first stage a very important development of the evaporation processes took place, which was implanted mainly in those countries in which it was not possible to have another source different from seawater, fundamentally to supply populations. Later on, it evolved towards another situation in which desalination was not contemplated as an unique source, but as an alternative source of supply that, though more expensive, can be justified according to some circumstances of social type or be the first stage in the production of fresh water. Currently the development reached by desalination technologies allows the massive production of water facing a moderate cost, contributing with flexible solutions that are able to adapt to each type of necessity. The desalination of surface water or ground waters and seawater allows to satisfy the increasing necessities of water in three big areas:

– Municipal or domestic uses: The supply of water to big cities near the sea can imply serious problems since it is necessary to consider much more than the provisioning sources. The production of fresh water by means of desalination units avoids the transport of water, which can be expensive and it also preserves the existent resources for the interior regions.
– Industrial uses: Industry uses water as raw material or as a mean of production. Great metallurgical industry, chemical industry and, mainly, thermal and nuclear power stations are large consumers of water. The dual plants that combine a thermal central with desalination units can be very interesting from the technical and economical point of view in certain areas.
– Agricultural uses: Agriculture consumes great quantities of water. The drop of production costs in high capacity plants (250,000 m^3/day) will allow to apply desalination for agricultural uses (particularly for greenhouse hydrophonic cultivations) if relatively cheap energy sources are available (Hairston, 1999).

2 GENERAL SITUATION OF WATER DESALINATION

In recent years the increase of the desalination capacity worldwide has exponentially passed from 1 hm^3/day in 1969 to 9 hm^3/day in 1982, and it was doubled ten years later. At the end of 1991, the installed capacity of desalination processes was 15.58 hm^3/day, distributed in 8,886 plants. At the end of 1993, the number of plants reached 9,900, with a total installed capacity around 19 hm^3/day. At the end of 2001 the desalination capacity installed worldwide was about 32.4 hm^3/day (about 11,600 m^3/year) being the number of installed desalting units, with a unit size of 100 hm^3/day or more, higher than 15,223 (Wangnick, 2002). Figure 1 shows this evolution for units with a production higher than 100 m^3/day. About 60% of this capacity corresponds to seawater desalination.

2.1 Geographical distribution

The production of fresh water is distributed among about 120 countries. However, more than 65% of the world production is concentrated in the Gulf countries and the USA. The rest of the capacity is distributed among a great number of countries with very small market percentages.

It is estimated that about 52% of the total capacity is located in the Middle East, and the six members of the denominated Gulf Cooperation Council (Kuwait, Bahrain, Qatar, Oman, UAE and Saudi Arabia) are producing about 38%, which represents 11.4 million m^3 of desalinated water per day (Wiseman, 2001).

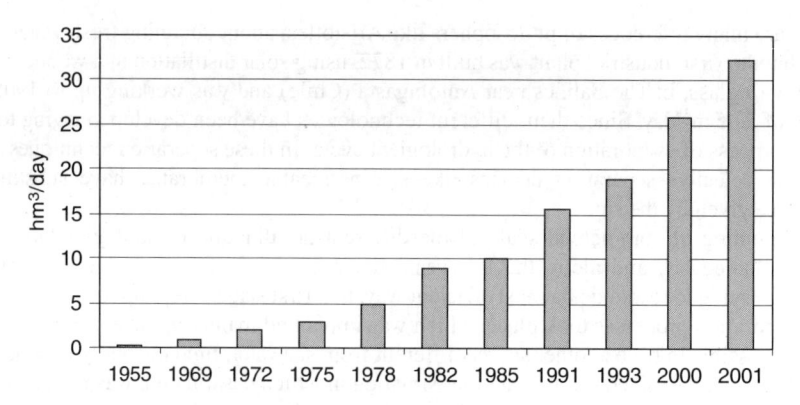

Figure 1. Evolution of total installed capacity of desalting facilities.

In relation to the type of technique used, about 76% corresponds to evaporative processes and only 21% to reverse osmosis. This fact is related to two simultaneous situations:
- The scanty rainfall in these arid countries, together with a high rate of evaporation, leads to deficits in their water budgets.
- Great oil availability and therefore very cheap energy sources.

On the other hand, the United States, with 15% of the total installed capacity and 20% of the existent units present the particularity of having RO units of small size, but it should be kept in mind that most of the facilities correspond to brackish water treatment (Birkett, 2002).

Desalination applied to urban supply continues being the most important application in relation to the size of the units and the installed capacity, and it is located in Gulf, South Mediterranean, and Caribbean countries. On the other hand, in the industrialized countries (North America, European Union and Japan) the desalination units used in the industry are more and more numerous, but only in some cases their capacity is greater than that corresponding to the urban water units above mentioned.

2.2 Distribution of processes

As for the installed capacity, approximately 52% corresponds to the evaporation technologies (with 93% for the MSF), 40% to reverse osmosis and the remaining 8% to other technologies (comprising ED). However, as for the number of plants, evaporation processes suppose 30% whereas RO and ED represent 58% and 12% respectively (Urrutia, 2001).

2.3 Capacity of desalination plants

About 25% of the installed desalination plants correspond to the range of 100–2,000 m^3/day, reaching 56% for plants producing less than 10,000 m^3/day. The group of plants up to 25,000 m^3/day of capacity reaches 14%. Within the rest of facilities there are plants with a capacity higher than 50,000 m^3/day. For example, in two locations like Al-Jobail and Al-Khobar, in Saudi Arabia, two desalination plants have a combined capacity of almost 1.3 million m^3 of water per day. In relation to the size, the MSF units are the biggest with capacities up to 50,000 m^3/day. MED and VC units can be competitive with 20,000 m^3/day, while the RO units have a maximum capacity of 7,500–8,000 m^3/day.

2.4 Cost of technology

Desalination projects around the world have demonstrated that they are a cost-effective source of water supply. Whereas 30 years ago water production costs were 2–3 €/m^3, this quantity has fallen

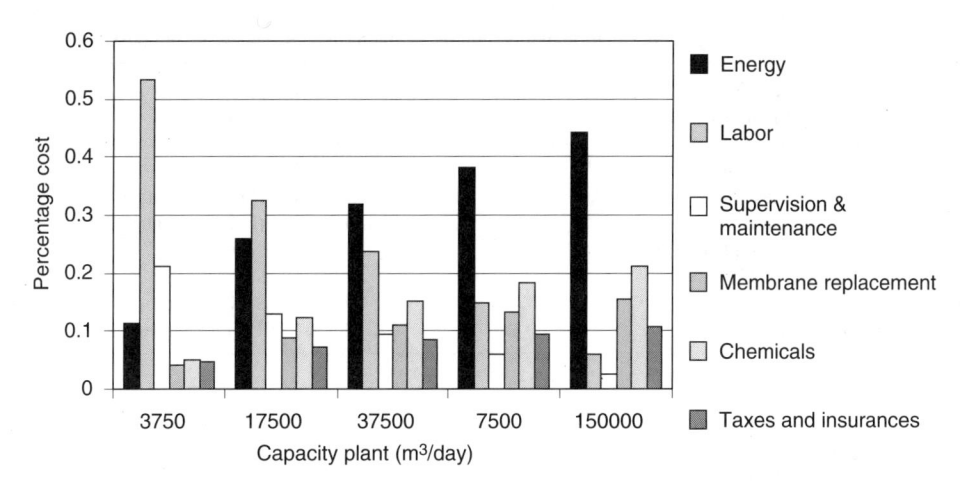

Figure 2. Water production costs distribution for RO desalination plants.

bellow 1 €/m^3. So, current prices referenced, including capital costs recovery, for desalinated brackish water are 0.25–0.60 €/m^3 in RO systems with capacities of 4,000 to 40,000 m^3/day. For seawater plants the prices are 0.75–1.25 €/m^3 including the costs for environmental protection (such as brine or concentrate disposal), distribution and losses in the storage and distribution system (Wiseman, 2001).

The specific investment costs for main desalination processes can be estimated in the following ranges: 1,100–1,600 €/(m^3 day) for MSF, 900–1,250 €/(m^3 day) for MED and 700–1000 €/(m^3 day) for RO (Allison, 2000). These costs for a MED plant are 10%–20% lower than for a MSF plant. The costs of RO units have come down to about 50% less compared with the 90s. The equipment for brackish water desalination by RO is usually less than 50% of seawater equipment cost. Concerning the operating costs, the scale economies has a significant effect in the final specific price as it is shown in the Figure 2.

2.5 Waste disposal

Desalination plants produce liquid wastes, generally a significant quantity of water, which may contain all or some of the following constituents: high salt concentrations, chemicals used during the pretreatment and cleaning processes, and toxic metals due to the contact of water with metallic materials used in the construction of the plant facilities. The disposal of this wastewater in an environmentally appropriate manner is an important part of the feasibility of a desalting facility. When the desalting plants are located near the sea, the problem will be considerably less important. Liquid wastes may be discharged directly into the ocean or combined with other discharges (e.g., power plant cooling water or sewage treatment plant effluent) before ocean discharge. Desalination plants also produce a small amount of solid waste (e.g., spent pretreatment filters and solid particles that are filtered out in the pretreatment process).

On the other hand, caution should be taken regarding the environmental impact on the receiving waters relative to the discharge of added chemicals, dissolved oxygen levels, and different water temperatures. The problems can be more significant when the desalination plants are installed in the inland, such as it is common for brackish water facilities. Much care should be taken not to contaminate any existing ground or surface water with the salts contained in the concentrate stream. Disposal may involve dilution, deep injection of the concentrate into a saline aquifer, evaporation, or transport by pipeline to a suitable disposal point. All of these methods add an extra percentage to the cost of the process and could drastically affect the economics of desalination depending as well on the disposal regulations in each country.

Table 6. The largest desalination plants running in Spain (m^3/day) (www.aedyr.com).

Las Palmas III	58,000 RO seawater
Bahía de Palma	53,000 RO seawater
Costa del Sol Occidental	45,000 RO seawater
Las Palmas-Telde	37,000 MED seawater
San Tugores	35,000 RO (brackish water)
Bajo Almanzora	30,000 RO (brackish water)
Mancomunidad del Sureste	25,000 RO seawater
Maspalomas II	24,000 RO seawater
Adeje-Arona	20,000 RO seawater
Maspalomas I	18,500 EDR (brackish water)

2.6 *Situation of the desalination in Spain*

In Spain, since it is a geographical area with important seasonal and local imbalances, facilities to produce fresh water from seawater were built more than 30 years ago, especially in places with no alternative sources, such as Canary Islands, Ceuta and Melilla.

The first desalination plants installed in Spain were located in Termolanza (Lanzarote island; 2,500 m^3/day) in 1964, Ceuta (4,000 m^3/day) and Alcudia (143 m^3/day) in 1966. In the 70s few desalination plants were installed in Canary Islands, and along the 80s a regular increase of the capacities installed was appreciated. During the 90s the biggest plants were characterized by being of municipal property. At the present time a new structure of management defined as build-own-operate/transfer (BOOT), which allows an approach to the participation of the private sector, is being introduced.

At the beginning, seawater facilities were used for urban and tourist supply, but in the recent years a change has taken place in this trend and, although the large constructions continue having this objective, the industrial facilities and especially those dedicated to agricultural watering are very numerous. As a consequence from the drought of the last years, Spain holds a predominant position in this sector in the world. At present, the approached distribution by uses can be the following: tourist and urban (54%), agricultural (30%) and industrial (16%).

The current capacity installed in Spain is about 800,000 m^3/day being distributed in 47% for seawater and 53% for brackish water (www.aedyr.com). Spain plays a leading role in Europe. The importance of desalination is reflected in the fact that Spain holds the fifth world position for installed capacity, and the predictions in relation to the facilities to be built in next years will allow to duplicate the above capacity. In relation to the type of technology used, more than 60% corresponds to RO systems.

The total quantity of desalination plants operating in Spain is higher than 700, 100 of which correspond to seawater. The most important of them are mentioned in the Table 6.

3 TECHNOLOGIES FOR WATER DESALINATION

The process to separate the salts that are dissolved in the water (desalination) requires a certain type of energy or driven force, and in fact, the form in which this energy is applied is one of the differentiating elements of the available processes. Table 7 shows the main desalination technologies, used nowadays, are shown in function of both the applied method and the type of used energy. The most typical operating range for these four major desalination processes is shown in Figure 3, indicating also the standard operating ranges of the different salt concentration expressed in part per million (ppm) of total dissolved solids (TDS).

3.1 *Processes operating by means of chemical bonds: Ion exchange*

Ion exchange resins are insoluble materials that have the property of exchanging their ions with the inorganic salts dissolved in the treated water. There are two types of resins: the anionic, that

Table 7. Main desalination technologies.

Technology	Method	Driven force
Ion exchange (IE)	Resin exchange	Diffusion-attraction
Multiple Effects Distillation (MED) Multistage Flash Evaporation (MSF) Vapor Compression (VC)	Evaporation	Steam
Reverse osmosis (RO) and Nanofiltration (NF)	Dense membrane	Hydraulic pressure
Electrodialysis reverse (EDR)	Porous membrane	Electric power

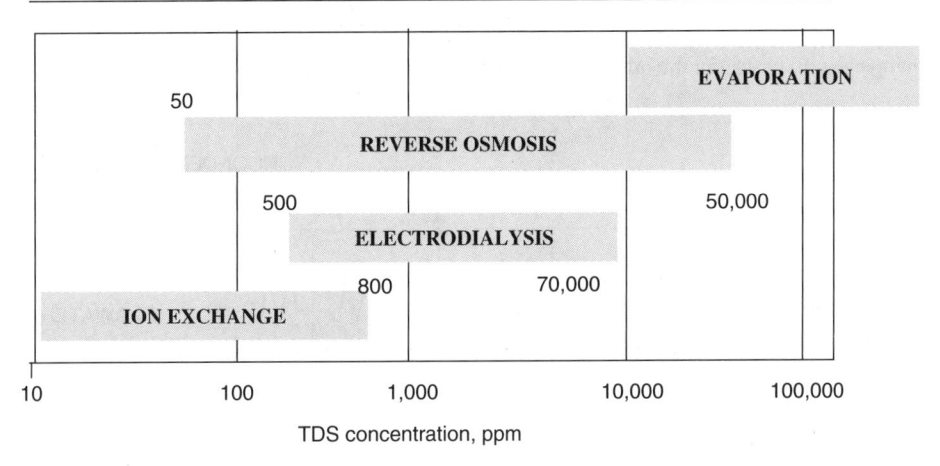

Figure 3. Operating range of the main desalination technologies.

allow replacing some anions from the liquid for other different anions; and the cationic, that allow replacing the cations present in the water. This technique gets a liquid of great purity if the treated water has few total dissolved solids (TDS). This technique is used in the industry to obtain feed water of boilers from waters with low salt content, or in the softening of waters enriched in calcium and magnesium. In order to replace the fixed ions in resins it is necessary a periodic regeneration process to recover the initial activity of the resins. The IE global process consumes not much energy, but the problem associated to a discontinuous process and the environmental management of the residual brines from the resin regeneration process make this desalination technique not the most attractive to be used.

3.2 *Evaporative processes*

Evaporation is a method that involves a phase separation in which the saline water is boiled to produce vapor of water that is condensed to fresh water. The different evaporative processes that produce drinkable water include the denominated multistage flash (MSF), multiple effects distillation (MED) and vapor compression (VC). All these techniques operate, generally, under the principle of reducing the vapor pressure of water inside the unit to produce boiling water at low temperatures with no additional heat. The evaporation units are designed with the purpose of conserving the thermal energy by means of exchanging condensation and vaporization heat inside the equipment. The major energy requirement in the evaporation process corresponds to the vaporization heat of the treated water.

3.2.1 *Multistage Flash Evaporation (MSF)*

Figure 4 shows a simplified scheme of a MSF evaporation unit. The incoming saline water crosses the heating stages and is pre-heated in the heat recovery sections of each subsequent stage. After crossing the last section of heat recovery, and before entering in the first stage, where the major evaporation happens, the feeding water is boiled even more in the boiler of the brine by using steam. At this point the water reaches its highest temperature, and later it passes through several stages where a sudden evaporation (flashing) takes place. The vapor pressure in each of these stages is controlled so the hot brine that enters each chamber with the appropriate temperature and pressure (lower and lower than in the precedent stage) can suffer an instantaneous and violent evaporation.

Fresh water is formed by condensation of water vapor, which is collected at each stage and passed on from stage to stage in parallel with the brine. At each stage, the product water is cooled and the surplus heat is recovered to preheat the feed water. Because of the large amount of flashing brine required in a MSF plant, a portion (50% to 75%) of the brine from the last stage is often mixed with the incoming feedwater, recirculated through the heat recovery sections of the brine heater, and flashed again through all of the subsequent stages.

By this way, the amount of water-conditioning chemicals that need to be added is reduced, and so happens with costs. On the other hand, it increases the salinity of the brine at the product end of

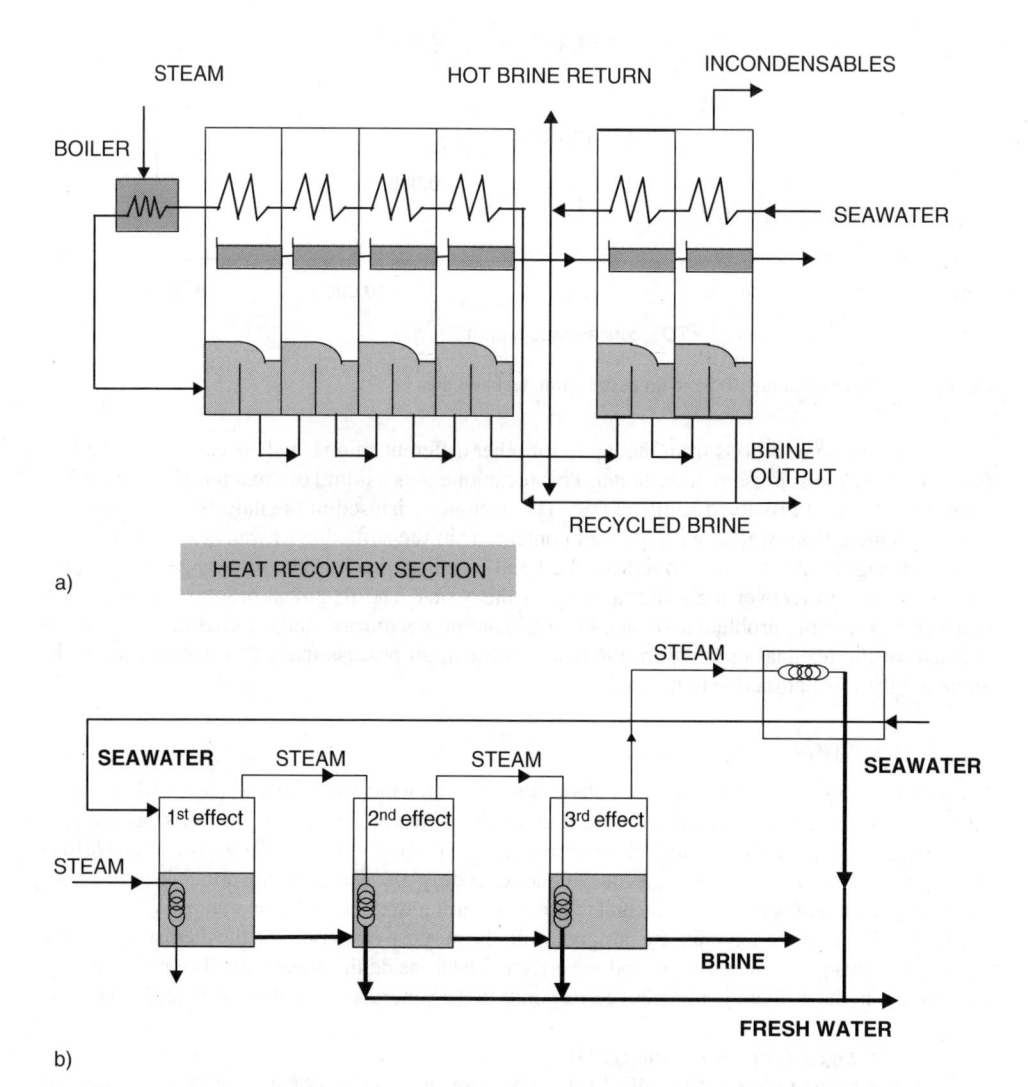

Figure 4. Evaporative processes for water desalination. a) Multistage Flash Evaporation unit b) Multiple Effect Distillation unit.

the plant; it raises the boiling point, as well as it increases the danger of corrosion and scaling in the plant. In order to maintain a proper brine density in the system, a portion of the concentrated brine from the last stage is discharged into the ocean. The discharge flow rate is controlled by the brine concentration at the last stage.

3.2.2 *Multiple Effect Distillation (MED)*

The evaporation or distillation in multiple effects (MED) was the pioneer technique in obtaining fresh water from seawater. The first plants started in 1930 in Saudi Arabia and they had two or three effects. In multiple-effect units vapor is condensed on one side of a tube wall while saline water is evaporated on the other side. The energy used for evaporation is the heat of condensation of the steam. Usually there is a series of condensation–evaporation processes taking place (each being an "effect"). The saline water is usually applied to the tubes in the form of a thin film so it evaporates easily. Although this technology is older than the MSF process described above, it has not been extensively used for water production. However, a new type of low-temperature, horizontal-tube MED process has been successfully developed and these plants seem to be both easy to operate and economical, since they can be made of aluminum or other low-cost materials.

3.2.3 *Vapor Compression (VC)*

In the 40s the improvement of the vapor compression was introduced in the evaporation units, passing from an efficiency of 20 kg of water evaporated per kg of fuel, to 200 kg of water per kg of fuel. However, scaling and corrosion continued being the most important problems. The method of vapor compression uses mechanical energy instead of the direct steam as source of heat (Figure 5).

The process of VC can be designed in a configuration of one single step or multi-effect. This process is characterized by low energy consumption and low operation costs. The VC plants have the advantage of being easily transported and installed. However, the initial investment and the maintenance costs suppose a disadvantage against other evaporation techniques. Furthermore, their capacity is somehow limited to 20,000 m^3/day.

3.3 Processes that use membranes of selective transport

3.3.1 *Electrodialysis (ED)*

ED was the first process of membranes developed for brackish waters desalination. An electrodialysis cell (Figure 6) consists on a series of narrow compartments through which the saline solution to be separated is pumped. Two types of membranes are assembled in the compartments in an alternative way: permeable membranes to the cations (CM) and permeable membranes to the anions (AM). In the opposed extremes of the cell the positive and negative electrodes are located.

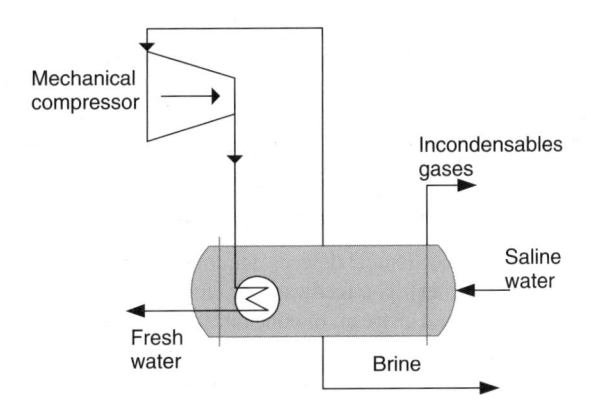

Figure 5. Diagram of the evaporation process by mechanical vapor compression.

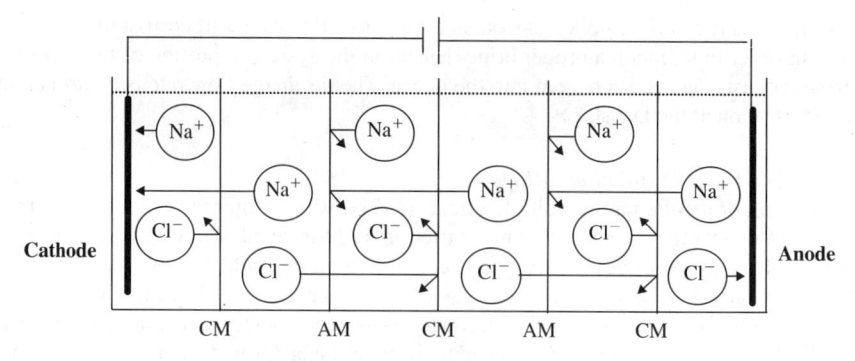

Figure 6. Schematic diagram of the electrodialysis process.

In a commercial cell with numerous compartments, the application of a continuous tension between the two electrodes located in both sides of the compartments causes a double migration: cations towards the cathode and anions towards the anode. As a consequence of this selective movement, a desalination process takes place in some compartments and a salt concentration in the adjacent ones. The reversible electrodialysis systems with a continuous change of polarity (EDR) is a competitive desalination technology to treat brackish waters with a 5,000 ppm TDS content or less, and reliability, economy and great operative experience are guaranteed (Figure 7).

3.3.2 Reverse Osmosis (RO) and Nanofiltration (NF)

In reverse osmosis and nanofiltration processes (Figure 8) the water in form of saline solution becomes fresh water when passing through a semipermeable membrane, by means of application a certain pressure. These technologies use synthetic membranes of several families of polymers, fundamentally aromatic polyamides and cellulose esters.

The RO membrane acts as a barrier against all dissolved salts and inorganic molecules, as well as organic molecules with a molecular weight greater than approximately 100 Daltons. Water molecules, on the other hand, pass freely through the membrane creating a purified product stream. Rejection of dissolved salts varies typically from 95% to more than 99%.

Nanofiltration refers to a special membrane process that rejects molecules around 1 nanometer (10 Angstroms) size. Dissolved salts are rejected in the range of 20%–98%. Also organic molecules with molecular weights greater than 200–400 Daltons are rejected. Monovalent anions salts (e.g., sodium chloride or calcium chloride) have rejections of 20%–80%, whereas divalent anions salts (e.g., magnesium sulfate) present rejections higher than 90%–98%. Transmembrane pressures are typically 3.5 to 16 bar for NF and 10 to 70 bar for RO process.

A reverse osmosis system consists of four main steps: pretreatment, pressurization, membrane separation and post-treatment. Figure 9 illustrates the basic components of a RO system:

- *Pretreatment*: The incoming feedwater is pretreated to be compatible with the membranes by removing suspended solids, adjusting the pH, and adding chemicals to avoid the scaling caused by calcium sulphate or carbonate.
- *Pressurization*: The pump increases the feedwater pressure up to the operating pressure according to both salinity of the feedwater and the type of used membrane.
- *Membrane Separation*: The semipermeable membranes inhibit the passage of dissolved salts, while allowing the water of desalinated product to cross it. In practice the membranes are assembled in various ways of mechanical devices denominated modules (plate and frame, tubular, spiral wound and hollow fiber). The feedwater driving to the membrane module produces a stream of fresh water product and a stream of concentrated brine at high pressure that can drive a power recovery device in order to recover part of the used energy.
- *Post-treatment*: The water product usually requires adjustment of the pH and degasification before being transferred to the system of distribution for the users. The product goes through an aeration column in which the pH increases from 5 to near 7.

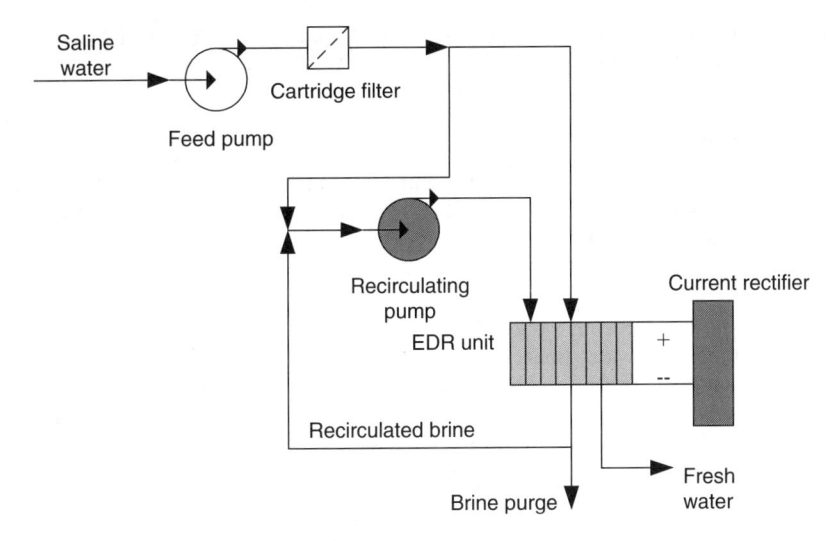

Figure 7. Diagram of a Reversible Electrodialysis (EDR) process.

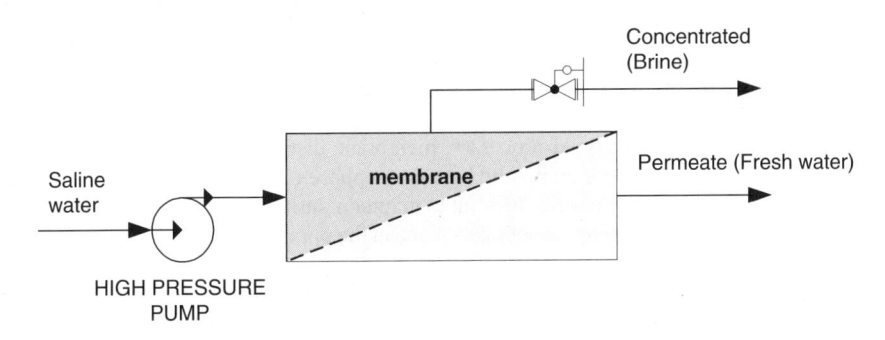

Figure 8. Diagram of basic RO process.

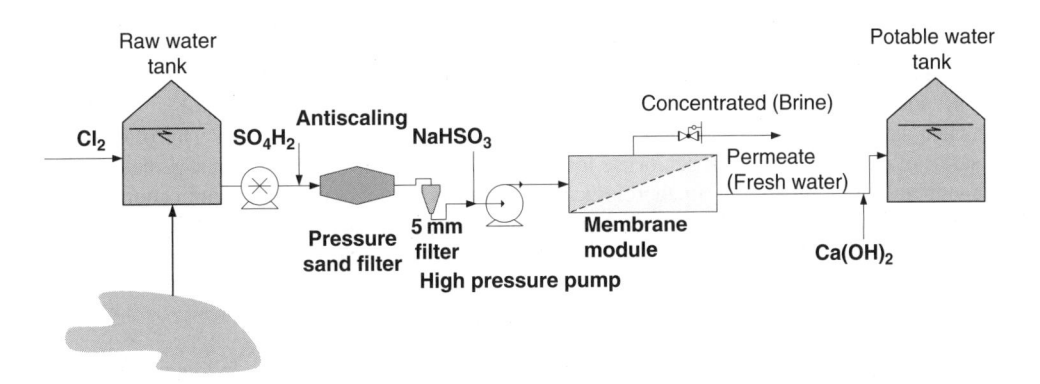

Figure 9. Diagram flow of desalination plant by RO membranes.

The reverse osmosis is a technology broadly used in the desalination of brackish waters and seawater whose main advantage is its low energy consumption. The nanofiltration is applied mainly in desalination for softening of waters (bivalent ions removing). Finally, it is important to emphasize that the improvement in the systems of energy recovery as well as the use of specific membranes have allowed to reduce the energy consumption from 4–5 kWh/m^3 to less than 3 kWh/m^3 with a recovery of 60% (Macharg, 2002).

3.4 Other processes

Other processes have also been used for water desalination, but they have not reached the commercial level of the evaporation, electrodialysis and reverse osmosis. However, on special circumstances they can be adapted to obtain fresh water.

3.4.1 Freezing

During the 50s and 60s freezing worked intensely in the general development of desalination. In this process, the saline water freezes and forms glasses of ice. Before all the mass of water freezes, the existent salts are removed by washing. Then, the ice is melted to produce fresh water. In order to obtain adequate quality water, successive freezing operations are necessary. Theoretically, freezing has some advantages against the evaporation process based on its low energy consumption, and lower corrosive and scaling effects. However, this process has not already passed to commercial scale due to the difficulties in manipulating mechanically the mixture of ice and water.

3.4.2 Membrane distillation

Membrane distillation is a relatively new process, having been introduced commercially only in the last few years. The process works by using a specialized membrane that allows passing vapor but not liquid water. This membrane is placed over a moving stream of warm water, and as the water vapor passes through the membrane it is condensed on a second surface which is at a lower temperature than that of the feedwater.

Compared to the most commercial processes, membrane distillation requires more space and uses a considerable energy to both pump and heat at the phase change. The membrane acts as an equilibrium stage whose performance is difficult to maintain during large periods of time. Usually, in a few days the condensation of the vapor takes place in the pores of the membrane, which blocks the normal performance of the application.

The main advantage of the membrane distillation is its easy cleaning and the use of small thermal differential to operate. Probably, its best application in water desalination is when it can get cheap thermal energy of low temperature.

3.4.3 Solar evaporation

The use of direct solar energy for water desalination has been investigated and used for some time. These processes imitate one part of the natural hydrological cycle, since the saline water is heated by the solar rays producing vapor of water.

Then, the vapor is condensed on a cold surface and collected as water product. This process has a series of difficulties that restrict its use for large scale production: great surfaces are needed and its vulnerability in relation to meteorological variations is very important. An application for this type of processes can be found in desalinating saline waters to small scale for a small population in which the solar energy is abundant and electricity is not available.

3.5 Brief comparison between the desalination technologies

The basic characteristics of the main desalination processes are shown in the following Table 8.

When making comparisons between the different desalination techniques from the economic point of view we should keep in mind different key points. In the first place thinking that these techniques are already used in more than 120 countries around the world is difficult to make an exact comparison. As the International Desalination Association (IDA) notes, "there are many different factors that enter into the calculation of capital and operating costs for desalination: capacity and type of plants, plant location, feed water, labor, energy, financing, concentrate disposal, and plant reliability" (IDA, 2002). On the other hand, the final cost of the desalinated water is not

Table 8. Advantages and drawbacks of desalination processes.

Technique	Advantages	Drawbacks
Evaporation	High production capacity High purity of the water product No salinity dependence on energy consumption Production of water and electricity	High energy consumption High initial investment High operation costs Low efficiency Difficult to control the water quality Great spaces are required
Reverse osmosis	Suitable for sea and brackish water Low investment costs Flexibility in quantity and quality of water Low energy consumption Modular configurations Water disinfection	It requires a good pretreatment for feed water High operating costs Needs high pressures Limited by the presence of specific ions
Electrodialysis	Low capital and operating costs Low energy consumption Simple operation	Applicable to low salinity waters Low production capacity It only separates ionic species Installation cost superior to RO

only a question of euros or dollars, since in certain cases the other alternatives of water supply in countries with problems of shortage are much more expensive.

For treatments of low salinity water (range of 500 to 5,000 mg/l in TDS), EDR and RO can be considered depending on the presence of particles that can cause some problem in certain type of membranes. For higher salinities, in general, the RO allows to obtain fresh water with smaller cost.

In the case of seawater desalination, four technologies are competing: MSF, MED, VC and RO. All these processes have years of operative experience and it can be considered that they have the same reliability. For small plants, with a production from 100 to 400 m^3/day, RO and VC present more advantages since the MSF and MED processes require more than double specific electric consumption, in the case of not having a close electric power generation plant. When choosing a process for further productions, the cost of the necessary energy is the decisive factor.

4 ENERGY REQUIREMENTS OF THE DESALINATION PROCESS

As we have already indicated, the desalination is a process that requires energy. The thermodynamics allows to establish the following relationship $dW = R \cdot T \cdot \ln(\gamma) \cdot dn$ that expresses the necessary work (dW) to eliminate (dn) moles of water from a saline solution. In Table 9 the values of the minimum work are shown for water desalination in function of both the degree of pure water recovery (product/feed ratio) and the solution salinity.

However the current desalination processes use more energy quantities to carry out the salt separation showing the low energetic yield of this type of separations (between 5% and 25%) based on surface phenomena from the point of view of both mass and heat transfer.

Nowadays the brackish water desalination is in exclusive competition of the membrane techniques (RO and EDR) using between 1.0 and 1.7 kWh/m^3 when underground brackish waters from 2,000 to 3,000 ppm of STD are treated. For salinities higher than 5,000 ppm consumptions are increased up to 2 and 2.5 kWh/m^3 with majority use of RO. In the case of seawater (30,000 to 40,000 ppm of STD) the energy consumption is about 5.6 kWh/m^3, but the recuperation of the pressure in the brine side allows saving energy between 35% and 40%, reaching in this way 4.5 kWh/m^3 including all steps of the process: caption, desalination and the supply of the water product. In all mentioned cases the consumption is exclusively in electric power form for the requirements of the pumping systems.

Table 9. Minimum work for water desalination.

Pure water recovery	Minimum work (kWh/m^3)
Seawater (35,000 ppm TDS)	
25	0.817
50	0.986
75	1.355
100	3.090
Brackish water (5,000 ppm TDS)	
50	0.19
90	0.28

The evaporative desalination systems are only used for seawater with energy consumption of 1.3 kWh in electric form and 48.5 kWh in form of heat, for each m^3 of desalinated water (3% of electricity and 97% of heat). The process of vapor compression allows reducing this difference to 8.5 kWh of electricity and 13.2 kWh of heat (40% electricity and 60% of heat).

The energy costs represent an important part of the real operating costs of operation whichever it is the used technology or the design of the desalination plant. Therefore, obtaining fresh water from any saline solution is a question of energy consumption. For this reason, the progress of the desalination techniques is also the intention in the research of systems whose combination allows both low energy consumption and high efficiency processes. In relation to the users of these systems, the decrease of the energy costs is one of the most important objectives since this represents the most important component in the total cost of the water.

So, nowadays the following procedures are used with the purpose of reducing the energy consumption in the desalination systems at industrial scale:

– Recuperation of energy from the brines.
– Dual systems desalination.
– Co-generation systems.

4.1 Energy recuperation

The conditions of high pressure and relatively low recovery that are given in RO make the option of installing a system of energy recuperation from the brine stream particularly interesting. These systems can be of several types:

– Rotary systems connected mechanically with the pump of high pressure.
– Rotary systems operating independently of the high pressure pump.
– Systems of accumulation and restitution of pressure in tanks.

In practice, the most usual systems are the first ones: a turbine that, coupled to the other end of the motor that moves the pump, receives the rejection stream with high pressure (50 bars) and restores this stream to atmospheric pressure. The second system, also known as booster, is similar to the turbo of the explosion motors (two solidary buns to the ends of an axis each one): on one hand, at the turbine, the rejection stream arrives with pressure; and in the other hand, a pump, inserted between the drive of the high pressure pump and the membrane module, allows to increase the pressure of raw water. According to the composition of the brine water in the rejection stream, it is very important the selection of the materials due to the corrosion.

The plants up to 300–400 m^3/day of capacity usually do not use systems of energy recovery, since their cost comes to compensate for greater productions than these values. Although evidently, the local cost of the energy is which puts the limits of this application, it is not conceived nowadays a plant greater than 1000 m^3/day without recovery system (Macharg, 2002). In Figure 10 the electric power consumptions are shown by each m^3 of desalinated water from seawater according to different recovery values in the process and using a turbine of energy recuperation from the concentrated brine.

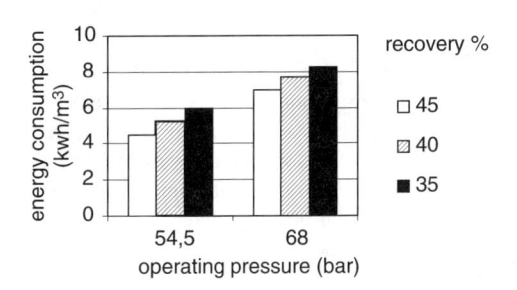

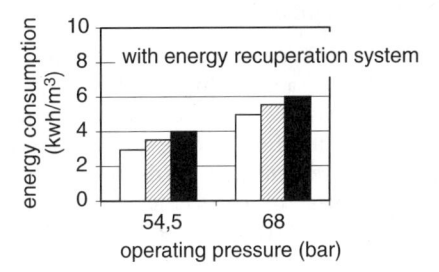

Figure 10. Specify energy consumptions for different recovery in RO desalination.

Table 10. Advantages of the use of RO versus evaporative processes in plants of dual configuration.

Less complex design	Modulated construction
	Flexible water and energy production
	Higher recovery ratio (less feed and brine)
	The produced water does not need cooling before consumption
	Shorter construction time
Facilities easier to operate	Start up and stop in less than 15 minutes
	Lower energy consumption (25%–50%)
	Minor number of components (<repair and maintenance)
Easier maintenance	Simpler operation problems solving
	Higher use of plastic components (<corrosion)
	Lower space is required
Easier proper place selection	The energy generation plant can settle down far from the urban centers
Smaller environmental impact	No particles or smoke emissions
	Water discharges at environment temperatures

In the field of brackish waters, desalination plants use the systems of energy recuperation with pressures above to 20 bars. For lower operation pressures their use is not economic yet (Steven, 1999).

4.2 The dual proposal

The energy production and the use of the secondary energy that is lost in different industrial processes are other aspects to keep in mind for saving the energy. The combination of these systems to produce both electric power and water is known as dual systems.

The electric power generation industry, when comes from thermal energy, has the basic task of increasing a value added to a certain fuel (fossil, gas or nuclear), by means of its transformation into energy. This process entails an important energy cost, because it is necessary to use between 2.5 and 3 energy units of fuel to produce one single electric unit. Nowadays, the combined gas cycles allow reducing that relationship below 2. The unused energy is wasted by the refrigeration water in form of low temperature heat. To take advantage of this heat, a system of evaporation of low temperature (like the one corresponding to MED of vertical tubes for seawater desalination) can be incorporated.

The coexistence of the Rankine cycle for power generation and the evaporative technologies like MSF for water desalination has provided economic benefits for both processes, mainly in Middle East countries, where this type of facilities has prevailed in the last years (Moch, 1997). However, at present, technological advances on the energy production processes (gas turbines and combined operation cycles), and the concern on the environmental impacts makes that RO present a series of advantages versus the evaporative processes, that favors its use in the dual systems in relation to the next future (Table 10).

4.3 Co-generation systems

Co-generation systems are other alternative technique for the effective supply of the necessary energy in the desalination process. In co-generation, a single energy source can perform several different functions. We have already commented that the proportion between the energy from fuel and the electricity that finally arrives to the consumer is about 31.5%. If this yield is measured to the output of the power station, it can rise to 35% in the current Spanish electric system. But if, apart from taking advantage of the electricity, we could do the same with the heat rejected, the yield could increase up to 85% (Chiou, 1997). The main advantage of the co-generation system is that it can significantly reduce the consumption of fuel when compared to the fuel needed for two separate plants. Since energy implies the larger operating cost in any desalination process, this can be an important economic. There are several possibilities when applying co-generation to water desalination:

- Evaporative desalination plant using the lower temperature steam in the low pressure side of the turbine after having used its energy to generate electricity. Then this steam is driven through the brine heater of evaporation plant, increasing the temperature of the incoming seawater. So, the condensate from the steam is returned to the boiler to be reheated for its use in the turbine.
- Desalination plant together with a small autonomous electric power station, using the electricity for RO pumping operations or the heat residual to a MED unit improving its thermal efficiency.
- RO desalination plant using the heat to increase the raw water temperature. So it is possible to improve the membrane permeability about 1% for each °C.

One of the disadvantages is that units are permanently connected together and, for the desalination plant to operate efficiently, the steam turbine must be operating. This permanent coupling can create a problem with water production when the demand of electricity is reduced or when the turbine or generator is down for repairs.

4.4 Hybrid process

At present, the idea of using in a combined way different types of desalination technologies as another method of reducing the overall costs of desalting is being consolidated. Such hybrid systems are not applicable to most of desalination installations, but they can prove to be an economic benefit in some cases (Tonner, 2002). An example to be considered could be to combine in some

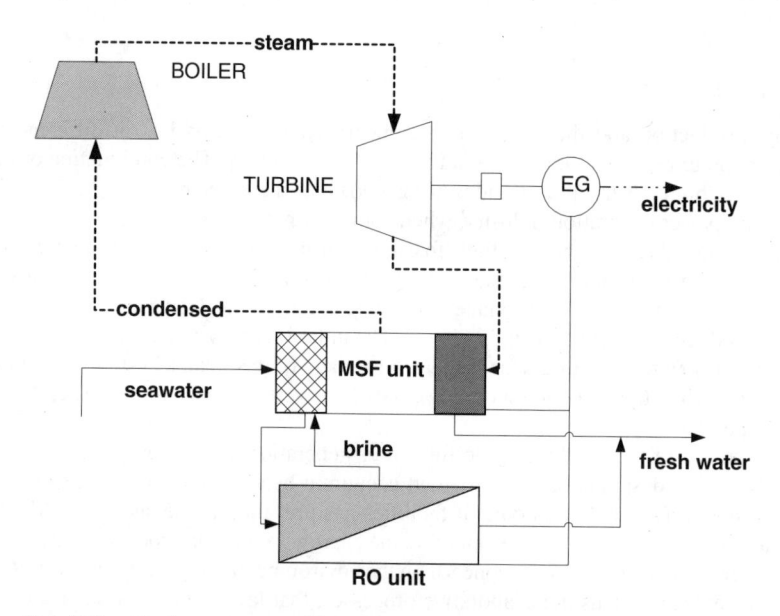

Figure 11. Hybrid desalination system combining RO and MSF technologies.

installation both MSF and RO processes with the purpose of conjugating their characteristics efficiently. Besides including a dual proposal (electricity and water) the system could provide a better agreement between the necessities of water and the electric power production. Figure 11 shows that type of hybrid desalination system.

The steam is used in a MSF plant to desalt seawater returning the condensed to the boiler. On the other hand, to stabilize the load of the generator it could be run a unit of RO during the periods of low electric consumption.

In this case, the seawater would be preheated in the MSF unit before entering the RO unit and the produced brine would constitute part of the feed in the evaporative process. Since the water produced by MSF has a very low content of TDS (30 ppm) it could settle down some operating conditions in the modules of RO to get a permeate with more content of salts that could be blended with the distilled water to provide right fresh water (240 ppm). This whole of proposals would allow getting important savings in operating costs from both energy and productivity of water points of view.

5 CONCLUSION

After the previous description, it seems clear that a continuous increase in the use of different technologies is currently happening to obtain water of quality due to the continuous deterioration of the hydrological resources. Therefore, it is necessary to keep in mind two aspects to face the future. On one hand, the protection and maintenance of the scarce existent resources and, on the other hand, the research and development of new ideas as well as new methods of financing.

Several changes have created the possibility for municipalities and private entities to consider desalination as a solution for their water management needs. Cost-saving innovations are transforming the desalination technology as financially viable alternative for municipalities. For example, membrane costs have dropped more than 10% per year for the past 20 years, and they are even expected to drop 29% within the next five years (Wolfe, 2000).

One of the key aspects of the desalination in the upcoming future is, of course, the energy. So, hybrid desalination systems, which combine reverse osmosis with the thermal process, and large-scale dual purpose systems will continue helping to reduce energy consumption. An example of it is given by the recent announcement of agreement between the Kingdom of Saudi Arabia and the three major oil companies to build an estimated 2,271,000 m^3/day of water and 4,000 MW of power (Andrews, 2001). In relation to the co-generation proposals different North African countries (Algeria, Egypt, Libya, Morocco and Tunisia) are considering the use of the nuclear energy (based on small size reactors) with RO units. In this way, the energy consumption of seawater desalination will follow a tendency to drop passing from the current 4–4.5 kWh/m^3 to 2–2.5 kWh/m^3.

The technology of membranes will bet for the design of new membranes with higher indexes of salts rejection and better water permeability, offering at the same time a superior resistance to the oxidizing agents used in water treatments. On the other hand, the manufacturers of membranes have improved their aim in the assembling modules which allows to increase the productivity of the facilities increasing the energy savings. Integrated membrane systems employing microfiltration or ultrafiltration can be used to pretreat water before RO, which replaces conventional pretreatment processes.

The costs of desalting water are decreasing while the movement towards privatization continues. There are possible cost advantages for a water utility when developers are allowed to do their own financing, design, and construction and they are essentially paid to deliver water to a customer. These BOOT systems are based on paying for delivered water and at present include contracts for a MSF facility in Abu Dhabi to deliver water (40,000 m^3/day) at about 0.70 €/m^3, a seawater RO facility in Larnarca (Cyprus) to deliver water (40,000 m^3/day) at about 0.73 €/m^3, and a 100,000 m^3/day seawater RO facility near Tampa (USA) 0.46 €/m^3. This fact will increase the expectations of using desalted water for agriculture (Kappaz, 2002).

The number of desalting plants of seawater will have a smaller increase. However, the large plants will be built in the Gulf countries, Singapore, Spain and Caribbean countries. In Mediterranean countries, there will be an important increase in the number of plants to treat brackish water with different problems (nitrates, heavy metals, excessive hardness, etc.) that will allow guaranteeing, at competitive prices, the water supply to coastal populations.

Finally, it is necessary to point out that all the scientific fields involved in desalination should join efforts to find solutions that allow to preserve the scarce resources and to economize the treatment of such a fabulous valuable natural asset like water. Even and daring to wish a utopia, we believe that it would be possible to solve the problems related to the shortage of the water if the dialogue between the main stakeholders (industrials, researchers, politicians, government, etc.) could surpass the national scale in order to analyze the problem at a world level, including all the affected regions.

REFERENCES

Allison, P. 2000. *Newsletter of Middle East Desalination Research Center Issue* 10 p.7.
Andrews, W. 2001. *Desalination & Water Reuse.* 11/3 p. 6–8.
Birkett, J.D. 2002. Desalination regional Report. *Desalination & Water Reuse.* 12/1.
Chiou, J.S. 1997. Desalination and Waste Treatment by a Co-generation System. Proced. IDA. *World Congress on Desalination and Water Reuse.* Vol 4 pp. 331–343 Madrid.
Hairston, D. 1999. Seawater gets fresh. *Chemical Engineering.* pp 32. March (1999).
IDA. 2002. International Desalination Association. *ABC of Desalting.* www.ida.bm/PDFS/Publications/ABCs.pdf.
Kappaz, M.H. 2002. *Desalination & Water Reuse.* 12/2 p. 12–22.
Lund, J.R. 2002. *Regional water system management; water conservation, water supply and system integration.* Cabrera, E.; Cobacho, R. & Lund, J.R. (Eds.) p. 5–11 ISBN 905809377B.
Macharg, J.P. 2002. *Desalination & Water Reuse.* 11/4 p. 48–52.
Moch, I. 1997. Project economics of dual purpose plants with RO technology. *Proced. IDA World Congress on Desalination and Water Reuse.* Vol 5 pp. 355–364 Madrid.
Sancho, et al. 2002. *Desalación de aguas.* SPUPV. Universidad Politécnica de Valancia.
Shiklomanov, I.A. 1998. *World Water Resources.* UNESCO.
Steven, J. 1999. *Desalination & Water Reuse.* Vol 8/4 pp. 34–40.
Tonner, J. 2002. *Desalination and Water Reuse.* sources directory pag. 4.
Urrutia, F. 2001. *Noticias AEDyR.* http://www.aedyr.com/
Wangnick, K. 2002. *Wangnick consulting GMB.* http://www.wangnick.com/market.htm.
Wiseman, R. et al. 2001. Desalination 138 7–15.
Wolfe, P. 2000. *Water & Wastewater Internacional.* (15)/3 p4.

Economical aspects

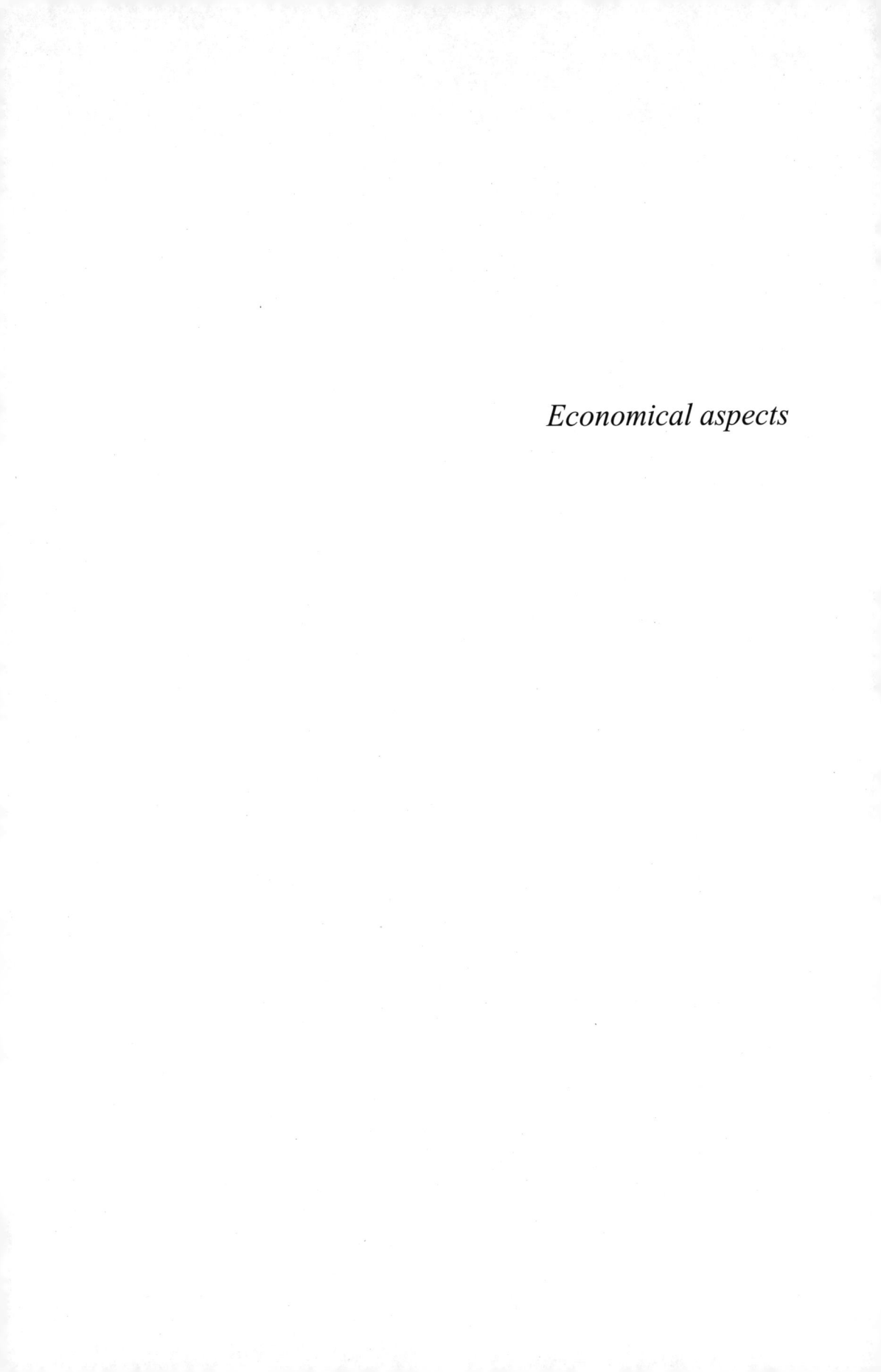

Water pricing: a basic tool for a sustainable water policy?

A. Massarutto
Dipartimento di Scienze Economiche, Università di Udine, Italy

ABSTRACT: Recent European legislation adopts the concept of full-cost recovery (FCR) as the guiding line to price water services. In this paper, the meaning of the FCR concept is analyzed and the economic rationale for its application is discussed. The basic idea here developed is that FCR is a misleading concept, since costs are by definition recovered, and the problem is to assess how it is covered. After presenting the various options available for cross-subsidization, an accounting framework for analyzing the structure of cost recovery of water services is proposed. Alternative ways of financing the service are listed according to their distance from the pure long run marginal cost, representing the theoretically optimum, though often practically unfeasible solution.

1 INTRODUCTION

A wide debate among water policy experts, economists and policymakers in the last decades has affirmed the importance of water pricing as a basic instrument for achieving sustainable water use (Rogers et al., 1998 and 2002; Dinar, 2000). Originally conceived as an economic principle aimed at efficient water resources use, water pricing has been later on understood as an instrument of environmental policy aimed at implementing the polluter-pays principle (PPP) and, more recently, linked to the idea of water as a social and economic good, one of the three corner stones of the definition of "water sustainability" contained in the Dublin Statement (Briscoe, 1996).

The political debate has often translated this principle into the more operational concept of full-cost recovery (FCR) as the guiding principle for setting prices for water services and financing the water industry. The European Water Framework Directive (Dir. 2000/60, hereafter WFD) makes explicit reference to this idea, by requiring (art. 9) that water pricing would be used in order to foster an efficient use of water resources and that each user is charged so as to guarantee an "adequate" recovery of costs, including those related to environmental externalities (European Commission, 2000). Similar recommendations have been repeatedly put forward by international entities such as the OECD and the World Bank (OECD, 1987 and 2000; The World Bank, 1993).

Many studies and debates have been carried on in recent years in order to discuss methodological and empirical aspects of FCR. However, there is still some confusion and overlapping meanings between concepts. Therefore, FCR, marginal-cost pricing and incentive pricing are often used as synonymous and proposed unresponsively as the "optimal" water pricing rule.

This substantial unclarity is probably in relation with the different perspectives from which water policy issues can be considered. Thus, the "individualistic" perspective of standard economists tends to stress the concept of individual marginal cost pricing that, in a well-functioning market, allows an efficient resource allocation (i.e., guarantees that, at the margin, each quantity of water generates the same value in all alternative uses, and collective welfare could not be improved by subtracting a certain quantity to a use and allowing it to another). The main problem for economists is to ensure that available economic (scarce) resources are allocated in the most beneficial uses (Spulber & Sabbaghi, 1994). The adoption of the polluter-pays principle (PPP) requires each user of resources having an economic value to compensate the society for the subtraction of these resources to other beneficial uses in the present or in the future. The internalization of the full opportunity cost of economically-valuable resources ensures that the best resource allocation is guaranteed from a microeconomic point of view.

On the other hand, the "hierarchical" perspective typical of engineers has been keener to consider the issue of cost recovery, intended as a balance between revenues and total costs. The main problem that is perceived here is how to finance construction and maintenance of water infrastructure in order to guarantee its long-term development, conservation and operation. The crucial point here is the willingness to ensure that water supply utilities dispose of *adequate financial resources* in order to sustain the costs in the frame of a declining role of the public sector in the water economy. In this meaning, FCR is functional to privatization policies and to the gradual retirement of the welfare state from the water sector: the money of taxpayers would then be spent in other public policies, less "mature" and demanding. To put it in other words, this purpose of FCR has more to do with politics than economics.

In turn, the "egalitarian" perspective is more concerned on the issue of price affordability and is in generally quite suspicious with respect to water pricing in general, and cost-recovery in particular. It sees water basically as an essential resource to be guaranteed to everyone regardless the economic condition. Water is seen as a source of socially relevant environmental functions, and the pricing issue should be inspired by the idea of "fairness" and equity.

Finally, the "environmentalist" point of view tends to emphasize the need to use water prices – and whatever other instrument – in order to promote the achievement of ecologically sustainable water management, in order to curb down pressures for consumptive water demand. The cost recovery argument here remains in the background, while the main concern is to avoid overuse and depletion of existing "natural" capital. Water pricing is seen as a tool for achieving an exogenously given environmental policy target.

This article tries to clarify the differences and to understand to which extent it is possible and useful to integrate them within a more general perspective based on the concept of sustainability. It adopts an economic perspective and therefore might be "biased" by the economic way of perceiving scarcity, efficiency and optimality. On the other hand, it explicitly refuses the point of view of "standard" economics and adopts an ecological economic perspective, whose main feature consists in the attempt to integrate the economic way of reasoning within a broader interdisciplinary understanding of environmental policy problems (Farber et al., 2002; Faucheaux et al., 1998; Green, 2003).

Section 2 provides a definition of the basic concepts whose understanding is crucial for discussing the pricing issue.

We first deal with the concept of "cost", defined as "opportunity cost" according to the economic theory. The crucial distinction between "industrial" (or "financial") and "external" costs is developed, before discussing issues related to the correct measurement and assessment of costs.

We move then to the understanding of the concept of "price", seen as one of the channels through which the economic balance of the water cycle is achieved. We argue that in the reality we rarely have "prices" in microeconomic terms, but rather a complex mixture of financial instruments (tariffs, specific ear-marked taxes and charges, general taxes). Each of these financial instruments may involve different degree of compensation among areas (territorial cross-subsidies), among consumers (e.g., productive uses vs. households) or among sectors of the economy. For this reason, what is mostly important to assess is not "cost recovery" as such – since it is always achieved in a way or the other – but rather the *financial structure* of the water economy and the related set of *incentives* that any financial structure provides to water users and other actors located along the value chain.

Finally, we discuss the concept of *value*, first by considering it only in its economic dimension, then by enlarging it to a broader and encompassing concept of sustainability.

Our conceptual model is now ready: water prices – intended as a particular way of financing the water sector through direct payments made by users – should be evaluated with respect to sustainability of water management, that involves the efficient allocation of natural water capital, the conservation of "critical" natural capital, the appropriateness of investment and depreciation patterns, affordability and satisfaction of basic water needs.

Section 3 develops the main arguments brought forward in order to justify the use of water prices. We can distinguish at least 3 different rationales for water pricing that are in some way linked to the definition of sustainability discussed above.

The first argument is based on the concept of allocative efficiency. It refers to the definition of "water as a social and economic good" and aims at an efficient allocation of water to its most productive uses (and, also, to an efficient allocation of money spent by public and private actors for water supply and sewerage infrastructure).

The second argument sees prices as an instrument of environmental policy. Given an environmental policy target that the government aims at achieving, environmental taxes – including water prices among others – might be helpful in achieving any given environmental policy target. The scope of pricing here is with respect to the objective of maintaining the critical natural capital.

The third argument is concerned on the economic balance of the water sector. FCR allows operators to achieve financial stability and to address safely the private capital market. The scope of pricing with this respect is to ensure that sound economic criteria are employed in the evaluation of costs and enough economic resources are devoted to investment (while public transfers are more risky and irregular).

While being potentially helpful for the sake of sustainability, water pricing has also some potential shortcomings that should be discussed. Section 4 discusses the limits of water pricing by analyzing the most important drawbacks, arising from low elasticity, transactions costs, public goods components, and perverse effects that can originate given the complex structure of the value chain of water services. Equity issues (affordability and accessibility) are also discussed.

In the end, the main thesis of this article is that the debate on pricing has somewhat overlapped different and alternative concepts. Water pricing allows an efficient resource allocation only provided it is based on marginal costs. Environmental policy objectives require the use of prices (or any other kind of environmental taxation) regardless the cost, provided that some proportionality with the undesired externality is maintained. In turn, cost recovery does not necessarily require any correspondence to marginal costs nor too complex scheme of volumetric pricing.

Once marginal cost pricing is left out (and this occurs in general because it is too costly and complicated to adopt), it becomes relatively indifferent from a microeconomic point of view if costs are paid by water users (through direct charges), by taxpayers (through the fiscal system) or by other economic actors (through cross-subsidies of different kind). What should be assessed is the set of incentives that the financial structure provides to the actors along the value chain of water services.

2 SOME USEFUL DEFINITIONS AND CONCEPTS

2.1 *The cost of water: industrial costs and external costs*

The standard economic definition of "cost" of any given good is based on the concept of "opportunity cost", namely the economic value of the sacrifice that is required in order to provide that good. Therefore, the cost of a certain good (whose production requires a certain amount of inputs such as labour and capital) is equal to the value of alternative goods that could have been obtained if labour and capital would be allocated elsewhere. Economic value is often defined as "willingness to pay" (WTP): it measures how much another user would be ready to pay in order to use those inputs in an alternative way.

A first relevant dimension is obviously the "industrial" (often referred to also as "financial") cost, namely the cost of services and infrastructure that are required in order to provide water to users.

Financial costs can be further divided in categories: operational and investment cost, labour and capital, etc.

In a perfectly competitive economy, market prices would reflect precisely the opportunity cost, and could therefore be used in order to evaluate it. On the other hand, non-perfectly competitive world, market prices might correspond only partially to opportunity costs, and this will pose some difficulties.

Environmental economists have long understood that when dealing with natural capital the relevant cost is not only the "industrial" one – namely, the opportunity cost of labour and capital used in order to make the natural resource available. There is also an "external" cost to be accounted for, namely the opportunity cost of the resource itself (Pearce & Turner, 1989; Spulber & Sabbaghi, 1994).

The cost of water is not just the cost of economic inputs that are necessary to use in order to make water available, but also the costs that some other users have to bear by means of reduced opportunities of using the "natural capital" in alternative ways, and the costs that are necessary for maintaining and improving the quality and quantity of the "natural capital" itself up to a level that is considered sufficient in terms of long-run sustainability.

The category of external cost can as well be further detailed by considering scarcity costs (the opportunity value of water in alternative economic uses), economic externalities (e.g., positive or negative effects for other economic actors that are not accounted for by users) and environmental externalities (regarding "environmental functions" that do not belong to the economic sphere), all evaluated in monetary terms (Rogers et al., 1998).

Further categories could be as well used. External costs might be intra-generational (for example, water pollution obliges downstream users to sustain extra costs for treating water up to satisfactory quality standards) or inter-generational (for example, aquifer contamination due to industrial or agricultural pollution). On the other hand, external costs can be intended in static as well as in dynamic terms. Permanent depletion of the resource can occur under specific circumstances: for example, aquifer overexploitation could lead to saline intrusion, and certain sources of underground pollution or eutrophication operate in the long run and can be recovered only in the very long run. Dams and reservoirs cause a permanent modification of landscape and outflow patterns. Wetlands disruption may well be reversed in the future by creating new ones, but in the meanwhile some species might have gone extinction.[1] But leaving apart these special cases, externalities are more typically occurring on a seasonal base, given that water is a renewable resource. In other terms, externalities occur because some uses of water prevent other environmental functions that the same water might offer. In principle, such externalities could be eliminated through a different pattern of resource allocation.

Finally, not all environmental functions that are sacrificed can be expressed in monetary terms (and thus easily traded-off with each other). Non-monetary externalities should also be considered: in this last case, of course, they cannot be accounted for as "costs" since they are unmeasurable in monetary terms, yet they might be figured in satellite accounts (Ekins, 2000).

It is important to stress that externalities may or may not arise – and their dimension might be greater or smaller – depending on local circumstances. In fact, an externality occurs only if water – or better to say environmental functions based on water – becomes "scarce", in the sense that some potential users are left unsatisfied. Figure 1 can clarify this point better.

In order to consider both internal and external costs and benefits, we can consider for each eventual water abstraction a "demand" – represented by the net value to the potential users; and a "supply", which is a marginal cost represented by the marginal value of alternative uses that might be sacrificed.

In Figure 1 we have stylized the typical allocation problem that occurs when a water user wishes to abstract water from a given point. On the horizontal axis, we measure the actual water availability at that point; we assume that on average it corresponds to OO'.

Curve DD represents the WTP of the water user. It has a negative slope, since we suppose that each additional quantity of water generates a diminishing improvement of users' benefit. The curve represents a net value, that is to say, the value of the additional product that can be obtained with water use, net of the industrial cost that has to be sustained in order to have that quantity of water available.

Curve NN' represent the WTP of other social actors (e.g., downstream water uses and instream uses) for maintaining the resource unused. The shape of the curve can be justified on the same

---

[1] In similar cases, the concept of "user cost" applies, meaning the present value of future uses and of alternative means of satisfying them in the future. Yet in most cases, the relevant dimension is the actual one: externalities are mostly caused by actual water uses to other potential actual water uses or actual environmental functions of water.

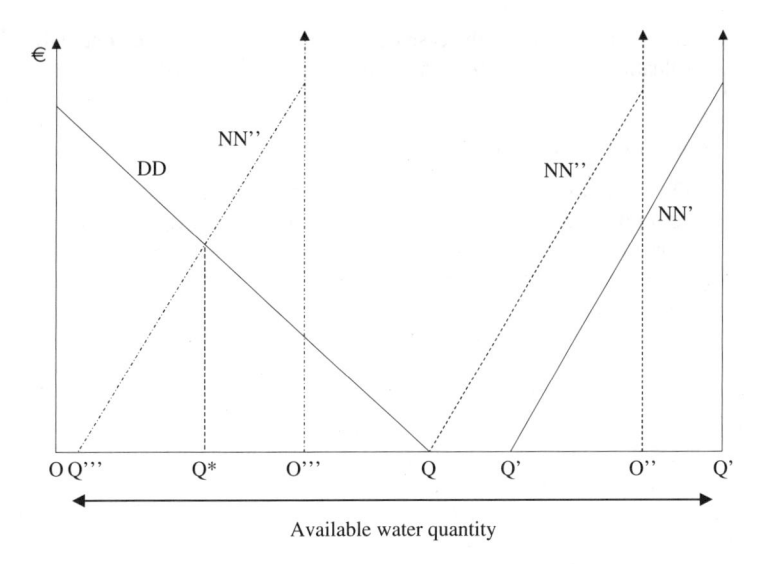

Figure 1. Optimal allocation of water, water scarcity and external costs.

grounds: the additional quantity of water left in the watercourse has a diminishing value; for example, the WTP for environmental instream uses falls to zero for additional quantities that exceed the "minimum acceptable flow". Curve NN' can therefore be interpreted as the marginal scarcity cost of water abstraction at that point.

Each user wishes to position on the point on which the marginal value will become negative: point Q for users and point Q' for non-users. If the available resource is OO', users would use OQ, non-users would require the quantity Q'O' be left instream, and a further additional quantity QQ' would remain unused. In this case, there is no allocation conflict, or, in economic terms, there is no external cost implied by the water use.

If the available quantity falls to O", the NN' curve is shifted to the left and becomes NN". In this case, both users' and non-users' demand use up all the available water: OQ is used upstream, QO" is taken by other users, and no free water is left. For available flows below O" (for example, O'''), a conflict arises: if water is used up to the level that satisfies users, an external cost arises (represented by the area below NN''' between Q''' and O'''). If water is left unused in order to generate no external cost, users would lose the value measured by the area under the DD curve between Q''' and O''' (plus the area between O''' and Q that would be left in any case due to water scarcity). An economic optimum could be found in point Q*, where the two marginal values are equal and the total value is maximum.

From this model, we can see that external costs may or may not arise, according to actual availability of water.[2] In case it arises, in order to comply with the requirement of full cost recovery, water users should compensate the society for the external cost (and therefore, they should be asked to pay for the equivalent of the area that is sacrificed, below the NN curve). Measuring this externality, however, is definitely not a simple task. While for "productive" uses it is relatively easy to calculate the

---

[2] In the long run, allocation rules decided by water authorities normally consider the average availability of water and make a compromise between potential stakeholders, including non-users. However, even if the long-run allocation has been made correctly, short-term variability might well produce temporary scarcity, even in river basins whose average flows are, on average, well above users' request. In any given location, therefore, external costs might arise in some years and not in others. For this reason, in more recent times flexible allocation rules (e.g., expressed in terms of percentage of available flow and not in absolute value) are being introduced.

additional value generated by water, in the case of instream uses this can be done only with the use of environmental valuation techniques, what normally requires expensive ad-hoc studies.[3]

2.2 Prices, charges, taxes: alternative ways of recovering the cost

Correspondent to the water *cost* we have the water *price*, that is, the amount of money that is paid by a water user in order to purchase water services and/or to achieve the right to use water. The water price requires a transaction between a "supplier" and a "user"; in fact, a water price might even not exist, provided that the user is able to self-supply water and no taxes or charges are requested. In fact, "cost recovery through price" could be assessed by comparing how much users pay for water respect to the total cost of water as defined above. In practice, this is not always easy.

We can individuate 4 levels representing the actors that might be present along the value chain of the water service (Figure 2). Transactions among levels and/or between each level and third parties supplying inputs correspond to monetary transfers.

The first level is represented by the owner of water resources, which in Europe is typically the state. Water resources uses are administered and governed by public authorities in order to achieve water policy objectives. The allocation of use rights between alternative users is one of the most important tasks that are fulfilled at this level. This level is normally financed by the public budget, even if water users might be required to contribute by paying various forms of charges, taxes, redevances, that might or might not be correlated with administrative costs and/or external costs.

The second level is represented by operators of large water storage and supply schemes, very often built under planning of public authorities and administered by dedicated agencies, either independent or controlled by the central or regional government.

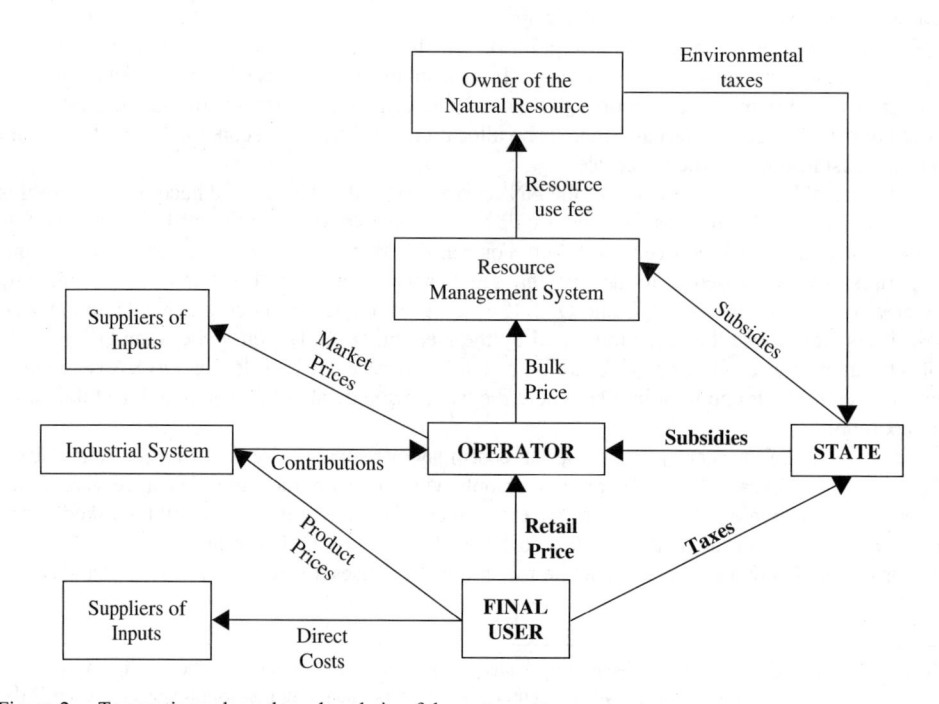

Figure 2. Transactions along the value chain of the water sector.

---

[3] For a survey of applicable methodologies either for productive and environmental water uses, see Fontana & Massarutto (1995), Gibbons (1986) and Tihansky (1975).

With various forms (public or private law; with the involvement of water users or directly expressed by the state; with or without the participation of the private sector, etc.), these entities build and/or operate the infrastructure; in the economy of water services, they act in practice as bulk suppliers. In many cases, these facilities serve multiple uses (namely, flood protection, flow regulation, energy, irrigation, public water supply), according to rules and priorities that are determined politically or agreed by users. In some countries (e.g., in the US) some degree of "water market" is envisaged at this level: owners of water rights might have some freedom to sell water to other users and/or to make their infrastructure available to other users.[4]

It is quite typical that financial resources are at least partially supplied by the public budget. Bulk supply nonetheless might, and in fact often does, involve a price paid by users, covering at least some fraction of the corresponding cost.

The third level is represented by the water supply and distribution system: it corresponds therefore to the retail sector. In Europe, the retail sector is normally functionally distinguished between separate networks for public water supply and sewerage (sometimes operated under the same system and by the same operator, sometimes separated), irrigation and sometimes industrial uses. While urban water services are normally under the control of municipalities or inter-municipal associations (with the only exception of the UK, where responsibility has been put on the central state in 1973 and privatised in 1989), retail supply for irrigation is normally managed through associations of farmers, sometimes voluntary and private, sometimes compulsory and/or governed by public law arrangements.[5]

Of course, farmers and industrial users might purchase water also by other water supply systems, and namely by public water supply operators. This choice is in absolute not common for irrigation, if not for very particular uses such as small horticulture and livestock breeding. The reason of the presence of a dual supply lies in the fact that irrigation normally requires water of largely inferior quality than human consumption and in far larger quantities, and therefore the involved cost in case of purchase of water from the PWS would be prohibitive. Industrial uses, in turn, are normally self supplied when water consumption is high, while they rely on the PWS for sanitary uses. Industrial sewerage often entails collective bodies with some public-private partnership.

Retail suppliers sometimes purchase water from bulk supply schemes (second level), sometimes produce water directly. In some cases, neighbouring networks trade water among themselves: a certain unit might produce water for its own necessity and also sell bulk water to other units. In a few cases (e.g., the British water companies or the Acquedotto Pugliese in Italy) retail suppliers operate on very large territorial units and manage complex water production schemes (dams, reservoirs, interbasin transfers).

All institutions at the third level are normally working on a cost-recovery base, even if the public budget, again, might, and often does, contribute in various ways, especially for capital expenditure.

Finally, we have water users. Even when all of the other three levels are present, water users provide themselves a number of activities, including the purchase and operation of equipment (e.g., pumps for irrigation, preparation of land). Unlike individual consumers, agriculture and industry are likely to produce at least some parts of the service value added directly. For example, an industrial polluter might wish to treat water internally before discharging it into the public system, in order to benefit from lower charges. Farmers may wish to manage their own boreholes instead of connecting to a collective irrigation system. Therefore, while the first and the fourth level are always present, the third and the second are represented only in Southern Europe. Sometimes there might be overlappings and division of competences is not always totally clear-cut. In other cases, the presence of multiple uses in one or another stage creates the scope for cross-subsidies and compensation of costs among public and private uses.

---

[4] In Europe, only in the UK there is some debate about the possibility to introduce common carrier obligations to owners of storage and transport facilities; the scope for similar policies seems nonetheless modest (Massarutto, 2001b).

[5] These institutions often date back to previous centuries or even to the Middle Age; they are very common in Europe for most of the collective actions entailed by the agricultural activity, not only for irrigation but also – for example – for land drainage, flood protection and rainwater management. Industrial uses for the most part are self-supplied, yet in some areas there are dedicated supply networks.

Consumers do usually pay an amount of money to operators of public utilities in exchange for the service they receive (retail price). These payments are called with different names (prices, taxes, charges, fees, etc.) and might also have different juridical nature (e.g., they might be compulsory or not; they might be charged directly by the operator, or by the public authority first, and then paid to the operator according to contractual arrangements); whatever the name or the nature, however, they correspond, very broadly, to the cost of the service, in the meaning that they are charged by the service operator and represent (one of) the revenues accruing to the operator.

The juridical nature of subjects located along the value chain can influence the nature of the payment. For example, in some cases public authorities responsible for providing the service obtain a revenue that belongs to the category of taxes (e.g., because it is compulsory) even if it is in fact calculated on a cost-recovery base (e.g., the wastewater charge in many European countries). In other cases, payments have the juridical nature of tariffs or prices, even though they maintain fiscal components (like the British "water charge", that in fact is very similar to a property tax).

As these examples show, the concept of water price has very different meanings in different parts of the world, and it is to some extent simplistic to discuss on "cost recovery" by simply comparing the final "water price" to the average cost. What we should assess in turn is the overall financial structure of the system, following the conceptual model underlined in Figure 2, where we have highlighted the basic elements of the public utility system: utility operators, final consumers and the state.

It is important to note that state contributions (subsidies) are in fact financed by taxation. In other words, the existence of subsidies does not mean that the cost is left "uncovered" – what is in fact impossible – but rather that individuals pay as taxpayers instead than as water users. Only in case there is an *intergenerational externality* (for example, because the present generation fails to cover the whole cost and transfers a part of it to the next one through public debt or depreciated natural or artificial capital) we could say that the cost is not "recovered". Of course subsidies financed through taxation introduce many potential distortions in the allocation of resources: the correct balance between subsidies and prices – or, better to say, between fiscal and endogenous financial sources – should be assessed having in mind the set of incentives that alternative financing structures will provide to water users.

One last opportunity that should be considered is the involvement of the private industry in the provision of the service. The concept of extended producer responsibility has been widely applied in environmental policies, though not too much in the case of water except in the case of sewerage. If the private sector is asked to accept certain costs within the product costs, this means that a part of the "cost of water" is in fact internalized by manufacturers of goods and later transferred on the price of those goods.

This means in practice that the burden of the cost will finally be passed on to the individual citizen, yet this time as a consumer and not anymore as a service user or as a taxpayer.

The same occurs, even with no incentive effects, when commercial and industrial customers are receiving the same public service as households and pay a higher price, therefore generating a cross-subsidy. This cost will in fact be reflected in the general cost of the commercial premise and be transferred on prices.

In the end, it is not possible to trace a precise line between "prices" (paid by individual consumers) and "subsidies" (paid by the public budget). The financial structure of water services (that is, the share of payments that individuals make as water users, consumers of goods and/or taxpayers) usually entails some degree of cost-sharing. This might occur on a territorial basis (e.g., the water cost is paid by water users, but costs are equally divided among a large territorial unit) as well as with cross-subsidies among water uses (e.g., industrial use bears part of the costs that would belong to households or agriculture) or among different services. Or, the state can contribute with direct or indirect payments financed by taxation.

Figure 3 contains a temptative taxonomy of the different kind of subsidies, starting from those that simply occur among water users (on a geographical base, through unification of management units or through ear-marked taxes); direct subsidies paid to some of the levels involved in the supply chain

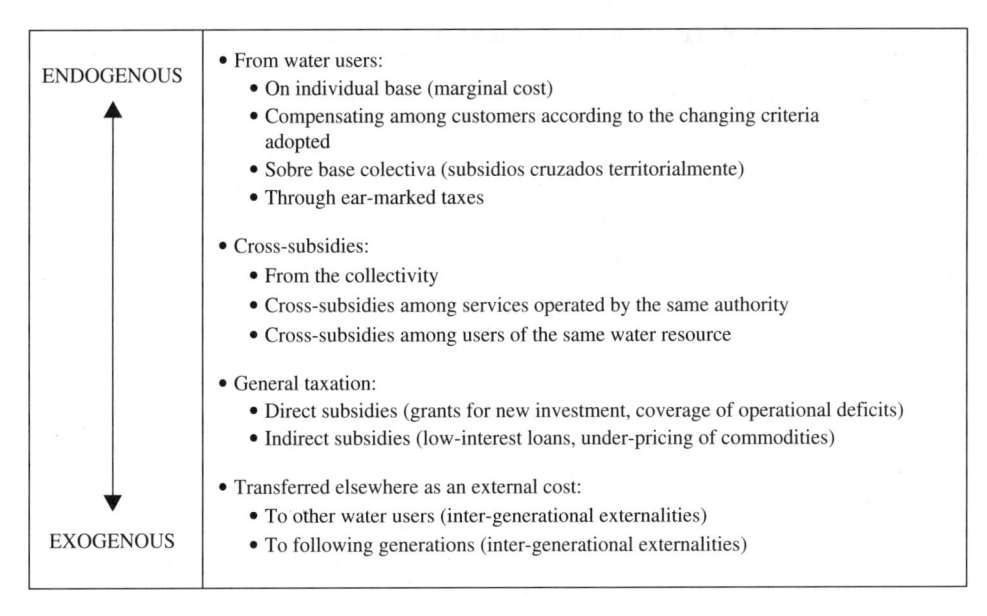

```
ENDOGENOUS     • From water users:
    ↑             • On individual base (marginal cost)
    |             • Compensating among customers according to the changing criteria
    |               adopted
    |             • Sobre base colectiva (subsidios cruzados territorialmente)
    |             • Through ear-marked taxes
    |
    |          • Cross-subsidies:
    |             • From the collectivity
    |             • Cross-subsidies among services operated by the same authority
    |             • Cross-subsidies among users of the same water resource
    |
    |          • General taxation:
    |             • Direct subsidies (grants for new investment, coverage of operational deficits)
    |             • Indirect subsidies (low-interest loans, under-pricing of commodities)
    |
    |          • Transferred elsewhere as an external cost:
    ↓             • To other water users (inter-generational externalities)
EXOGENOUS         • To following generations (inter-generational externalities)
```

Figure 3. Alternative ways of financing water services.

by a public authority; indirect subsidies (e.g., tax allowances, low-interest loans); and externalities (costs that are simply discharged on some other water users, of the present or the next generation).

The reason for this is either technical (since water resources and water infrastructure are consumed "collectively", and it is not always possible nor useful to determine the exact individual cost) or lays on ethical grounds (aiming at reducing the total cost of water to certain categories of users).

2.3 The economic value of water

The third concept is that of value, and this is crucial either in order to understand the reactions of users to water policy or for assessing the desirability of a given policy. We can distinguish for this purpose an individual value (for the individual user) and a social value (benefits that accrue to the society as a whole in terms of public goods and/or positive externalities).

A water user might be willing to purchase water (either for off-stream abstractions or for in-stream uses) as long as the price paid, plus the direct costs, is lower than the value that the same user attributes to water. In economic terms, value is synonymous with utility: a good is valuable as long as it provides utility, and this can be measured by means of the utility of alternative goods. If we abandon the strict utilitarianism of neoclassical economics, other dimensions should be considered. In the case of water, it is particularly important not to neglect the dimension of human needs, as well as values that belong to non-economic spheres (e.g., symbolic, religious, etc.).

Limiting our analysis to economic values only (that is, those related to the sphere of utility), these are in any case quite larger than the pure and direct utility, since the concept of "economic value" entails many dimensions: direct and indirect values, use and non use values, economic and non economic values (Turner & Postle, 1994; Fontana & Massarutto, 1995). The sum of economic values that an individual is ready to attribute to water is summarized by the concept of willingness to pay (WTP), that is, the maximum amount that a person would theoretically accept to pay – the maximum amount of alternative welfare opportunities it would be ready to sacrifice – in order to have water available while remaining better off.

Productive uses, such as irrigation, normally entail a use value only. In such cases, WTP can be evaluated by calculating the additional income generated by an additional quantity of water. This is clearly influenced by the productivity of water. For example, in the case of irrigation different sets of crop prices and other market features will influence the economic performance of the farm and therefore the desirability of irrigation.

The evaluation of individual WTP is not always easy, since the contribution of water in general does not depend simply on the marginal improvement of production. Other aspects, often difficult to individuate and quantify should be considered as well (e.g., reduction of risk, possibility to engage in more risky and profitable land use choices).

In case of final uses, in turn, this procedure cannot be used, and other techniques should be adopted. For example, individuals may be asked to express their willingness to pay through the use of appropriate surveys such as contingent valuation. In alternative, we can imagine to derive the WTP from the behaviour of individuals (e.g., if they are ready to spend a certain quantity of money for visiting a fishery, we can assume that their WTP for that fishery is higher). As far as use values are concerned, these techniques can reveal magnitudes of some importance. Gibbons (1986) reports cases in which the indirect use value of instream uses (especially recreational ones) expressed by individuals could even be higher than the use value of consumptive uses.

In case of environmental externalities, other values should be considered; those are not directly linked to the actual use, but could as well be expressed in terms of monetary WTP at least in part. Conceptually, this WTP could be measured if a stakeholder for some non-economic environmental function would accept a given monetary compensation in exchange of the sacrifice of that environmental function. If the sacrifice is not acceptable at all (i.e., it is considered as "inestimable"), a monetary value could be as well be measured in terms of the cost of the cheapest alternative for satisfying the same environmental function (Merrett, 1997 and 2002).

Individual WTP is an important component of value. However, when dealing with environmental resources such as water, we should include also collective dimensions of value that belong to the category of "public goods" or "merit goods".

In the first case, we are dealing with values that are not accruing to a specific individual. In technical terms, a good is "public" when it can be consumed contemporarily by many users without causing extra costs (a new user can be added without a marginal cost) and the exclusion of any individual consumer is impossible or too costly. Flood protection, drainage of rainwater, health benefits associated with centralized water supply, collective compliance with environmental regulation can be cited as examples. Environmental values (e.g., conservation of wetlands and of the river landscape, creation of natural parks) are other examples.

The second case occurs when there are value dimensions that are not expressed by individuals, but are instead revealed through the political process. Affordability of water to low-income families is an obvious example. Another example can be made in the case of irrigation: it does not only contribute to the farmer's income (individual WTP), but produces also other benefits that are valuable from the society as a whole (e.g., it may help to improve one country's ability to be self sufficient in food production; it may help to slow down the abandonment of countryside; it can allow employment opportunities, and so on).

2.4 *Sustainability*

According to the now commonplace definition agreed in the Dublin International Conference on Water and the Environment of 1992, the concept of sustainability should involve the ecological dimension (water as a finite and vulnerable resource), as well as a social (accessibility of water seen as an indispensable social need; participatory and democratic approach in water policy decisions) and an economic one (water as an economic good to allocate efficiently). This three dimensions, appropriately specified in terms of indicators, provides the evaluation basis for water use.

The concept of sustainability, defined in this way, allows to enlarge quite a lot the simple rationale of neoclassical economics, according to which a given allocation of water resources is "efficient" – and therefore optimal – as far as all users obtain the same net marginal benefit (resulting as the difference between individual value and the sum of price and individual costs, possibly including external costs and benefits in the calculation) and public money is spent up to the level that equals the marginal returns to society.

The sustainability concept allows us to consider further dimensions that are not appropriately dealt with in the neoclassical economic framework, namely *equity* (defined as a human right to access to basic needs) and *conservation of natural capital*.

Water is seen then either as a source of irremissible values or as a "critical natural capital", providing basic functions for which there are no substitutes. Nonetheless, the concept itself remains to a certain degree ambiguous and requires an appropriate specification that can occur only through the political process and on a case-by-case basis.

As a critical natural capital, sustainable use of water cannot be assessed at an overall scale, but rather on a local scale, where all relevant environmental functions should be preserved and water use should be kept below natural recharge of the renewable resource (Faucheaux & O'Connor, 1998; Ekins, 2000). The relevant territorial scale for the water balance can be a larger one only if resources are "averaged out" in a larger territorial unit by physical man-made infrastructure. Yet in order to do so, an economic as well as a social dimension of sustainability should also be considered: the cost of infrastructure should be fairly shared among generations (i.e., the next generations should not be charged for benefits that are enjoyed by the present generation); at the same time, the price charged on users for this purpose should not exceed a critical limit that excludes those who cannot pay (Barraqué, 1999).

The equity argument, in turn, legitimates water policies aiming at a "socialization" of the cost of large water schemes, in order to allow "each" user – regardless its geographical location – to satisfy at least "basic" water needs. This principle of equity in the allocation of water resources is well agreed (Correia et al., 1999). However, the exact definition of "each user" and "basic need" is of course a delicate political issue; in Mediterranean countries, the usual approach in the past has been to intend the geographical size at least at the regional or district level (municipal for public water supply); and to include irrigation among the "basic needs". The result has been a significant degree of socialization of the cost of providing enough water for satisfying irrigation needs in the driest regions. In Spain and in Italy, as well as in the rest of Southern Europe, water resources planning has meant for a long time assessment of "water needs" at the desired territorial level and provision of supply systems that would be able to meet the expected need, whatever the price – or with just a very rough cost-benefit analysis, very much over-estimating social benefits (Vergés, 2002a; Barraqué, 2000a).

As a general criterion, widely agreed in spite of slightly different formulation and terminology adopted, we can assume that relevant functions of the critical natural capital ("water needs") should be effectively satisfied without creating prejudice for the integrity of the natural resource (quantitative and qualitative water balance between available renewable resources and uses) while "socially relevant" water uses should be affordable for anybody regardless income levels and social conditions. Therefore indicators of sustainability should reflect available environmental functions, appropriateness of investment in and depreciation of natural and man-made capital in order to achieve a satisfactory trade-off between them, affordability and accessibility issues, allocation of water among competing uses according to its "value", and finally an efficient use of economic resources in order to ensure that water services are supplied so as to avoid the creation of monopoly rents of any kind (Figure 4).

Transforming these general concepts into proper indicators is not straightforward. We cannot enter deeply into the complex and rich debate on the practical definition of indicators for assessing sustainability of water use (see de Carli et al., 2002 for a discussion). We can nonetheless note that definitions and underlying principles maintain some degree of fuzziness.

The concept of "prejudice", for example, is ambiguous. It can be intended as an absolute prohibition of alteration of the natural capital stocks (in terms of quantity and quality) or, in a more elastic way, keeping in mind that even a degraded resource could continue to be used though with higher treatment costs and technological equipment. In other words, since there is some possibility for substituting natural capital with man-made capital, a certain degree of diminution of the natural capital stock could be considered as sustainable provided investment in man-made capital can compensate for this.

The concept of "water needs" is as well quite a slippery one, since its very meaning should be assessed through a political evaluation. Just to make an example: we could assume that water for

Environmental sustainability *Discourage depletion of critical natural capital*	Equity *Guarantee that "merit uses" have due access to water resources under fair and equitable conditions*
• Guarantee ecological functions of water natural capital • Minimize the recourse to "supply side" • Minimize the alteration of natural outflow patterns	• Identify "water needs" (⇔basic environmental functions) • Keep level and dynamics of prices below the threshold that makes it unaffordable for some users • Achieve an equitable and democratically accepted way to share the cost of managing water resources
• Guarantee financial stability of water systems • Compensate adequately economic resources that are used as inputs • Cash flows should guarantee the conservation of value of physical assets • Each new infrastructure binds the next generation to cover its cost in the future⇔Minimize the creation of artificial capital **Financial sustainability** *Guarantee long term reproduction of physical assets*	• Allocative efficiency: water should be allocated in order to privilege uses with the highest social value • Allocative efficiency: the cost of provision of water services (to non-merit uses) should be confronted to their value • X-efficiency: costs should be as close as possible to the minimum (intended in dynamic terms) • Not encourage over-capacity, over-staffing, gold-plating, etc. • Cost coverage should be intended as for efficient costs only • Regulation should ensure an optimal allocation of risks among shareholders, users and taxpayers **Efficiency** *Guarantee that water is allocated to most beneficial uses and economic resources are not wasted*

Figure 4. Sustainability in 4 objectives.

human consumption belongs to the "water needs" to be satisfied, but is the same true for all household uses? Should it also apply in the same way for second houses?

For this very reason, it has been argued that the choice of indicators and their evaluation should not be imposed "ex ante" by means of a general definition of sustainability encompassing all relevant criteria and to be applied everywhere; rather, it would need to result from a participatory political process, in which relevant environmental functions are identified, trade-offs verified and measured, alternative solutions are examined (de Carli et al., 2002).

3 THE ECONOMIC RATIONALE FOR WATER PRICING AND FCR

3.1 *The economic optimal pricing rules: allocative efficiency*

According to the economic theory, optimal pricing of any private goods or services should reflect their long run marginal cost (LRMC). Each customer should pay for any *additional unit* according to the *additional cost* that her demand requires. In this case – and only in this case – the price functions as a signal of economic scarcity. Each user will decide to purchase an extra amount of water only if its price is greater than its value. As far as the price includes all relevant cost dimensions, we can be sure that society as a whole improves its welfare: the utility obtained is greater than the disutility created.

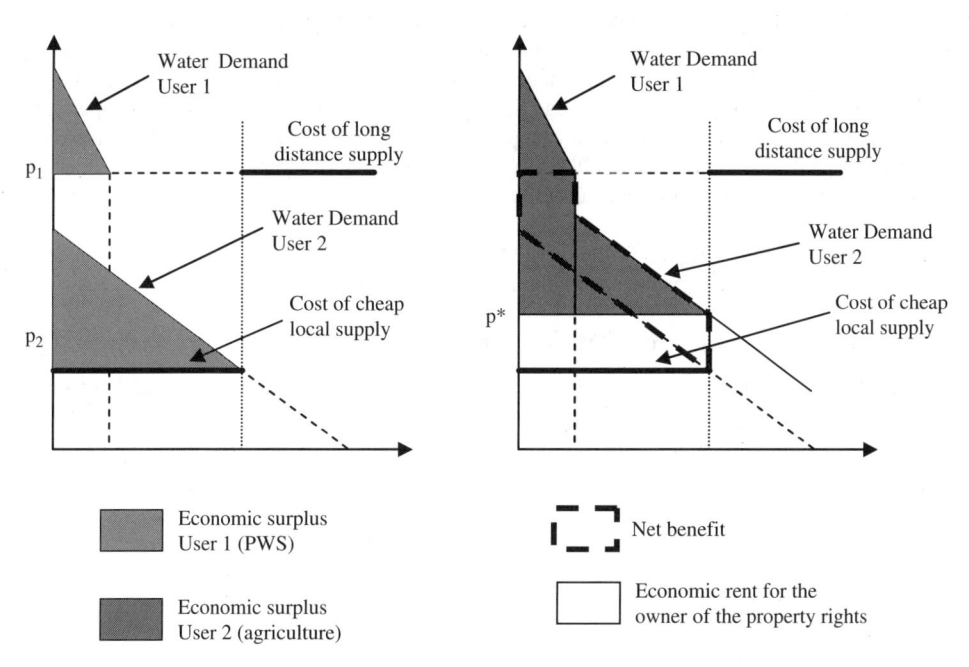

Figure 5. Pricing as an allocation instrument.

It is quite intuitive to understand why this happens. If marginal value is greater than marginal cost, there would be a potential welfare gain by supplying an additional quantity. Vice-versa, if marginal cost would be higher, there would be a welfare gain from reducing supply.[6]

Therefore, if for some reason prices diverge from marginal cost, water users will receive a distorted signal. With water price lower than MC, for example, users will be encouraged to demand more water than actually efficient.

This conceptual scheme can be applied for example in order to evaluate the opportunity to build a new water supply scheme (with higher marginal cost) in order to supply additional quantities of water. There are many concrete examples of similar questions in Europe: for example, when agriculture uses the cheapest local resources, forcing public water supply to rely on long-distance transfers.

As we can see from Figure 5, from the pure point of view of economic efficiency, it is preferable that existing water is shared so as to satisfy all water request of household supply first, and then leave the remaining to agriculture without building the new facility. The total welfare (sum of the benefit of farmers and households) is higher in this case, rather than in case household supply is obtained from long distance transfers, and all cheap local water is used by agriculture. In other words, domestic users would be ready to buy water from farmers paying a price that is higher than the agricultural income that would be otherwise obtained, and lower than the price of long-distance water.

While the application of the marginal cost rule is universal, in certain cases – namely, when costs are sub-additive and/or marginal costs are difficult or complex to calculate – the willingness

---

[6] Considering long-run costs instead than short-run ones means that investment costs should be considered as well. This is appropriate as far as new investment decisions are concerned. For example, if the additional value created by an irrigation project is lower than the investment cost plus the marginal operational costs, the project would not be economically efficient. On the other hand, when dealing with already existing investment, short-run costs might be more appropriate as a term or reference. For example, if a reservoir is already existing, the decision whether to use if for supplying water or not should be based on the short-run marginal cost only (since the fixed cost has been already paid).

to recover fixed costs has suggested second-best alternatives. As a matter of fact, when average costs are decreasing because of economies of scale, the marginal cost is lower than average cost.

We can note from Figure 6 that collective welfare continues to be maximized by the application of marginal cost, but this causes a loss to the service operator.[7] In such cases, we can either imagine that the state could cover this deficit through money transfers, or accept a minor welfare loss by allowing the operator to charge the average instead than the marginal cost. Alternatively, if the operator is able to apply different prices to different categories of consumers (and this "price discrimination" is legally accepted), the same effect could be obtained by charging different groups of consumers the marginal cost plus a quota of the fixed cost that is inversely proportional to their demand elasticity (Ramsey pricing).

FCR in the strict meaning is a further relaxation of both concepts: it requires a correspondence between total costs and total revenues, without going too precisely into the issue of which customers pay for which costs.[8]

The limits to the application of LRMC are of four types:

- The marginal cost principle is an optimal pricing rule only if monitoring individual consumption is not difficult or costly and demand is elastic enough.
- Recovery of sunk costs in the long run might not be guaranteed by the strict adoption of LRMC nor by AC and RP.

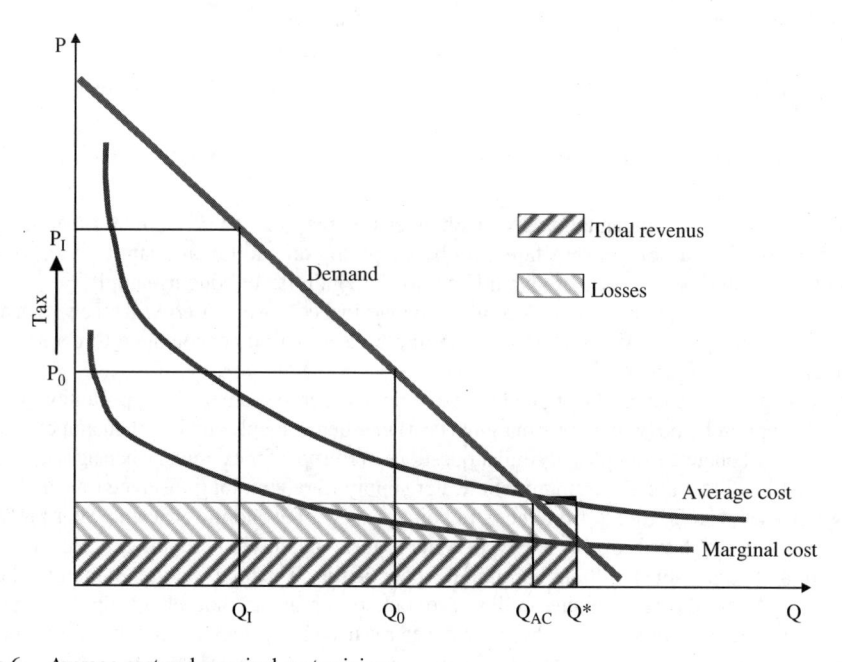

Figure 6. Average cost and marginal cost pricing.

---

[7] If the price was equal to average cost, demand would be Q_{AC}. Collective welfare could be improved by increasing output levels up to Q^*: social benefit (the area below the demand curve between Q_{AC} and Q^*) is greater than social cost (the area below marginal cost curve in the same reach).

[8] Normally, the FCR is specified as if the full cost be matched by charges on each territorial unit in which the service is supplied by an independent operator. Yet this very criterion leads to very different outcomes in Europe: to make only an example, England and Wales have only 10 large water supply and sewerage systems, while Italy or France count the separate undertakings in the order of 10,000. It is clearly not the same thing to require balance of costs and revenues for each individual undertaking, or for larger aggregates.

- Water services usually entail a public good dimension as well as a private good one; therefore, we should include among the benefits also components of collective utility that individuals do not consider (or consider only partially).
- Allocative efficiency is not the only goal of public agencies; once the allocative benefit of the marginal cost pricing is lost, the price paid acquires a fiscal nature, and there is no reason but a political one to use the PPP instead of a progressive criterion for allocating the tax burden. Distributive issues might be relevant as well: in other words, society is not necessarily indifferent with regard to the concrete distribution of costs and benefits.

3.2 Water prices as environmental policy instruments

Independently on marginal cost issues, water prices might be used by water authorities as demand management tools, and therefore maneuvered in order to achieve given water policy targets.

The rationale for this use of prices lies in the well-known theory of environmental taxes starting from the work of Pigou in the 30s until the work of Baumol & Oates (1989) that is now considered as the standard reference.[9] It is based on the idea that demand reacts negatively to price variations, for the same reason seen above: users will purchase an additional unit of water until the additional expenditure (the price of the additional unit) is lower than the economic value (utility). Since utility decreases with quantity (additional quantities produce diminishing additional utility) this results in a demand curve with a negative slope, such as the one represented in Figure 6. At lower prices, users demand more water; at higher prices, demand is reduced.

From the same figure, it is easy to understand the mechanism through which water prices can provide an incentive to curb down water demand: supposing that the demand target is Q_1, this is not achieved at the current price P_0 (since at that price, demand is Q_0). The water authority could then impose a water tax (an additional water price) so as the final sum paid by the water user becomes P_1.

We can assume then that each increase of the water price, however motivated, would have the same effect. It is important to note that the effect occurs because users pay an additional quantity of money for additional quantities of water, and not because their total expenditure for water is higher.

In other words, it is important not to make confusion between cost-recovery and incentive pricing, since the latter requires that prices are some function of the quantity that is actually demanded.

Full-cost pricing of water is perfectly compatible with a poorly incentivating system (e.g., if rules for allocating water and the pricing methods will continue to be based on flat rates); on the other side, an incentivating pricing system could be designed even if the total revenue does not recover the full cost. In any case, the economic literature invites us to be at least cautious in retaining that pricing water alone could be sufficient in order to promote water sustainability; nonetheless, it can be a very useful policy instrument provided that it is structured in an incentivating manner. The incentive effect depends on the shape of the demand curve: the more elastic the demand curve, the more effective is the price signal. Since what is important is the cost of the last quantity of water purchased (and not the total or the average cost), it can be believed that an increasing-block tariff structure could reach this effect more easily (above a certain quantity, the additional cost of a new unit grows higher).

3.3 Financial equilibrium of operators and the public budget

As we have seen in previous sections, allocative benefits of water pricing occur only if it follows the marginal cost rule, while environmental incentivation simply requires that prices are in some relation with quantity (hopefully, a monothonic growing function of quantity such as in the increasing-block model). In both cases, there is no strict requirement for FCR (in the sense that allocative and

---

[9] The main innovation introduced by Baumol & Oates (1989) lays in the fact that Pigou considered that taxes should be calculated in such a way to internalize completely externalities in order to foster an optimal allocation of environmental resources. Baumol & Oates (1989) recognize the difficulty to measure externalities and therefore advocate the use of taxes even if the target has been determined exogenously.

environmental objectives can be pursued even if FCR is not achieved), while, on the other hand, the achievement of FCR per se does not necessarily imply allocative nor environmental benefits.

The last argument spent in favour of FCR has nothing to do with incentives, rather it is based on the need to guarantee that water management systems can be self-sufficient from a financial point of view. This argument is either motivated by public budget restraints, or by the belief that independent service-oriented water management systems are keener to behave efficiently rather than state agencies.

The first motivation can be considered as a *de facto* statement rather than a *normative* judgment. Regardless political preferences in favour or against state intervention in the economy, the supply of state-financed goods requires public spending and therefore taxation or creation of public debt. Most western economies are facing a crisis with this respect: the growth of the size of the public sector has already reached a critical point; fiscal policies are thus constrained and better concentrated in those areas of public intervention that require redistribution of resources. In this sense, expenditure for water services should be faced through the direct involvement of users, even when it maintains some elements of taxation, which will likely incur into less opposition from users.

Water pricing can thus be seen as an indispensable ingredient of any privatisation policy, since it is the only way water management agencies can have access to the private capital market. This model of financing is more suitable for "ordinary" management of water systems (requiring smaller but continuous investment, instead than single and concentrated in time), being less conditioned by the tendency of public expenditure to depend on economic cycles.

Once the initial investment is done, direct responsibility for service operators on long term maintenance would decrease the risk of opportunistic behaviour, encourage the adoption of good management practices, stimulate cost reduction; otherwise, a concrete risk of running into a "vicious circle of public spending" would be faced: if ordinary management is not able to generate enough financial resources for keeping the system in good conditions, sooner or later a new public investment will be needed in order to replace the infrastructure.

The second motivation, relies on the fact that independent service-oriented agencies, being forced to provide value for money to their customers, will not be able to incur into deficits and will thus be stimulated to adopt cost savings and not to invest in overcapacity. On the other hand, users will develop a cost-conscious attitude towards the service. This motivation is also linked to a political preference – well rooted into the sustainability debate – for user-governed self-regulating systems for managing environmental resources for the sake of a better achievement of an equitable sharing of environmental resources.

While these arguments lead in the direction of "cost recovery" rather than of economic pricing, it should also be noted that in many cases adopting full-cost recovery – thus obliging each system of users to repay the 100% of the total cost – might lead to pricing levels that are likely to exceed the threshold of affordability. This is particularly the case when investment in new infrastructure is foreseen.

For this reason, some authors have developed the concept of "sustainability cost" (or "quasifull-cost recovery") meaning that prices should not necessarily cover the total cost, but rather be able to finance the long-term maintenance of the capital infrastructure, including natural capital. In other words, there might well be a public contribution for the initial investment, provided that later on users will be able to continue financing operation and depreciation of the system in order to prevent its value from reducing (and therefore preventing the need for new public contributions in the future) (Tardieu & Préfol, 2002).

4 SHORTCOMINGS AND DRAWBACKS OF WATER PRICING AND FCR

4.1 *Assessing the cost: financial costs*

From the above discussion, "financial costs" represent the cost of inputs (labour, capital, etc.) that have been used for providing the water service. In principle, these costs should result from the accounting of operators, provided that they have been appropriately calculated. In fact, this task is not always easy for many reasons.

First of all, it is reasonable to assume that at least some fractions of the value added of water services originate from non-competitive markets: this is due to the territorial monopoly in which utilities operate in each local market, but also to the economic rents that are contained in segments such as good quality water resources (when privately controlled) or technological equipment. These complications should be considered in order to assess the true economic cost. In the European context, however, this complication is not too much important, since water resources normally belong to the state and allocation rules are governed politically.

Other inputs, conversely, might be supplied by the public sector for free or at a subsidized price, and their value should be included as well. In all countries, the state supplies in this way a lot of inputs (research, education, vocational training, but also, for example, waterworks aimed at the provision of public goods such as flood protection or territorial planning).

A further source of divarication between market prices and true costs lies in the cost of capital, given the very long depreciation schedule of fixed assets. Artificial transformation of the water environment cannot be easily distinguished from the "natural capital" itself;[10] even if only water supply and sewerage infrastructure is considered, equipment life, especially in the case of reservoirs and pipelines, have economic lives of many decades or even a century.

In the case of long-lived infrastructure such as water supply and sewerage facilities, however, private investors usually require much shorter repayment schedules (and thus higher interest rates) in order to achieve risk/return profiles that are comparable with the rest of the economy, while the public sector can – and according to some, should – be less pessimistic and prudent. This myopic behaviour of private capital markets will mean that capital purchased on the market will have a higher price than its opportunity cost (measured by the social discount rate).

The adoption of a market interest rate instead than with respect to public accounting procedures that do not apply market interest rates causes dramatic increases of the final price. Barraqué (1999) estimates a range of 200%–300% for a French case study; a still higher range results in the Italian case studies analyzed by de Carli et al. (2002).

Should the market interest rate or a social discount rate be applied to such investments is a long-debated and never definitively resolved issue in public sector economics (Florio, 1991) and more specifically in environmental economics (Bromley, 1996; Ekins, 2000). In theory, the use of market interest rates is appropriate only if private investors discount for the future correctly: many environmental economists argue that this is definitely not true in the case of natural resources given the long term horizon and radical uncertainty about future development.

It seems justified, therefore, to consider water artificial capital by the same standards as reproducible natural capital, given the sustainability issues that are associated with it; and therefore, apply the same discount rate that is applied to choices affecting natural capital.

It is also important to note that the financial cost can also entail an inter-generational dimension: who has to pay for the infrastructure and for its maintenance over time? Supposing that at time 0 a new facility is built and there are no public subsidies, the capital cost will be repaid during time: depreciation quotas will be calculated in order that the financial cash flow thus generated will allow to repay the debt. At the end of the economic life, the value of infrastructure will be zero, and a new investment will be made, and so on. When this simple mechanism is not applicable because public intervention is necessary, some complications may arise. For example, if the initial investment is financed through public debt and this will remain constant over time (that is, it will be never reimbursed, while bond owners will continue to receive a perpetual rent), this means that the cost will be shared among the present and the next generation. If the present generation fails to set aside enough resources for reconstruction, it will "consume" the value of something whose cost will continue to be paid by taxpayers of next generations; on the other hand, these will have in addition to create another debt for financing the reconstruction.

---

[10] For example, the territorial landscape of Northern Italy has been modeled during centuries with a long series of public works that definitely altered the "natural" pattern of water flow: the Po plain was actually an immense wetland. The conservation of the "natural environment" of the lagoon of Venice would entail the complete disruption of the city.

Three general approaches can be found in the international experience with respect to the assessment of capital costs.

The approach used in Germany consists in a systematic revaluation of assets at their reinstatement value (that is, as if they would have to be reconstructed again), and to calculate depreciation according to that value. This model allows that investment is constantly sufficient to cover the real depreciation of capital: in other words, the value of infrastructure remains constant over time. On the other hand, this model causes potential distortions if the service operator would use depreciation cash flows in order to finance other investments, and can place an undue burden on the generation that first realizes the infrastructure (in fact, it would have to pay twice, once for the initial building of the infrastructure, once again for maintaining its value during time).

In turn, the approach used in the UK consists in calculating only investment that is required for keeping the infrastructure value during time, what allows operators to fulfil their obligations with customers. In principle, the water service operator could receive infrastructure for free (the cost having been covered, for example, through public debt), with the duty to maintain it in good operation and to finance the necessary investment for ordinary and extraordinary maintenance as well as reconstruction, from a certain point in time onwards. This model is more fair to the generation that first realizes the infrastructure (since its capital cost has been socialized), but requires on the other hand that investment is actually sufficient. Since the operator is the only subject who really knows the efficiency condition of assets, there might be a risk of underinvestment and consequent devaluation of infrastructure along time. At a certain point, the operator will not be in the condition to continue operating it, and public investment will be required in order to rebuild the assets.

A further possibility, that is more often practiced in the lease contracts, accounts only for ordinary and extraordinary maintenance. The first time construction will be financed apart (with a direct contribution of users through direct charges and taxes or more often through the public budget), and operator will use it until the end of its economic life. Users would pay water charges covering operational and maintenance cost, while the cost of infrastructure will be paid separately by them as taxpayers or as service users (with an additional charge). At that point, a new investment will be requested, and the process will start again.

Finally, a further intergenerational issue should be considered, given that part of the infrastructure costs are transferred onto the next generations (e.g., because of technological lock-in, or bad maintenance of long-lived assets). Artificialisation of the water system in this meaning represents a mortgage on the future generations, since they will be "forced" to pay the cost of capital infrastructure built by the present generation in order to compensate for the disruption of natural capital. A sort of an "option value" should therefore be considered as a further component of the industrial cost. This component is nonetheless very difficult to quantify.

4.2 *Assessing the cost: external costs*

With respect to external costs, again the concept is quite straightforward, but many empirical difficulties arise when we try to calculate it.

Very little applied research has been conducted in recent years in order to assess the magnitude of external costs in a systematic way: a few insights from the available literature suggest that there is a great variability according to local conditions and to the value of environmental functions requiring the non-use of water (IVM-EFTEC, 1998, Turner & Postle, 1994; MacMahon & Postle, 2000).

In fact, what most studies have assessed the *value of alternative non consumptive uses*, rather than the *marginal cost of water abstractions*. Yet the linkage between the two is not everywhere obvious: the impact depends on the local water balance, to which water abstractions may contribute positively or negatively depending on where the abstraction and discharge points are located. It can be argued that the most critical cases are those in which massive uncontrolled abstraction from aquifer occur, rather than those using large and regulated surface water transfers; but even this very general statement requires to be carefully assessed on site.

Despite the spotlight nature of these studies and the impossibility to be conclusive at the present state of the art, it seems quite clear that the magnitude of external costs is very much site-specific and dependant on seasonal aspects. When important recreational and landscape conservation dimensions are present, the value of this external cost might well be high enough so as to overcompensate the value of "productive" uses; yet this occurs only in special cases, while in normal ones, once there are no critical natural capital losses involved, the magnitude can be supposed as far lower.

Merrett (1997) adds to these remarks the idea that once ecological aspects are at stake, monetary estimates based on the WTP might severely underestimate the "cost", and argues therefore that the external costs should be evaluated through environmental impact assessment rather than in monetary terms. The external cost should in other terms be assessed on the basis of the actions that would be necessary in order to recover the natural capital up to the level that guarantees environmental functions.

In fact, the literature on monetary evaluation is quite unanimous in believing that empirical estimates of the "total economic value" at best can capture indirect use values (such as recreational use of water or landscape amenity), while non-use values and obviously non-economic values cannot be captured by available techniques although they can be defined on a theoretical base (Green, 2003).

In the end, while the theoretical definition of external cost is clear-cut, it is difficult to use it in practice for an assessment of the true cost in specific cases. Most applied evaluation studies in fact address the problem of measuring specific non-use values (e.g., the recreational and landscape value), yet it is not straightforward to associate this "value" to a "cost" that is generated by another specific water use (e.g., abstractions for irrigation), since the negative externalities for specific uses normally arise from a complex and interrelated set of other water and soil uses, and a specific ad-hoc hydraulic model is usually required (Fontana & Massarutto, 1995; Merrett, 1997; IVM-EFTEC, 1998).

For this reason, some authors have suggested a different approach, based on the idea of "cost of sustainability" (Barraqué, 1999). According to this methodology, the "full cost" of water should be defined as the (theoretical) cost that should be encountered if water use was ecologically sustainable – that is, if all relevant environmental functions of water resources are guaranteed to the present and the next generation, *and* enough investment is put in place in order to maintain over time the value of physical water assets.

The application of this assessment methodology to some case studies of European urban water supply and sewerage has provided many surprises, since it has enlightened that the "true" cost is very often unaccounted for (Correia et al., 1999; de Carli et al., 2002).

Other approaches refuse to impose a pre-determined evaluation criterion: analysis should start from the identification and measurement of environmental functions that are relevant for different stakeholders groups and of "governance issues" that emerge from the eventual clash between them (Faucheaux & O'Connor, 1998). The "external cost" should therefore better considered as (one among the) possible indicators of the existence of a governance issue to be solved through a political process, rather than simply as a number to be added to the total cost.[11]

4.3 *The real effectiveness of volumetric prices: price elasticity*

From the previous section we should be aware that optimal pricing rules require customers to pay a price that corresponds to individual marginal cost or – second best solution – to average cost, in order to achieve the most efficient resource allocation compatible with the structure of the water industry.

The consequence is that sub-optimal pricing rules will deviate from maximum efficiency. With inefficient water prices, water demand will be too high or too low. The society could be better off if some inputs would be transferred from the water sector to another one.

However, we have seen in section 2 that this is not always the case. Water prices are often not correlated with costs, and in many segments of the water industry there are direct or indirect subsidies paid by the government and financed through general taxation. Can this be justified in some way?

---

[11] An attempt to apply this conceptual model to the case of water is taking place through the GOUVERNe project, sponsored by the European Commission and actually close to the end (O'Connor, 2003, for further details).

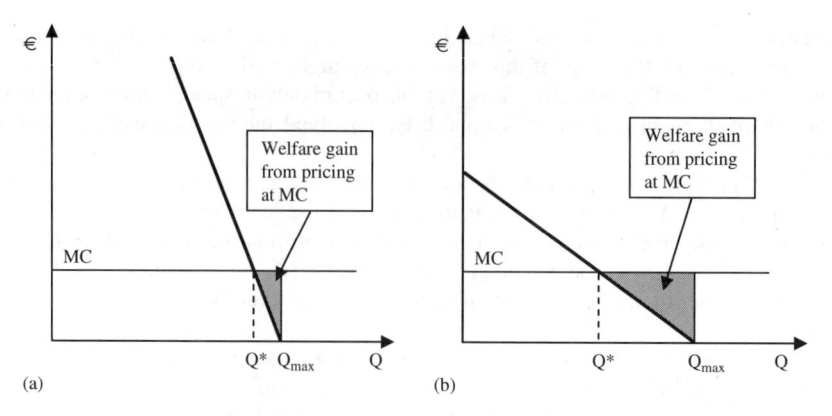

Figure 7. Allocative benefits of pricing depend on elasticity of demand.

A first issue to be raised concerns the size of welfare improvements that optimal pricing allows with respect to free provision. It can be easily shown (Figure 7) that these improvements are linked to water demand elasticity, namely the way water demand reacts to price variations.

Empirical estimates of water demand elasticity have been carried on in a great number of contexts. Results disagree quite a lot; although, some general statements can be made. As a general introductory remark, we can note that applied studies show the necessity to separate different segments of water demand, since they exhibit quite different behaviour.

As far as household demand is concerned, empirical evidence does not provide much support to advocates of water pricing. Evidence for a significant impact of water pricing and consumption has been sometimes provided (for a survey see OECD, 2000), but in most cases elasticity is reported to be quite low.

The elasticity of household water demand has been reviewed in a number of recent studies (OECD, 2000; Dalhuisen et al., 2001). In general, applied studies report elasticity values that are quite low on average (in the reach of 0–0.7). The existence of metered demand and increasing block structures appear as the most important element in determining the reaction of water users. Country-specific aspects should be considered as well, since practical possibilities to reduce water demand occur particularly for some uses (e.g., garden watering, swimming pools, car washing, reduction of leakage within private properties) than for others, and depend very much on aspects such as the actual level of water bills. With this respect, it is not particularly surprising for example that Northern American studies report in general far higher elasticity values (US water consumption is 4 times higher than European average).

Nonetheless, the introduction of metering is reported to have allowed significant reductions of demand in nearly all of the case studies analyzed in the literature, with figures ranging from 10%–15% to 30%–40% of initial demand. Again, the use of incentive pricing structure with rapidly increasing block rates or supplementary charges in critical periods has been as well quite successful as a demand management tool even in European experiences (OECD, 2000).

These results should nonetheless be carefully interpreted. For example, in England and Wales it is apparent that metered per capita consumption is much lower than non-metered one. However, since the choice of the meter is left to the individual consumer, the correlation might be explained very simply by assuming that only those consumers knowing that their water consumption is low had asked for the meter, and this cannot be considered as an example of "price elasticity".

Other authors have emphasized the need to better understand the social and behavioral mechanisms underlying water demand, since the brutal price-quantity relation is neither credible nor sufficient as an explanatory variable (Merrett, 2002).

On the other hand, it should also be noted that in many cases the result has also been that of putting pressure on the economy of water supply systems, whose revenues have fallen short of the coverage of fixed costs.

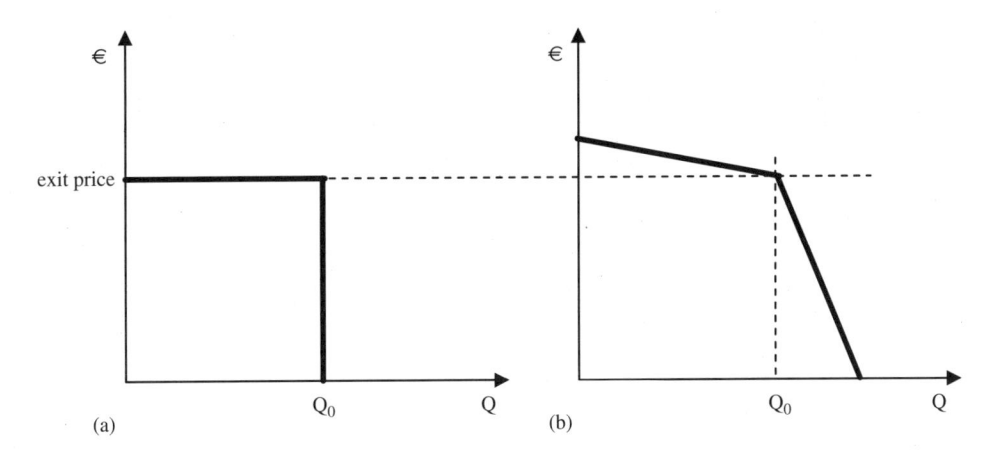

Figure 8. The typical shape of the water demand curve for productive uses.

As far as productive uses are concerned, two different aspects should be emphasized:

– The ex-ante decision to use water as an input of a certain productive process.
– The ex-post decision on how much water to use and how to use it.

The first decision is basically a decision whether the (total) cost of the water input is or is not overcompensated by revenues from the sale of outputs. Since water requirements of most water using processes is given, the decision is whether to start the water consumptive activity or not, and not how much water to use.

The convenience to engage in a water consumptive process in a particular place (e.g., irrigated agriculture, industrial processing requiring water for cooling) is thus not very much dependent on the water price *structure*, but much more on its *level*. As Tardieu & Préfol (2002) put it, this decision depends on the "strategic" value of water. In extreme, the demand curve could assume a shape like in Figure 8a: the water quantity consumed will be Q_0 until the price becomes higher than a threshold ("exit price"), above which demand suddenly drops to zero. More frequently, demand has some marginal reaction to prices, though very asymmetric: very low elasticity below the exit price, and very high elasticity above it (Figure 8b).

The second decision can be in principle influenced by water price structure, provided that technology allows some alternative options. Again, we can distinguish between short and medium term decisions.

For example, in the case of agriculture, a crucial distinction should be made, depending if we consider the very short term (e.g., the occurrence of a drought during the agronomic season once cropping choices have been made and most costs have been already sunk); the short term (e.g., at the beginning of the agronomic season, when farmers decide which crops to produce) and the long term (when the issues at stake can be whether to adapt the farm to irrigation (e.g., by investing in irrigation facilities or in greenhouses), or the choice of irrigation system (e.g., spray or drip irrigation).

In the very short term, elasticity can be very small: all costs have been already sustained, and the risk of losing the crop means that the value of water for the farmer corresponds to the total crop value. In case of a water shortage, therefore, prices would have to be substantially higher if they had to be used for the purpose of reducing demand or allocate water to the most productive uses first. This is what Tardieu and Préfol (2002) call the "tactical" value of water, arguing that it reaches a maximum in the phase of pollination and ripening (Figure 9).

As a result, water pricing can have rather different impacts if used as an instrument for orienting long-term strategies of water users rather than as an instrument to face short-term water shortages and induce reallocation of water in case of a drought.

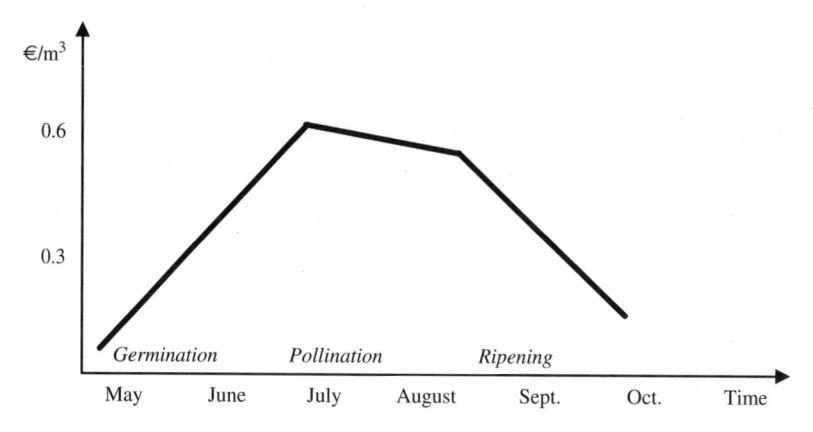

Figure 9. The "tactical" value of irrigation water as a function of the growing cycle.

In the case of agriculture, where the most significant opportunities for using prices as demand management tools are normally forecasted, again empirical evidence shows little sensitivity to water price variations, even with volumetric charges, and a much higher sensitivity to the overall water price level, especially once a critical threshold is reached (Garrido, 1999; Dinar, 2000).

In the Southern European context, many recent studies seem to confirm this statement. For example, a comparative study conducted on 8 European regions in different agronomic, environmental and geographic context show the relevance of the exit price threshold for the decisions concerning whether to engage in irrigated agriculture or not. The adoption of FCR would determine that most of the irrigated COP cultivations in Southern Europe would become unprofitable, while wherever the threshold is not trespassed, water demand increases rather than decreasing (Massarutto, 2002a). The same study shows on the other hand that the elasticity to marginal price variations is quite small, particularly in the short run.

This means that the decisive effect is not the incentive to reduce consumption at the margin, but rather the crude fact that the overall price is or is not higher than the threshold that makes irrigation convenient. As far as Southern European agriculture is concerned, this is likely to occur only in a few cases (continental Spain and Southern Italy; here and then Southern France and Northern Italy) where low-value crops (cereals, oilseed, grassland, etc.) are combined with costly water transfers with a water cost that exceeds 0.20–0.30 €/m^3. On the other hand, high value crops (Mediterranean products, horticulture, greenhouses, etc.) seem ready to pay prices that are even 10 times higher before giving up irrigation (Massarutto, 2002a).

Quite similar conclusions are likely to appear in the case of industrial water demand, although in this case there are two specific issues to be considered.

First of all, industry is much more "mobile" than agriculture. Investment decisions can be located in the most suitable areas more easily, and this could be a factor influencing location of water-intensive activities.

Second, the water cost normally accounts for a very little share of the total costs: this is likely to reduce the responsiveness of decisions to water prices, even if a wise pricing policy might be a welcome component of an action plan aimed at promoting water saving.

Third, industrial uses – particularly the most "water intensive" ones – are quite often self-supplied through direct abstractions rather than using collective facilities. As some experiences show the optimal strategy in these cases is to create collective facilities to which individual industries should connect, even compulsorily, and give up individual abstractions. In order to facilitate this, charges for the collective facility should better be flat rather than proportional, thus contradicting the principle of LRMC.

Finally, and probably most important, industrial uses are normally much more critical from an environmental point of view for what concerns sewerage and treatment of discharges rather than for water

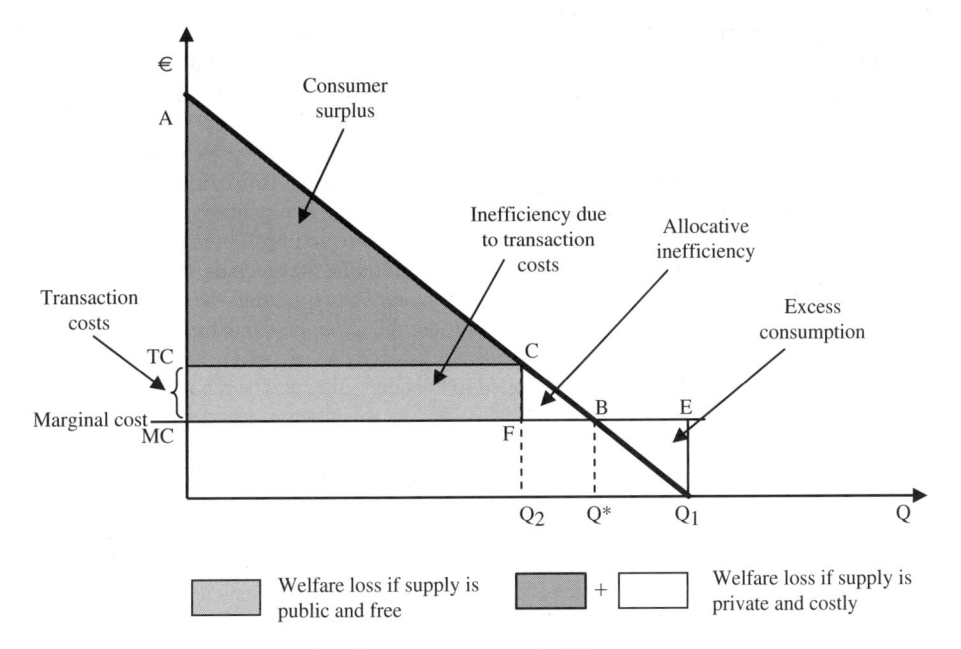

Figure 10. Transactions costs and welfare improvements of pricing.

consumption. The empirical literature does not support very much the idea that economic incentives such as wastewater taxes, sewerage charges, etc. can have a major effect on pollution loads, at least not following the "marginal" impact at the base of the theory of environmental taxation (OECD, 1997). On the other hand, it is the absolute value of costs what probably stimulates industry to engage in the search for cleaner technologies. It should also be noted that wastewater facilities entail normally fixed costs rather than variable costs: once facilities are built, therefore, the incentive for reducing pollution is quite small, at least as far as the capacity load is not trespassed.

In the end, what empirical literature on water pricing suggests, is that while water pricing could be a useful demand-management tool, yet the simple idea that "optimal" water prices based on LMRC can automatically achieve a socially optimal outcome should be rejected (Merrett, 1997).

4.4 *Marginal cost pricing and transactions costs*

The argument developed in the previous paragraph is further reinforced, if we consider that in order to charge "optimal" prices, water service operators should incur into extra costs for the sake of measuring individual consumption. In economic terms, these are "transactions costs", that would not appear if water was supplied for free or unmetered.

The economic theory demonstrates the superior allocative efficiency of providing the service through the public sector when the allocative benefits of the adoption of market-clearing prices are overcompensated by transactions costs (Stiglitz, 1988). This is particularly true in case demand is inelastic, since the allocative welfare loss of the excess demand is in this case lower.

Figure 10 helps us to understand the problem. We assume that water supply requires a positive marginal cost – that is, an extra quantity of water costs the society a certain amount of resources (MC). With the optimal price, P = MC, water demand would be Q^*. Social benefit results from the difference between total consumer utility (the area below the demand curve between 0 and Q^*) and the total cost (the area below MC in the same range). Net benefit is therefore the whole triangle MC-A-B.

If the service was supplied for free, customers will be willing to consume an extra quantity until the marginal utility for them is positive (namely, until Q_1, where the demand curve meets the horizontal

231

axis). This originates a welfare loss for the society: the total additional cost (the area under the MC curve between Q* and Q₁) is larger than the additional benefit (the area below the demand curve in the same range). The difference corresponds to the triangle Q₁-E-B, which can be interpreted as the welfare loss with respect to the social optimum.

On the other hand, if we wish to charge the optimal price, we need to monitor individual consumption – better to say, individual marginal contribution to the total cost, that is not only a function of quantity of service consumed but also of the period of time in which consumption occurs, given that marginal costs are a function of resource availability and level of use of infrastructure. This means to install individual meters in any single household, that include either a capital cost (the metering equipment) or an operational cost (e.g., personnel reading the meter, billing systems, expedition, management of complaints, etc. A similar situation occurs in the case of sewerage (e.g., chemical analysis of effluents).

These transactions costs (TC) should be added to the other costs and be recovered as well. Thus the optimal price becomes in this case P = MC + TC. At this price, water demand is Q₂. In this case, collective benefit is now the difference between the area below the demand curve in the range OQ₂ and the new cost, corresponding to the triangle TC-A-C. Welfare loss is represented by the transactions cost (area in green), plus a further loss corresponding to the triangle C-F-B.

In both cases we have a welfare loss, and therefore in order to decide we have to confront both areas. Their relative size depends on the magnitude of TC (in relation to MC) and on the slope of the demand curve. It is easy to show that if TC is small and demand elasticity is high, it is more efficient to run into the transactions cost and charge consumers directly; vice-versa, if TC is relatively high and demand elasticity is small.

The question can be resolved only empirically, since it ultimately depends on (i) demand elasticity, (ii) size of transactions costs relative to total costs.

In Table 1 we have tried to estimate the break-even level of transactions costs under some alternative hypotheses on water price elasticity and starting level. We have assumed that elasticity is 0 up to a price level of 0.5 €/m³ and that at that price water consumption would be 200 l/(capita day). We have then evaluated, for different levels of water price needed to equal marginal production costs (in a range between 0.75 and 1.5 €/m³). As we can see, when water production costs are low enough (0.75–1), metering is convenient only if its costs are very small (not higher than 0.01–0.03 €/m³ corresponding to no more than 4% of the production costs). With higher water prices, the break even threshold increases quite rapidly. It should be noted that a figure of 1.5 €/m³ for production and distribution cost is quite high, and is reported in Europe only in the least fortunate circumstances. Wherever the water system can count on cheap local resources, the lower values seem more appropriate.

This simulation seems to suggest that metering can be useful only when it is coupled with costly supplies. This case arises for example when locally available cheap water is limited, and the system should rely on more costly water imports to cover excess demand.

Moreover, it should be considered that marginal costs of sewerage depend more on effluent quality rather than on water quantity. Further transactions costs should be incurred into in order to provide an appropriate measure of pollutants contained in discharges.

Table 1. Welfare loss of free supply and metering with an isoelastic water demand curve.

Price = marginal production cost €/m³	Break-even marginal transactions cost €/m³ (% of marginal cost)			
	$\eta = -0.1$	$\eta = -0.2$	$\eta = -0.3$	$\eta = -0.4$
0.75	0.01 (1%)	0.02 (2%)	0.02 (3%)	0.03 (4%)
1	0.03 (3%)	0.06 (6%)	0.11 (11%)	0.17 (17%)
1.25	0.06 (5%)	0.16 (13%)	0.33 (26%)	0.63 (50%)
1.5	0.14 (9%)	0.35 (23%)	0.81 (54%)	always

4.5 Water services as public goods

As we have argued before, water services might include – and in fact often do include – components with the character of the public good, in the sense that they are non-rival and non-excludible. Since in this case no marginal cost is occurring, the full cost results higher than the LRMC.

The standard economic literature on this subject has recognized that public goods are better financed through the public budget because of the free rider problem (the individual has no incentive to contribute since it is expected that others will do). On the other hand, the public supply of public goods is also problematic (the state has no correct information on the true social willingness to pay and there is no guarantee that any democratic decision would ever find out the "optimal" quantity of public goods.

When public goods are supplied by the same economic activity that also provides private goods, it is not too inefficient to leave the market provide on its own: if the private good is valuable enough, the individual will pay for it in any case and thus pay for the public good also. In the water sector, this could be for example the case of public hygiene and sanitation (public goods), that are inevitably correlated with water supply and disposal of wastewater (both private goods for which the individual has a WTP that is alone sufficient for having the water service purchased anyway).

However, in the water sector there are as well many examples in which valuable public good might be associated with insufficient demand for the related private good. The most relevant cases regard large water storage and river management facilities (for the sake of flood protection or restorement of natural habitats), rainwater management and sewage treatment.

Most of the ongoing new projects for water resources development in Europe, for example, fail to pass the cost-benefit test unless a public good dimension is also considered in the valuation (Barraqué, 2000b; Verges, 2002b). Uncertainties in the estimation of variables such as the potential damage caused by a flood and the way to consider alternative means to provide the same public goods are the most critical points. This is usually a very delicate point since it often occurs that in the valuation process public goods are overestimated in order to justify government subsidies to the new infrastructure (Barraqué, 2000b).

In these cases, as a long-time established theory in the public sector economics shows, the optimal solution is to provide the service through the public sector and finance it out of general or local taxation. At least, once the public good dimension is clearly distinguished, a two-part tariff should be applied, in order to recover the fixed cost out of a flat-fee assimilable to a tax and the variable cost out of a variable charge proportional to consumption.

The problem of government failure in the identification of the "real" demand for public goods can by no means be neglected; nonetheless, it might be improved through a more transparent decisionmaking and stekeholder participation, probably with better chances for success than through the search of automatic and self-revealing procedures like the cost-benefit analysis (Faucheux & O'Connor, 1998).

4.6 Water infrastructure as a semi-public good

Water infrastructure is a typical example of a semi-public good subject to congestion (like a bridge or a road). That is, until a critical threshold (in terms of capacity) is reached, the cost of adding a new consumer or the consumption of an extra quantity is low and limited with the few dimensions having a non-fixed cost (e.g., energy, reagents). Fixed costs, that represent the largest share of the full cost (70%–90%) should then be considered apart, since in economic terms it is inefficient to exclude consumers that are not ready to pay that price but still have a positive willingness to pay associated with consumption.

When the capacity threshold is close, however, this conclusion is not anymore valid: extra consumption would require an extension of the infrastructure and therefore a new cost. This applies of course to any new investment decision.

This point can be better understood with the aid of Figure 11, in which we have assumed for simplicity that the marginal cost is 0. Suppose that the infrastructure capacity is Q_m and water demand is WD. If the supplier would charge consumers in order to cover fixed costs (e.g., at price P)

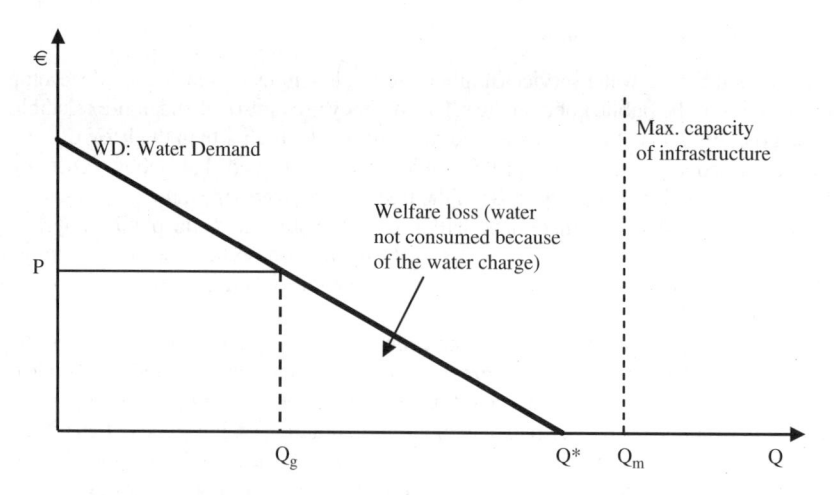

Figure 11. Welfare losses from charging cost-recovery prices when exclusion is not desirable.

some consumers would decide that the price is not worth paying, since the value of water for them is lower for quantities higher than Q_g. However this leads to an inefficient resource allocation, since the infrastructure would be able to supply extra water at 0 cost. If water was supplied for free (e.g., if the fixed cost was covered through taxation), the cost for society would be the same, but water consumption would be higher (Q^*) allowing therefore the extra benefit represented by the highlighted triangle.

In turn, if the infrastructure capacity Q_m would be lower than Q^*, supplying water for free would mean that some demand could not be satisfied and new investment is required (or a scarcity cost for someone would emerge). In this case the cost of capital cannot be considered as fixed, but should be included as a marginal cost in the evaluation.

In these cases, it could be efficient to charge the marginal cost only. The fixed cost should be covered through some kind of taxation. We should note once again that what is important here is not the fact that something is called "tax" or not, its payment is compulsory or voluntary, but rather the idea that (a certain group of) people are asked to share a cost among themselves according to a fixed-rate scheme, regardless the actual consumption of the service. Of course, if a free rider problem exists, it might be required that some kind of imperativeness is introduced.

The fixed part of a two-part tariff, with this respect, is equivalent to a personal tax whose condition is being connected to the service (in other words, it looks like a "club admission charge", where the "club" is represented by the community of customers that have decided to be connected to the service).

Without this arrangement, the fixed cost would be divided according to the total quantity consumed (ending up in a higher marginal price and maybe in lower consumption, with the risk that the system will fail to recover costs).

The argument can be further developed by distinguishing new facilities and already existing ones. In this last case, that occurs very often in Europe, charging the LRMC could entail that the infrastructure loses its economic utility for customers, and no one will ever use it in the future; however, this might not result in an optimal outcome, since there are sunk costs to be considered (barriers to exit, decommissioning costs, etc.). This might occur because infrastructure was subsidised in the past, but also because of planning mistakes and unanticipated future events conditioning the economic viability of a project.

In other terms, once sunk costs have been actually incurred into, pricing at the LRMC would mean that the economic risk of the investment is fully assumed by the public budget: customers will buy the service only if the value (i.e., their willingness to pay) is higher than the charge. Otherwise, the service will not be bought: the sunk cost will not be repaid and this will entail an economic loss for the supply operator, that is usually the state or some private establishment whose contract with the public sector normally entails that the risk is sustained by the public sector.

On the other hand, there are many long run costs whose occurrence is actually unforeseeable at the moment when the initial investment is decided; therefore, they cannot be accounted for at the time of the evaluation of the project. The same occurs for the estimation of the willingness to pay of future users, that might be lower than initially predicted (e.g., thanks to technological innovation).

In fact, full-cost pricing in the general theory of management and accounting, is suggested as a good solution when demand is inelastic. This is in fact what we just argued above about environmental utilities, yet this might not always be the case if we consider a particular water resource instead than "the demand for water" in a larger meaning. This is particularly true if we consider the level of bulk supply – where, indeed, the most of the public subsidies are usually concentrated: large distance transfers are normally justified from the water user point of view only if the state covers part of the cost. If the long distance supply was charged at the full cost, the users' WTP could be insufficient and demand would drop.

In other words, while the individual consumer can hardly be believed as sensitive to marginal price variations, the application of the full cost of a complex infrastructure might impact so greatly on the price so as to make it absolutely undesirable for the user to use that infrastructure.

It might be assumed, for example, that domestic consumers have no alternatives for receiving water supply but to connect to the distribution system, therefore their demand is captive and not too much price-elastic. The same is not true for the distribution system itself, that might have different alternatives for obtaining bulk water to distribute (e.g., connect to a large water transfer scheme or invest in technology for cleaning up and maintaining local underground resources). If the cost of the large transfer scheme rises suddenly, the local distributor might find it cheaper to give up buying bulk water and start producing it from local resources, or engaging in a cleanup of local aquifers, or introduce water saving measures.

In the case of irrigation, the decision whether to irrigate depends on the differential value created by irrigation. If this value is completely absorbed by the price, the farmer will stop using water and convert to dryland farming, or possibly enter into some other land management strategy. The already built supply system and irrigation facility would then be left unused and the sunk cost abandoned to the agency that financed the investment.

In all of these cases, if water prices were suddenly raised up to the LRMC (including the cost of investment) the result would be simply that the facilities are abandoned. Charging at the short run marginal cost (that is, at the operational cost plus the cost that should be met in any case even if the facility was unused) would alleviate part of the burden on the state budget.

Of course, the same does not apply to new water projects, where the rationale of LRMC is much more robust. However, an equity problem might arise in this case, since users that could benefit from the new project might consider this asymmetric treatment unfair.

4.7 *Perverse effects of water prices as environmental policy tools*

We have discussed above that the effectiveness of water prices as environmental policy tools (in order to promote water saving or prevention of water pollution) is largely dependant on demand elasticity. While this is usually quite low, an appropriate design of water tariffs (e.g., through a sharply increasing block structure) might help with the achievement of demand management targets.

On the other hand, perverse effects could also originate, depending on the complexity of the value chain.

A first potentially harmful effect of pricing (and especially of incentive pricing) depends on the availability of alternative supply options.

Many water users need not connect to collective facilities, since they could rely on self-supply (or self-treatment of discharges). These practices are still very common in many rural areas as well as in intensely industrialized regions, and are often causing environmental problems (e.g., overexploitation of aquifers). In such cases, the connection to a public network should be considered also as an environmental policy instrument, and charging policies should not discourage the abandonment of environmentally unfriendly practices.

Much the same occurs for many commercial and small industry wastewater: in many European countries, local authorities find it preferable to compulsory connect these users and apply a flat price, in order not to encourage illegal practices. Again, this can be avoided if effective monitoring and punishments are foreseen, yet with a substantial increase of transactions costs.

Given the high costs for monitoring and ensuring compliance, many countries have chosen to provide collective services to which users should compulsorily connect; in order to reduce the incentive to free ride, tariffs paid to these systems are often flat or lump-sum, and very weakly correlated with actual consumption. In this way, users are not encouraged to use freely the natural resource. Lump-sum charges could well be raised up to the cost recovery level, in order not to create a burden for the public finance; it is nonetheless obvious in this case that the allocative benefit would be lost. A fixed charge paid for a compulsory connection resembles in fact very closely a form of taxation, with the only difference laying in the fact that the service operator – instead than the public administration – raises it directly and allots its revenues to the service budget.

A second potentially perverse effect might occur with respect to water operators themselves, depending on the possibility that water prices and charges are shifted up and down along the value chain. While water prices might have an influence on final consumers, for sure they represent an incentive for water service operators to sell more water, since their revenues will depend on the quantity sold. Unless an appropriate charge is raised on them, they will have very little incentive to adopt actions aimed at improving water supply efficiency (e.g., reduce distribution losses) and to encourage water saving measures on the demand side.

For this reason, water price regulation should at least be able to "decouple" the incentive effect, by making the operators' revenues independent of the quantity sold. This could be done for example by allocating the variable part of the tariff (the increasing block structure) to the water authority (and not to the operator), while the operator's revenues would be composed by the fixed part of the tariff and a lump-sum allotment paid back by the water authority itself. In alternative, water abstraction charges could be introduced with a substantial value and an increasing block structure, with the possibility to transfer the tax to consumers only up to the amount that corresponds to water effectively sold (in this way, water losses would remain as a financial loss for the water company).

A third negative effect from water pricing might occur in specific circumstances, depending on the structure of water demand. In the case of irrigation, for example, it has been shown that water pricing has the effect of increasing the overall efficiency of irrigation (since water is allocated to more productive crops), but not necessarily of reducing water demand (Massarutto 2002a). This occurs because more valuable crops have also a less elastic water demand and require higher water quantities.

More generally, in some cases the objective of more efficient water allocation and demand reduction are in conflict with each other. When this occurs, until the water price reaches the exit threshold, the effect of pricing will be more to achieve the former rather than the latter objective.

4.8 *Which cost to recover? Budget equilibrium vs. economic efficiency*

A further perverse effect that prices might have in terms of efficiency regards the economic behaviour of water supply and sewerage operators.

So far we have assumed that "the cost" we are dealing with is the cost actually sustained in order to provide the service. However, water utilities are natural monopolies, in which the beneficial effects of competition do not occur. As a long-lasting tradition of economic studies has shown, the presence of an unregulated monopoly is source of at least three kinds of inefficiencies.

The first one is of allocative nature: since the monopolist behaves in order to maximize profits, it will supply lower quantities at higher prices than the theoretical optimum. In other terms, it will charge a price that is higher than marginal cost, supplying the quantity that corresponds to the point in which marginal revenues equal marginal costs. This problem can be in principle resolved, if an external regulator (or a public owner) obliges the monopolist to charge the marginal cost (or, as a second-best alternative, the average cost). This scheme has inspired for example the so called "cost-plus" pricing method, based on the ex-post calculation of average operational costs plus a given rate of return on investment.

The second one is more "political": the monopolist could use its power in order to exclude certain subjects from consumption or discriminating among them according to its own preferences. In principle, also this problem can be solved through public regulation or property, through the establishment and enforcement of appropriate antitrust norms.

The third inefficiency, in turn, cannot be solved in the same way, since it is common to private and regulated monopolists: namely, since no competitor threatens its position, the monopolist is not forced to the same cost discipline that would have occurred in a competitive environment. In fact, in a competitive market, if some operator has higher costs than the lowest possible, some other operators will supply the market and it will be thrown out. The same occurs if it fails to invest enough resources for innovation or if it chooses too high an economic scale. In a monopolistic market this pressure does not occur, and actual costs can be significantly higher than the theoretical minimum. At the same time, if operators are allowed to recover costs without any risk, they will tend to invest in over-capacity (this is the so called "Averch-Johnson" effect, typical in case of cost-plus tariff regulation).

In order to cope with this problem, the only general solution is to introduce as much competition as possible (true or simulated), given the structural characteristics of the water industry. As a summary of the rich debate occurred in the last 20 years with this respect, we can assert that the water industry presents some opportunities for introducing some indirect forms of competition (e.g., competition for the market through open competitive bids; contestability of property of monopolistic utility; competition among the value chain for technology, equipment and construction activities); yet these are not sufficient nor resolutory, therefore requiring some sort of incentive regulation on costs or revenues (Rees, 1998; Massarutto, 2001b).

In simple words, it is not efficient to allow operators to charge the "actual" costs; rather, the tariff schemes should include some incentive effects that will force the operator to make efforts for increasing its efficiency.

In the economic literature, there are two general approaches that can be used, in various forms (Massarutto, 2001b).

The first approach is based on benchmarking against a "theoretical" cost function, that can be derived analytically or through an econometric estimate based on comparative analysis (e.g., through a regression on all water suppliers that operate under similar conditions). This effort generates a "standard cost", which could be used as a basis for setting charges instead than the actual cost. In this approach, a deviation of actual costs from the standard cost are supposed to be caused by inefficiency and not by specific circumstances that are typical of the concerned area. Of course, in order to be sure of this, the model specified for calculating the cost should be sophisticated enough so as to encompass all "exogenous" variables that are likely to influence costs.

Models of this kind are already used for example in the UK and in Italy. If the reliability of the model is not to be trusted at 100%, benchmarking could lead to less drastic measures. For example, in the UK the evaluative benchmarking models are not used for the purpose of setting charges, but rather to provide the regulator with better information on the efficiency of different water companies. In Italy the standard cost is used for the purpose of setting charges, yet a deviation of $\pm 30\%$ from the benchmark is allowed (Massarutto, 2002c).

The second approach is based on automatic incentive pricing structures such as the price cap mechanism. In this case, what is regulated is not the level of price (as a function of costs), but rather the future dynamics of prices. Starting from a point in time in which actual tariffs cover actual costs, the regulator might put an upper limit to future prices that incorporate the efficiency gains that operators are in the condition to obtain. Since the price cap remains fixed for a certain number of years (usually 5 to 10), the operator has the incentive to strengthen efforts aimed at cost reductions (since any reduction that is greater than the expected one will lead to a profit, and vice-versa profits would fall below the actual limit).

The experience made so far (particularly in the UK, where this regulatory model has been applied for 13 years after 1989) shows that the price-cap mechanism can be effective in promoting cost reduction; on the other hand, it requires an unambiguous specification and verification of quality, and it is not easy to use in case the industry is in a dynamic phase of the investment cycle. In fact, since quality and environmental policy are at present the most important cost drivers of the water industry, effective price-caps require a discretional evaluation of what new investment is required: this is a task that the economic regulator cannot perform too effectively, given the strong territorial specificities that characterize the water industry.

4.9 Water prices and affordability of water

Last but not least, the strict application of LRMC might have undesired social and distributive consequences given that consumption of water services is only weakly correlated with income and natural availability is very unequal even within the same country or region.

Since the access to basic services is often regarded as a "social right" to be guaranteed independently from the readiness to pay, it might be considered unfair to charge the full cost to any individual consumer. This concept has been developed in the literature on sustainability with respect to the "ethical" sustainability, according to which the water price should never be higher than a critical threshold, that can be measured for example in terms of the share of the household water bill on the family budget (Barraqué, 1999; de Carli et al., 2003).

Again, the cost-revenues correspondence might be searched for at a greater territorial scale, even when multiple suppliers exist (for example, through compensatory lump-sum payments). If this compensation is done, however, the allocative benefits would be once more lost, and there are no allocative reasons for preferring this solution to another one based on general taxation.

This argument was decisive at the beginning of the 20th Century when ideas such as "municipal socialism" and "welfare state" were elaborated. Later on, economists argued that essential needs should be guaranteed regardless the efficiency considerations that are implicit in the market pricing; while Sen (1982) has developed the idea that "essential" needs are correlated with the satisfaction of the basic "functioning" requirements of any given collectivity.

While there is no doubt that water supply and sewerage belong to the category of essential services, this argument has probably lost some of its importance nowadays, given the declining weight that public services have in the average income of affluent societies.

Rogers et al. (2002) also point out, with reference to developing countries, that the price of water services should be evaluated against the costs that households would have to sustain anyway in order to purchase water from private vendors. If there is no alternative to develop collective water systems than to ask households to pay, a payment that guarantees a reasonable cost recovery could be preferable since the price asked by private vendors in the market is also very high.

It must also be stressed that water can be seen as an "essential good" only for a part of total consumption, and that in fact people spend on bottled water figures that are one order of magnitude higher than the cost of tap water. Moreover, irrigation and industry cannot be considered as "essential needs" at least in developed world.

Yet recent privatization policies in many countries such as the UK have revealed that once prices have been brought up to the full cost recovery level for an increasing number of services, their aggregate weight is not anymore negligible, even in developed countries; *a fortiori* this argument is valid for developing countries, where the application of full marginal cost of water would simply make the connection to the public system unaffordable for most families.

Particularly during the expansive phases of the investment cycle, when new infrastructure is developed, the adoption of the FCR may cause dramatic increases in charges, especially caused by the financial cost of capital.

These negative effects could be nonetheless compensated without sacrificing completely the idea of LRMC, for example by foreseeing compensative payments or other equivalent measures (e.g., vouchers) in aid of the low-income families (OECD, 2000). These solutions face nonetheless political difficulties in many countries and leave open the question about how to single out households that have an entitlement to access to the subsidy.

In the case of irrigation, the issue of rules for water allocation cannot be separated from that of social justice and rural development policy. Particularly in developing countries, but also in the

EU Mediterranean countries, the access to water is a prerequisite to any agricultural production and consequently on the economic viability of entire regions. In this sense water is "essential" to them, even it might be obviously questionable whether productive agriculture itself is essential in those regions. Again this argument loses some of its importance if there are alternative economic activities or at least alternative agricultural models (e.g., based on rural tourism). If agriculture is the only concretely practicable way to guarantee the existence of a local economy, however, the problem changes radically: while it is necessary to avoid a wasteful use of water or improductive expenditure in waterworks, it is also evident that other considerations (e.g., the social cost of unemployment, emigration and land abandonment; alternative subsidies that should be paid in any case in order to maintain agriculture) are also important aspects to evaluate.

5 CONCLUSIONS

Water pricing is now widely considered in the water policy community as an important instrument to be used with the aim of achieving a sustainable water use. In this paper, we have discussed the basic reasons that lay behind this assumption and tried to single out advantages and shortcomings relative to water pricing. In the end, we have the impression that, while remaining useful and important, water pricing has probably been mythicized.

As an instrument aimed at the efficient allocation of water resources among users and as an instrument of environmental policy, water pricing can be (moderately) effective but it will hardly be sufficient. Although the reaction of water demand to prices should be better understood and although many recent studies have shown the importance of high marginal costs (to be achieved through increasing block structures), water demand remains rather inelastic, particularly for those uses that are recognized as the most consumptive. Important shortcomings also emerge, arising from the public good nature of some components of the economic value of water services, from transactions costs in the application of volumetric prices. In turn, volumetric prices do not necessarily raise equity issues, at least provided that these effects are carefully examined and avoided, and a socially acceptable way of defining prices is used. The design of pricing mechanisms should be perceived as fair.

On the other hand, the greatest importance of pricing lays probably in the fact that it allows water systems to be managed under decentralized and independent institutions, therefore increasing cost consciousness of users and avoiding the distortions potentially created by government intervention. To reach this objective, prices do not necessarily need to be based on marginal cost nor on effective consumption. In turn, the imposition of strict cost recovery constraints could mean that prices could rise at unacceptably high levels. Some way of redistributing the impact of water prices on families, with the creation of cost-sharing devices as well as through appropriately conceived subsidies should not be refused a priori, but rather should be evaluated with respect to the set of incentives created.

It has also been shown that in reality the exact meaning of "price" should be better clarified. Between individual marginal cost pricing (the theoretical optimum, though not always easy nor cheap to achieve) and totally subsidized schemes, the reality shows a wide number of solutions that are more or less directly and explicitly involving water users. Most of these mechanisms can be considered as "water pricing", although not always having the same allocative and incentive effect. The use of prices as incentive instruments, as allocative instruments and as cost recovery instruments does not require the same approach to pricing and, to some extent, could even be in conflict with each other.

Finally, water pricing can have perverse incentive effects that go towards the opposite direction with respect to sustainability, encouraging overcapacity, increasing water demand and indirectly promoting environmentally unfriendly substitutes of water demand. These potentially negative effects should be considered as well while designing pricing mechanisms.

For sure, water pricing – regardless its relation to sustainability – is a necessary outcome of the present phase, in which the public finance is put under great pressure and it is simply not possible anymore to conceive state-financed water systems whose costs are charged, in a way or the other, to taxpayers. This should not prevent from the use of public subsidies: despite "radical" interpretations of FCR, there are many good reasons to believe that water finance can well deviate from the orthodox and literal textbook pricing. The Dublin definition of sustainability in fact contains no explicit requirement for FCR intended in its literal meaning. It requires instead that patterns of allocation of water and of economic burdens that are linked to the use of water are democratically legitimated through an open participatory process.

As Colin Green (2003) puts it: "economists usually assume that prices always work but nothing else does ... In practice, prices do not seem to work particularly well in water management as a means of changing behaviour, and other tools can be more effective. Water pricing remains one among other water policy instruments; progresses in economic research have allowed to better understand its functioning and potential usefulness, and for sure there is space for a wider use of economic instruments in the water policy field. On the other hand, by no means its use should be confused with the idea that water would become some "market good" that is traded against prices like any other commodity.

REFERENCES

Barraqué B., 1999, *Environmental, economic and ethical sustainability of water service industry*, in Correia F.N., Water 21: towards a sustainable European water policy, Report to the European Commission, Dg12

Barraqué B., 2000a, *Les demandes en eau en Catalogne: perspective Europeene sur le projet d'aqueduc du Rhone a Barcelone*, Report to the French Ministry of the Environment, Paris

Barraqué B., 2000b, *Privatising the water sector: the French experience*, in Holwarth F., Kraemer A., *Environmental consequences of privatizing the water sector in Germany*, proceedings of the conference held in Berlin, November 2000, Berlin, Ecologic

Baumol W.J., Oates W.E., 1988, *The theory of environmental policy*, NY, Cambridge University Press

Briscoe J., 1996, Water as an economic good: the idea and what it means in practice, International Commission on Irrigation and Drainage, Cairo

Bromley D., ed., 1996, *The handbook of environmental economics*, Basil Blackwell, Cambridge Ma., USA

Brown C.V., Jackson P.M., 1990, *Public sector economics*, Basil Blackwell, London

Correia F.N. et al. 1999, *Water 21: towards a sustainable European water policy*, Final Report to the European Commission, Dg12

Dalhuisen J.M., Florax R.J., de Groot H.L., Nijkamp P., 2001, *Price and income elasticities of residential water demand*, Discussion Paper TI 2001-057/3, Tinbergen Institute, Amsterdam (http://www.tinbergen.nl)

de Carli A., Massarutto A., Paccagnan V., 2003, "Water sustainability: concepts, indicators and the Italian experience", *Economia delle Fonti di Energia e dell'Ambiente*, forthcoming

Dinar A., 2000, *The political economy of water pricing reforms*. Oxford University Press: Oxford, UK

Ekins P., 2000, *Economic growth and environmental sustainability*, Routledge, London

European Commission, Dg Environment, 2000, *A Study on the Economic Valuation of Environmental Externalities from Landfill Disposal and Incineration of Waste*, Bruxelles, URL: http://europa.eu.int/comm/environment/enveco/studies2.htm#28

European Commission, Dg Environment, 2000, *Pricing policies for sustainble management of water resources*, COM(2000) 477 final, Brussels

European Commission, Dg Environment, 2001, *Sixth Framework Programme for Environmental Protection*, Bruxelles; URL:http://www.europa.eu.int/comm/environment

Farber S.C., Costanza R., Wilson M., 2002, "Economic and ecological concepts for valuing ecosystem services", *Ecological Economics*, n. 41, 375–392

Faucheaux S., O'Connor M., J.van der Straaten (eds.), 1998, *Sustainable Development: Concept, Rationalities and Strategies*, Kluwer, Amsterdam

Faucheaux S., O'Connor M., 1998, *Valuation for Sustainable Development: Methods and Policy Indicators*, Edward Elgar Publisher, 1998

Florio M., 1991, *La valutazione dei progetti pubblici*, il Mulino, Bologna

Fontana M., Massarutto A., 1995, *La valutazione economica della domanda di acqua: metodologie di stima e applicazioni empiriche*, Quaderni di ricerca Iefe, Università Bocconi, Milano

Garrido A., 1999, "Pricing for Water Use in the Agricultural Sector," in European Commission DGXI and Instituto da Água, *Pricing Water: Economics, Environment and Society*: Sintra

Gibbons D.C., 1986, *The Economic value of Water*, Resource for the Future, Washington DC

Green C., 2003, *A Handbook of the Economics of Water*, Wiley (forthcoming)

IVM – EFTEC, 1998, *External economic benefits and costs of water and solid waste investments*, Report R98/11, Vrije Universiteit, Amsterdam

Massarutto A., 2001a, Water pricing, the Common agricultural policy and irrigation water use, Report to the European Commission, Dg Environment. URL: http://www.europa.eu.int/comm/dg11

Massarutto A., 2001b, Liberalisation and privatisation in environmental utilities in the international experience, XIII SIEP Conference, Pavia, 5–6 ottobre

Massarutto A., 2002a, "Water pricing and irrigation water demand: efficiency vs. sustainability", *European Environment*, forthcoming

Massarutto A., 2002b, *Environmental, economic and ethical sustainability of the water industry: a case-study from Italy*, Working paper Series in Economics, 03-02, Dipartimento di scienze economiche, Università di Udine

Massarutto A., 2002c, *Regulation and institutions of water management in Italy*, Working paper series in economics 02_, Dipartimento di scienze economiche, università di Udine (http://web.uniud.it/dse/)

McMahon P., Postle M., 2000, "Environmental valuation and water resources planning in England and Wales", *Water policy*, n.2, pp. 397–421

Merrett S., 1997, *Introduction to the economics of water resources: an international perspective*, London, UCL Press

Merrett S., 2002, "Deconstructing households' willingness to pay for water in low-income countries", *Water Policy*, n. 4, 157–172

O'Connor M. (ed), 2003, GOUVERNe Final Report, European Commission Dg12, Bruxelles.

OECD, 1987, *Pricing of water services*, Paris

OECD, 1997, *Evaluating Economic Instruments for Environmental Policy*, Paris

OECD, 2000, *The price of water: trends in the OECD countries*, Paris

Pearce D., 1999, *Water pricing: investigating conceptual and theoretical issues*, Proceedings of the Conference "Pricing Water: economics, environment and society", Sintra, 6–7 September

Pearce D.W., Turner R.K., 1989, *Economics of natural resources and the environment,* Harvester-Weatsheaf, London

Prost T., ed., 1998, *R&D and water management systems in a perspective of sustainable development,* Reporet to the European Commission, Dg12, Bruxelles

Rees J., 1998, "Regulation and private participation in the water and sanitation services", *Natural Resources Forum*, vol, 22 n.2 pp. 95–105

Rogers P., Bhatia R., Huber A., 1998, "Water as a social and economic good: how to put principles into practice", Global Water Partnership background Paper n.2, Stockholm, (http://www.gwpforum.org/gwp/library/Tac2.pdf)

Rogers P., de Silva R., Bhatia R., 2002, "Water as an economic good: how to use prices to promote equity, efficiency and sustainability", *Water Policy*, n.4, 1–17

Sen A., 1982, *Choice, welfare and measurement*, Oxford, Basil Blackwell

Spulber N., Sabbaghi A., 1994, *Economics of water resources: from regulation to privatization*, Kluwer Academic Publishers, Dordrecht

Stiglitz J.E., 1988, *Economics of the public sector*, McGraw Hill, New York

Tardieu H., Préfol B., 2002, Full cost or sustainability cost pricing in irrigated agriculture: charging for water can be effective, but is it sufficient?, *Irrigation and Drainage*, vol. 51, 97–107

The World Bank, 1993, Water Resources Management, The World Bank, Washington DC, USA

Tihansky D., 1975, *A Survey of Empirical Benefit Studies*, in H.M. Peskin, E. Seskin (a cura di), "Cost-Benefit Analysis and Water Pollution Policy, The Urban Institute, Washington D.C., USA

Turner R.K., Postle M., 1994, *Valuing the water environment: an economic perspective*, mimeo, CSERGE, University of East Anglia, Norwich, GB

Vergés J.C., 2002a, *El saqueo del agua en Espa*a*, Barcelona, La Tempestad.

Vergés J.C., 2002b, The two-headed Ebro Hydrodinasaur, World Bank Water Forum 2002, Washington DC, 6–8 May 2002

Author index

Barraqué B. 39
Bogardi J.J. 17

Cabrera E. 3
Chocat B. 171
Cobacho R. 3
Connor D. 157

Fereres E. 157

Howarth D. 141

Lora J. 187

Martin B. 99
Massarutto A. 209

Mays L.W. 117

Sancho M. 187
Soriano E. 187
Szollosi-Nagi A. 17

Wolff G. 51